checked
DEC 2013 (CD)
(LT to keep)

581.4 ING

DIVERSITY AND EVOLUTION OF LAND PLANTS

WITHDRAWN

2

009079

DIVERSITY AND EVOLUTION OF LAND PLANTS

MARTIN INGROUILLE

CHAPMAN & HALL
University and Professional Division
London · Glasgow · New York · Tokyo · Melbourne · Madras

Published by Chapman & Hall, 2-6 Boundary Row, London SE1 8HN

Chapman & Hall, 2-6 Boundary Row, London SE1 8HN, UK

Blackie Academic & Professional, Wester Cleddens Road, Bishopbriggs, Glasgow G64 2NZ, UK

Chapman & Hall, 29 West 35th Street, New York NY10001, USA

Chapman & Hall Japan, Thomson Publishing Japan, Hirakawacho Nemoto Building, 6F, 1-7-11 Hirakawa-cho, Chiyoda-ku, Tokyo 102, Japan

Chapman & Hall Australia, Thomas Nelson Australia, 102 Dodds Street, South Melbourne, Victoria 3205, Australia

Chapman & Hall India, R. Seshadri, 32 Second Main Road, CIT East, Madras 600 035, India

First edition 1992
Reprinted 1992

© 1992 Martin Ingrouille

Phototypeset in 10/12pt Plantin by Intype, London
Printed in Great Britain by Clays Ltd, Bungay

ISBN 0 412 44230 2

Apart from any fair dealing for the purposes of research or private study, or criticism or review, as permitted under the UK Copyright Designs and Patents Act, 1988, this publication may not be reproduced, stored, or transmitted, in any form or by any means, without the prior permission in writing of the publishers, or in the case of reprographic reproduction only in accordance with the terms of the licences issued by the Copyright Licensing Agency in the UK, or in accordance with the terms of licences issued by the appropriate Reproduction Rights Organization outside the UK. Enquiries concerning reproduction outside the terms stated here should be sent to the publishers at the London address printed on this page.

The publisher makes no representation, express or implied, with regard to the accuracy of the information contained in this book and cannot accept any legal responsibility or liability for any errors or omissions that may be made.

A catalogue record for this book is available from the British Library
Library of Congress Cataloging-in-Publication Data available

Contents

Preface vii
Acknowledgements viii

1 The study of diversity 1
1.1 The names of organisms 1
1.2 Studying plant structure 5
1.3 Characters 7

2 The plant body: plant behaviour 24
2.1 Plants are different from animals 24
2.2 The environment of plants 33
2.3 Plant growth and behaviour 36

3 The first land plants: patterns of diversity 55
3.1 Plant fossils 55
3.2 Algal ancestors 58
3.3 An early community of land plants 59
3.4 Diversifying on land 63
3.5 The axis and its appendages 73
3.6 The age of gymnosperms: extinct seed plant groups 78

4 Sex and dispersal: gametes, spores, seeds and fruits 88
4.1 The alternation of generations 88
4.2 The sporophyte 90
4.3 The gametophyte 99
4.4 Heterothallism 107
4.5 Seed plants: the ovule 110
4.6 A classification of land plants 130

5 Flowers: evolution and diversity 133
5.1 The evolution of angiosperms 133
5.2 Cross pollination 138
5.3 Breeding systems 141
5.4 Evolutionary trends 148
5.5 A review of angiosperms 157

6 Trees: adaptations in woods and forests **187**
6.1 Wood anatomy 187
6.2 Tree architecture 197
6.3 Leaves 205
6.4 Bark and periderm 208
6.5 Epiphytes 210

7 Adaptive growth forms: the limiting physical environment **223**
7.1 Water relations 223
7.2 The bryophytes; non-vascular plants 227
7.3 Gas relations 234
7.4 Aquatic plants 236
7.5 Nutrient relations 242
7.6 Surviving environmental extremes 255
7.7 Life forms 257

8 Competition, herbivory and dispersal: the limiting biotic environment **263**
8.1 Reproductive strategies 263
8.2 Herbivory 267
8.3 Dispersal 279
8.4 Establishment 287
8.5 A behavioural classification of plants 287

9 Cultivated plants: conclusion **291**
9.1 Exploited diversity, the uses of plants 291
9.2 The genetic history of crops 293
9.3 Future prospects 294
9.4 Conclusion 294

Glossary **297**
References **316**
Index **327**

Preface

Such is the diversity of the plant kingdom it seems a hopeless task to try to describe and to explain it. There are probably more than 300 000 species of plants. Any single text is impossibly limited. The traditional systematic treatment of describing each group, one at a time, requires much more space than a single text if it is not to be hopelessly superficial. Anyway the systematic approach is dull to a modern audience not attuned to it. It is impossible to see the wood for the trees. The difficulty is compounded because systematics has become a specialist study. Morphology and anatomy are increasingly technical and abstruse and there is a plethora of taxonomic groups which can only be comprehended by the expert.

How is one to describe in a single short text the basics of anatomy and morphology of the land plants and convey an impression of the great diversity of structures and forms which exist? The approach adopted here is to describe variation not in terms of taxonomic groups but by relating diversity to evolutionary trends in structures. Differences between plants are related to the circumstances of the plants, their physical and biotic environments. Evolutionary trends as recorded in the fossil record are described where they are known, but even here comparison with living forms is made so that structures can be visualised more easily in functional terms, as adaptations.

My hope is that this text will provide a single readable introduction to the land plants and stimulate the student to more study. The text represents my own process of learning about land plants, which is still at an early stage. The more one knows, the more there is to know.

Acknowledgements

As well as the references cited in the text, I have been greatly aided, again and again, in my studies by a small number of sources, each excellent in their own way, which I here gratefully acknowledge. They may provide the next step for the interested student: for reference, Mabberley (1987); as a general text Bold, Alexopoulos and Delevoryas (1987); for tracheophytes only Gifford and Foster (1990); for anatomy Fahn (1982), Mauseth (1988) and Esau (1965); for adaptation, Crawford (1989) for bryophytes, Schofield (1985) and Watson (1967); for pteridophytes and gymnosperms Sporne (1974 and 1975); for angiosperms Cronquist (1981) and Stebbins (1974).

I would like to acknowledge the help given to me by my colleagues, and Martin Cullum, who helped prepare some of the material for me. In particular I would like to thank Prof. W.G.Chaloner for reading the manuscript, spotting some errors, and making several useful suggestions. The writing was greatly aided by a microcomputer bought with the support of the Systematics Association.

I would finally like to thank the following individuals and organizations who have given permission for the reproduction of illustrations (figure numbers in parentheses): R.Gornall and the BSBI (1.3); Ecological Society of America (7.10); G.Duckett and the Linnean Society (4.8, 4.9); B.H.Tiffney and Academic Press, Australia (4.16); A.C.Gibson and Harvard University Press (6.4); K.J.Niklas The New York Botanical Garden (8.7); J.P.Grime and Unwin Hyman (8.9).

The study of diversity

1

1.1 THE NAMES OF ORGANISMS
1.2 STUDYING PLANT STRUCTURE
1.3 CHARACTERS

SUMMARY

The study of biological diversity is the science of taxonomy or systematics in a broad understanding of those terms. The study, at its simplest, requires only a basic understanding of how groups are named and classifications are structured and the knowledge of some terminology. This is described here together with some of the basic skills required for looking at plant material.

The chapter, from section 1.3, then considers some more complex ideas in systematics (taxonomy) about the nature of variation and how it should be treated in classifications. Systematics is also the study of evolution and the way the characters of living organisms may reflect either evolutionary history or evolutionary processes and of how adaptation may provide a framework for understanding the significance of biological diversity. This is a complicated and contentious area but it makes the study of diversity a particularly interesting and exciting part of science. However this part of the chapter can be safely skipped without loss of understanding of the rest of the book.

1.1 THE NAMES OF ORGANISMS

The study of biological diversity and the placing of organisms in groups on the basis of shared relationship is called taxonomy or systematics. Groups of organisms are named in two ways. There are formal taxonomic names of groups, called taxa (singular = taxon), which are part of a hierarchical classification. Higher taxonomic names are distinguished by having a special set of endings indicating at which rank in the taxonomic hierarchy they belong. Species and genus names are latinized and italicized or underlined. Taxonomic names are invented as a means of labelling a taxon. Each species must have a different name. The names do not have to mean anything or refer to any particular characteristic of the taxon to which they are attached.

Alternatively there are informal vernacular names. Many taxa do not have a precise vernacular name. Different species can share the same vernacular name and different vernacular names are used in different areas. It is for this reason that the latinized species or the generic name is preferred for scientific accuracy. Many vernacular names of higher taxa originated as formal taxonomic ones in classifications which have now been superseded. However these vernacular names of higher taxa are often useful; because they have been around a long time, we feel confident of what they actually mean.

Vernacular names are of various kinds. They are commonly used names like 'fern', 'moss', 'horsetail'. Many record a particular characteristic like 'vascular plant'. Others are similar though the descriptive term has a Latin or Greek derivation like 'cryptogam' (crypto = concealed, gam = reproductive parts) and 'gymnosperm' (gymno = naked, sperm = seed). A very useful informal ending for vernacular names is '-phyte' which means plant hence 'bryophyte', moss plant or moss (bryo = moss). However note that '-phyta' rather than phyte is a formal taxonomic ending for the rank of Division. A list of vernacular names, a kind of vernacular classification of the higher taxa of land plants is presented at the end of Chapter 3.

1.1.1 Species, genus and family

Species, genus and family are probably the most important kinds of taxonomic groups. A species has a binomial name which includes the name of the genus to which the species belongs, written first, and then a specific epithet. Where there is no possibility of confusion the species name may be abbreviated to the first letter of the generic name plus the specific epithet. Specific and generic names are always indicated by being underlined or in italics. A genus name always begins with a capital letter and may be used on its own. The species name (the binomial, which of course includes the generic name) is the most important taxonomic name. Species are the most important taxa. They are the basic units of study in all parts of biology. Many species and genera were recognized long before taxonomy became a science. It was the works of the Swedish biologist Linnaeus, *Genera Plantarum* of 1737 and *Species Plantarum* of 1753 which are taken as the start of modern plant taxonomy.

Sometimes the same species name has been defined in different ways by different authors. To enable the reader to understand which concept of the species is being used, it is common for the authority to be indicated by including his or her name after the species name. For example *Pteridium aquilinum* (L.)Kuhn, was first described by Linnaeus (abbreviated to L.) as *Pteris aquilina*. Later Kuhn placed bracken in a different genus, that of *Pteridium*. Note that the ending of the specific epithet changes to correspond to the gender of the new generic name. There are many rules governing the naming of species (Stace, 1989).

Families of plants have been recognized for a long time. In the 18th century a French botanist called A.-L. de Jussieu who was the nephew of the gardener at Versailles, wrote a *Genera Plantarum* which divided plants into 100 'natural orders', which are now regarded as families. We still recognize many of these families today even though many others are also now recognized. They are 'natural' families, in the sense that they are easily recognized. Species, genus and family are probably the easiest taxonomic categories to recognize. Taxonomists try to base the definition of these taxonomic units on readily distinguishable features so that their names are useful to other biologists.

1.1.2 The taxonomic hierarchy

Classifications of living organisms are hierarchical. As one goes up the hierarchy successive levels contain ever larger numbers of lesser taxa. Not all the taxonomic ranks have to be used. The endings of the names of higher taxa change to indicate the rank of the taxon (Table 1.1).

Table 1.1 The taxonomic hierarchy (important ranks in capitals)

	suffix	
KINGDOM		Plantae
Subkingdom	-bionta	Embryobionta
DIVISION	-phyta	Tracheophyta
Subdivision	-phytina	Spermatophytina
CLASS	-opsida	Magnoliopsida
Subclass	-idae	Magnolidae
Superorder	-anae	Asteranae
ORDER	-ales	Asterales
Suborder	-ineae	Asterineae
FAMILY	-aceae	Asteraceae
Subfamily	-oideae	Asteroideae
Tribe	-eae	Astereae
Subtribe	-inae	
GENUS		*Bellis*
SPECIES		*B. perennis*

As well as the main ranks, sub- and super- ranks may be used which take their place in the hierarchy in exactly the same way as the full ranks. Some old and well established family names are allowed as synonyms, so for example the family Asteraceae can also be called Compositae. It is increasingly common to follow the practice of naming higher taxa by using the name of a lower taxon, ultimately a species, which represents it as a 'Type' so in the example in Table 1.1 the type of all the taxa from Tribe to Superorder is a species of the genus *Aster*. Above that the type is a species of *Magnolia*. This neat system of pigeon-holing is a kind of filing system which allows ready access to information about all kinds of groups, taxa at all ranks in the hierarchy.

Within a species individuals share a high degree of similarity. Species of a genus recognizably belong together. Different genera within a family share an overall resemblance but have many differences. The higher one goes up the hierarchy, the harder it is to define the taxon. The taxon contains more and more variation and fewer and fewer characteristics are held in common. In some cases the things in common are shared not because the species in that taxon share a recent common ancestor but because of parallel or convergent evolution.

1.1.3 Different classifications

The system of naming taxa in a classification is very orderly. It is governed by a set of internationally agreed rules for nomenclature though the rules for algae and fungi are slightly different from other plants. Algal and fungal taxa also have some different endings. However, many different classifications of land plants have been proposed. They are confusing to the non-expert because although the different classifications share many similarities they may differ in subtle ways. The potential for confusion is great, especially if extinct fossil groups are included because the classification of fossils is less certain than that of extant plants.

Differences are of two sorts. First, there are nomenclatural differences in classifications which arise from differences in the application of the hierarchy of names.

1. Synonymous names are allowable above the generic level and commonly used like the Dicotyledonidae (= Magnolidae) or Angiospermopsida (= Magnoliopsida). Different root words may be used. For example the flowering plants or angiosperms are the Magnoliophyta in Cronquist (1981) based on the plant *Magnolia*; they are Anthophyta in Bold *et al.* (1987) from the Greek *anthos* =flower, and they might also be called the Angiospermophyta from the Greek *angion* =vessel and *sperma* =seed. Such high-level taxa with different names may be identical.
2. Some old-fashioned names, without the helpful recommended standardized ending for different taxonomic ranks, are still allowed, and are in use though they are not recommended. For example the flowering plants when treated as a class are often called Angiospermae rather than Angiospermopsida, with the proper -opsida ending. The standardized endings of higher taxa above Order are not mandatory.
3. The rank in the taxonomic hierarchy at which a group is recognized may be different so that the ending of the name differs even though the same root is used. For example the flowering plants may be recognized as a Division, the Magnoliophyta (or Angiospermophyta), a Subdivision, the Magnoliophytina (or Angiospermophytina), or a Class, the Magnoliopsida (or Angiospermopsida).
4. Changes in rank are complicated because they may also conceal differences in the set of taxa included or excluded. For example the Class Angiospermopsida, for all flowering plants, in Stace (1989) is identical to the Division Magnoliophyta in Cronquist (1981) and not his Class Magnoliopsida, which includes only the dicotyledons. In this case we are reminded to look for a difference because a different root name is used but this is not always the case. Taxa with an identical name but used by different authors may include or exclude different taxa.

Despite all these difficulties having a proper hierarchical system of naming plants is very important because it allows the possibility of precision while the use of common vernacular names is fraught with unresolvable difficulties.

Scientists from different parts of the world, indeed from different ages, are able to understand each other if scientific names are used.

The other, and more important, differences in classifications are taxonomic. They are more fundamental differences in the way the hierarchy should be ordered, and which groups are closely related to each other. Some of the differences of opinion arise because there is no agreement on what a classification is for. There is a difference of emphasis by different workers. Some regard a classification primarily as an information retrieval system which may also, almost as a subsidiary feature, indicate evolutionary relationship. Others are more concerned with producing a classification which reflects evolutionary relationship. If the evolutionary relationship has been correctly determined the classification will then provide a useful information retrieval system.

There is also a difference between the methods used by taxonomists to detect groups and define taxa. This is a complex and contentious area, of how taxonomists work. Arguments centre around the nature of characters and patterns of evolution.

1.2 STUDYING PLANT STRUCTURE

The skills required for studying plant structure are relatively simple. The most important skill is using your eyes carefully. The shape and size are recorded in drawings and notes. Many budding botanists are put off because they think they cannot draw. However, botanical drawings are not artistic interpretations but precise records of what can be seen.

The drawings should include the minimum amount of work necessary to convey the information. In textbooks like this, where drawings have been reduced, stippling or shading is used to emphasize structures. However normally shading should be avoided because it is usually unnecessary and if not done properly obscures rather than clarifies the drawing. If an organ occurs repeatedly on a plant it is pointless to draw the same thing several times. The position of the repeats should be indicated only. The drawings should be large enough to make it easy to draw what you want to show. Information is conveyed by labelling and by the inclusion of short notes. The latter are useful for conveying information which is difficult to draw and not for discursive speculations.

The most important aid to botanical drawing is a sharp pencil!

The essence of studying structure is making comparison between different organs and different organisms. The most important requirement is that the size and relative sizes of parts is accurately recorded. A common mistake is to record size as a magnification factor. It is much better to include a scale line.

1.2.1 Sectioning

The internal structure of plants can be observed by making sections and mounting these on glass slides for examination under a microscope. There are microtomes which can produce professional quality sections or serial sections. For soft plant material or complex organs where the relative position of parts needs to be retained, the material is first embedded in wax. For most plant material, like leaves or herbaceous or semi-woody stems, wax embedding is not necessary and it is sufficient to clamp the material between two pieces of cork. Woody tissue requires special treatment. It needs to be softened by a jet of steam while it is clamped in the microtome.

For the student all that is normally required for taking sections is a very sharp razor-blade and sometimes something to support the tissue while it is being sectioned. Pith, polystyrene or cork are frequently employed. It is important to keep the cut surface wet at all times. Sectioning is carried out by a stroking motion with the blade. Elongated cylindrical structures are sectioned in three planes; a transverse section (t.s.), a radial longitudinal section (r.l.s.) and a tangential longitudinal section (t.l.s.). It takes a mental leap to reconstruct the three dimensional structure of the anatomy of a living plant from the views at different angles that different sections provide.

Sections can be mounted in water or 50% glycerine. The visibility of parts of a section may be improved by adjusting the microscope condenser or using a temporary stain: 1% aqueous methylene blue stains cell walls leaving the cuticle unstained; a drop of phloroglucin plus a drop of concentrated HCl turns lignin red. There are many other stains which are useful for different purposes (Cutler, 1978).

1.2.2 Using a microscope properly

It is important that the lenses are kept clean. The condenser must be focused. Place a sharp point like the tip of a needle on the light source. Bring it into sharp outline by moving the condenser up and down. The iris diaphragm should be opened/closed so that it just passes out of the margin of the field of view.

In examining a specimen you should start with a low power objective first and work your way up through higher magnifications. Measurements can be made using an eyepiece graticule calibrated with a slide micrometer. If these are not available you can gauge the size of objects by the fraction of the field of view they occupy. You can measure the actual size of the field of view under the lowest magnification by placing a clear plastic ruler under the microscope. Divide in proportion to the change in magnification for higher power objectives.

1.2.3 Technical terms

One of the greatest difficulties for the new student is the number of botanical terms which are in current use. They are required to describe the great diversity of structures precisely, and to name the multiplicity of organisms. This is a difficulty which has to be faced. DON'T PANIC. With experience the student will be able to apply terms with increasing confidence. In this book I have included a lot of specialist terms because in the long run they make it easier to describe plant variation. The index will point out other places where the term is used in context and also there is a glossary.

There are some easily learnt aids to understanding because many botanical terms are derived from Latin or Greek roots. If these are learnt a term can be taken apart and understood. So, for example, 'gynoecium' is the 'oecium', the room, where the 'gyno', the female parts are found, i.e. the female reproductive part of a plant.

Ignorance of the correct technical term is no barrier to understanding. There are ways of describing using ordinary English, though they are often more long-winded. A botanist can get a long way with a small vocabulary of adjectives describing shape, texture and arrangement. Many will be used in context later in the text. A few to begin with are shown in Figs 1.1 and 1.2 (overleaf). Many special terms, especially those describing reproductive or anatomical structures are described and figured over the following chapters. There are several specialized texts which provide a guide to terms for flowering plant morphology (Webberling, 1989; Bell, 1991).

1.3 CHARACTERS

The observable characteristics of an individual are its **phenotype**. The phenotype is the result of a developmental interaction between the genetic coding of the individual, the **genotype**, and the environment. Some variations prove to have a close and simple connection to the genotype. These are called genetic **variants** and are the result of genetic variation. Some characters although coded genetically are much modified by the environment. Many plant variants have been proved to be genetic variants by being taken into cultivation. If after growing in a range of environmental conditions and after many generations of vegetative propagation the variant remains distinct it is very likely to be genetically determined. The variant may be formally taxonomically recognized as a **variety** or **form** or **cultivar**. For example, there are hundreds of garden cultivars of heath, *Erica*, which have been collected from nature and maintain their beautiful shape and colour in cultivation. More rarely the genetic basis of a variation has been tested by a genetic analysis of the progeny of a cross between two different variants.

Most variation has not been scientifically tested either by cultivation experiments or by crossings. However, in nature most variants occupy a range of environments and if the variant maintains itself despite variation in the environment there can be some confidence that it is at least partly determined genetically. However, there are few absolutes. Many characters, though genetically determined, are **plastic** in expression.

1.3.1 Plasticity

A plant can be regarded as being constructed in a very regular way by the repetition of a basic module. Each **module** of a plant is genetically identical to every other in the plant, except in rare situations. However, a very regular structure is very rarely produced. In nature development is plastic so that

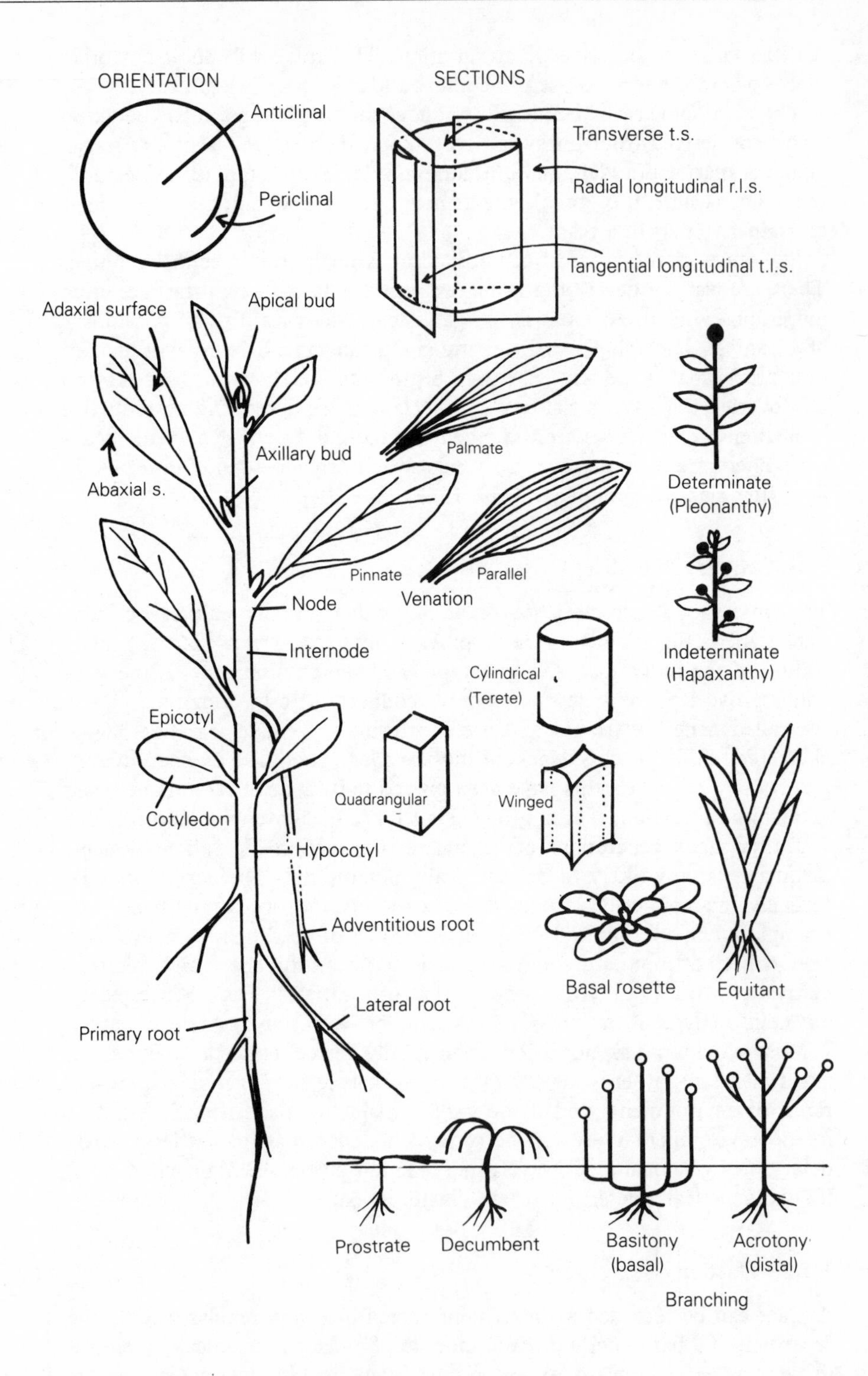

Fig. 1.1. Botanical terminology I.

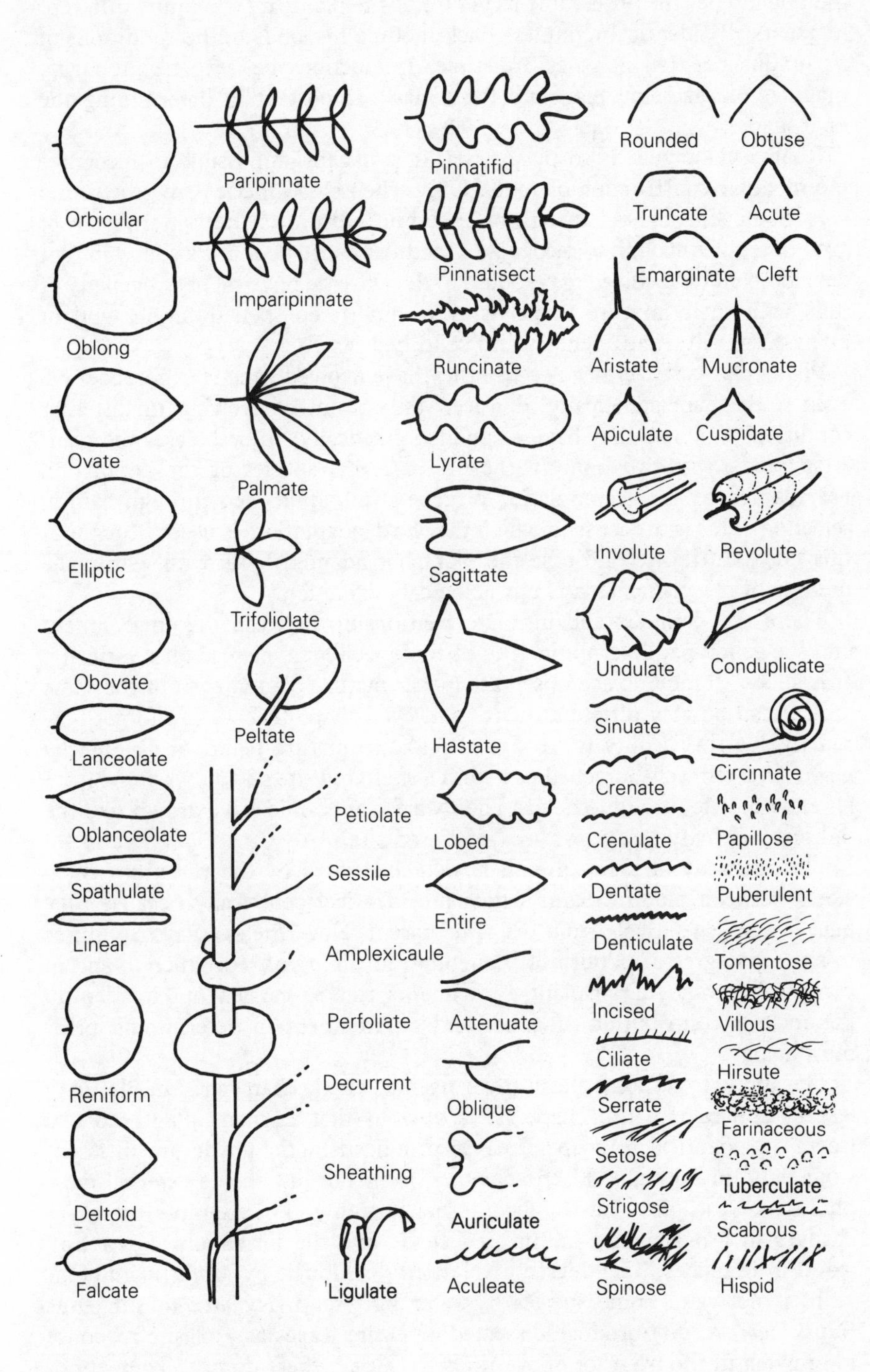

Fig. 1.2. Botanical terminology II.

the phenotype, the observable structure of the plant, may be quite different in genetically identical modules. Each module responds to the conditions of its unique position in space and time. Interactions between neighbouring modules of the same plant are particularly important in determining the final shape.

Root development is so plastic that it is usually impossible to recognize the modular construction of roots. Only when some species are grown in a very homogeneous soil matrix or in liquid culture is the pattern of root growth regular enough to recognize a modular construction. Plasticity in root development has evolved in response to the extreme physical heterogeneity of soils with their mixture of particles with different penetrabilities and of many sizes: silt, sand, pebbles and solid bedrock.

Plants may differ either because they have evolved genetic differences or, even if they are genetically identical, as a result of growing in different conditions to which they have responded plastically. In both cases the plant is adapted to its situation. In the first case the species or the population has adapted by the accumulation of genetic mutations and through natural selection. This is the sense in which the word 'adaptation' is used throughout this book. In the second case the plant has adapted by growth within the lifetime of the plant. This is physiological adaptation.

There is a complex and intimate relationship between the environment and the genotype, determining the phenotype. Plants may adapt plastically. It is such a commonplace observation that mature plants, even of the same species can be very different in size and even shape that we can forget how remarkable this facility is. In Fig. 1.3(a) two mature plants of *Gentianella amarella*, a tall well-branched one from a sheltered site and a dwarf one from an exposed site, are illustrated. The dwarf plant could have grown like the tall one if it had been grown in a sheltered site.

The plasticity of plant growth is well illustrated by the variation which exists between mature plants which are identical genetically because they have arisen vegetatively from the same parent. Nevertheless they may differ in size, in degree or pattern of branching, in timing of reproduction and in many other ways. Transplant experiments can be carried out to identify the relative contribution of genes and environment in determining plant morphology.

The ability to respond plastically is itself a genetic characteristic. Plasticity is very apparent in leaf shape. It is obvious that in many plants no two leaves are identical in shape. Leaves produced in the shade are different from those in the light. Plastic changes occur not just in the external morphology of the plant but in its internal anatomy. For example there are changes in leaf anatomy in sun and shade and the formation of **reaction wood** in branches. The degree of plasticity exhibited by *Morus nigra* (Fig. 1.3b) is not rare. Some species of water buttercup (*Ranunculus* subgenus *Batrachium*) readily produce dissected or entire leaves as a plastic response to growing in the water or above it (Fig. 1.3c). Others do not. They either produce ony dissected or entire leaves (Bradshaw, 1965).

Species differ in the extent of their plasticity or, alternatively, they respond

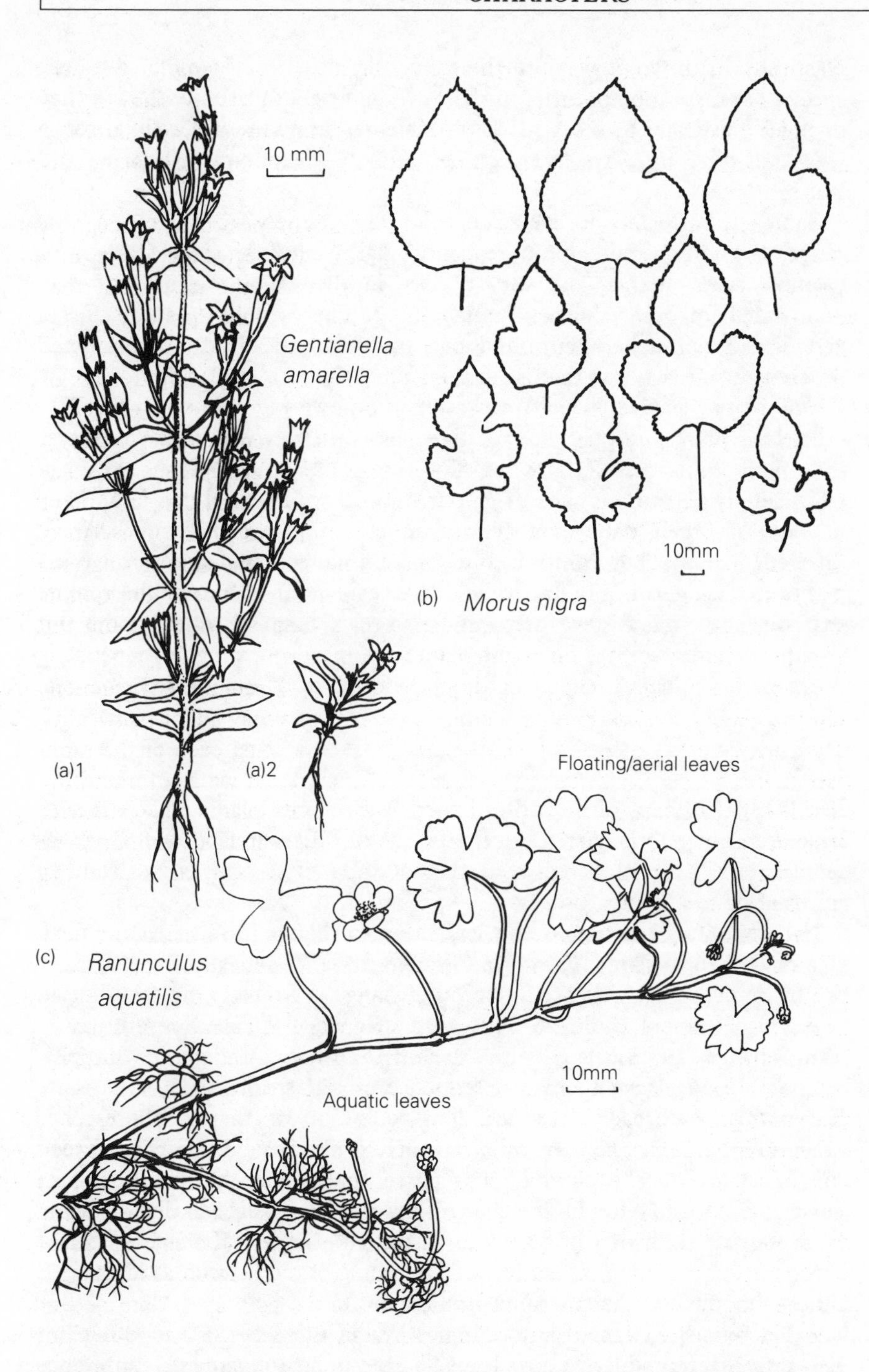

Fig. 1.3. Plasticity: (a) *Gentianella amarella*, plants from 1. sheltered chalk grassland, 2. exposed cliff; (b) *Morus nigra*, leaves from a single individual; (c) *Ranunculus aquatilis*, dissected submerged leaves and entire, aerial leaves.

plastically in different ways to the same stimulus. For example, different species of grasses do not differ in their overall level of plasticity though they respond plastically to the same environmental stimulus, different nitrogen levels, in different ways by adopting a unique phenotype (Robinson and Rorison 1988).

Some garden plants are well known to be extremely adaptable to a wide range of conditions and able to respond by growing differently. Others, for example some alpines, are very precise in their requirements and recommended for expert horticulturalists only. This is unlikely to be due to genetic differences between individuals but rather genetically programmed differences in plasticity; that is, in their physiological adaptability.

One of the features which marks out the flowering plants compared to other land plants, and perhaps partly explains their evolutionary success, is their marked plasticity. Many non-flowering plants are remarkably uniform from simply repetitious mosses to remarkably symmetrical conifers. They do not vary their pattern of growth either within the plant or between different plants. They conform to a simple architectural model. Flowering plants are also genetically diverse and have evolved mechanisms to promote that diversity. They have many more species than all other plants put together. Each species of flowering plant contains a wide range of genetically different individuals, each responding plastically to its unique circumstances.

1.3.2 Development

It is likely that many of the most important variations between plant species have arisen by subtle changes in the timing of different aspects of development. Some differences between the modules of a plant arise from an inherent developmental process.

For example, it is common for leaf shape to change from a juvenile form to a mature form as in *Chamaecyparis* (Fig. 1.4a).

This is particularly marked when the change is from a simple outline to one which is more complex. This kind of change is called heteroblastic. Many changes are associated with the shift from vegetative growth to reproduction. In *Phyteuma* there is a sequence of leaf shapes from the rosette leaves which are broadly ovate with long petioles to awl-shaped inflorescence bracts (Fig. 1.4c). In ivy, *Hedera helix*, there is a striking difference between the lobed leaf of the climbing vegetative shoots and the simple leaf of flowering shoots. Heteroblastic changes may be irreversible so that a cutting from the flowering part of an ivy maintains its simple leaf shape even if it is rooted and grows on vegetatively. The timing of the heteroblastic changes can be modified by the environmental conditions.

A heteroblastic change can occur not just in the vegetative modules but also into the reproductive structures. There is no fundamental difference between the vegetative and fertile modules. This is well illustrated by the gross similarity of fertile and vegetative fronds in many ferns. In the climbing fern *Lygodium* (Fig. 1.4b), the heteroblastic change occurs within the frond; the lower **pinnae** are vegetative and the upper fertile. In some rather primi-

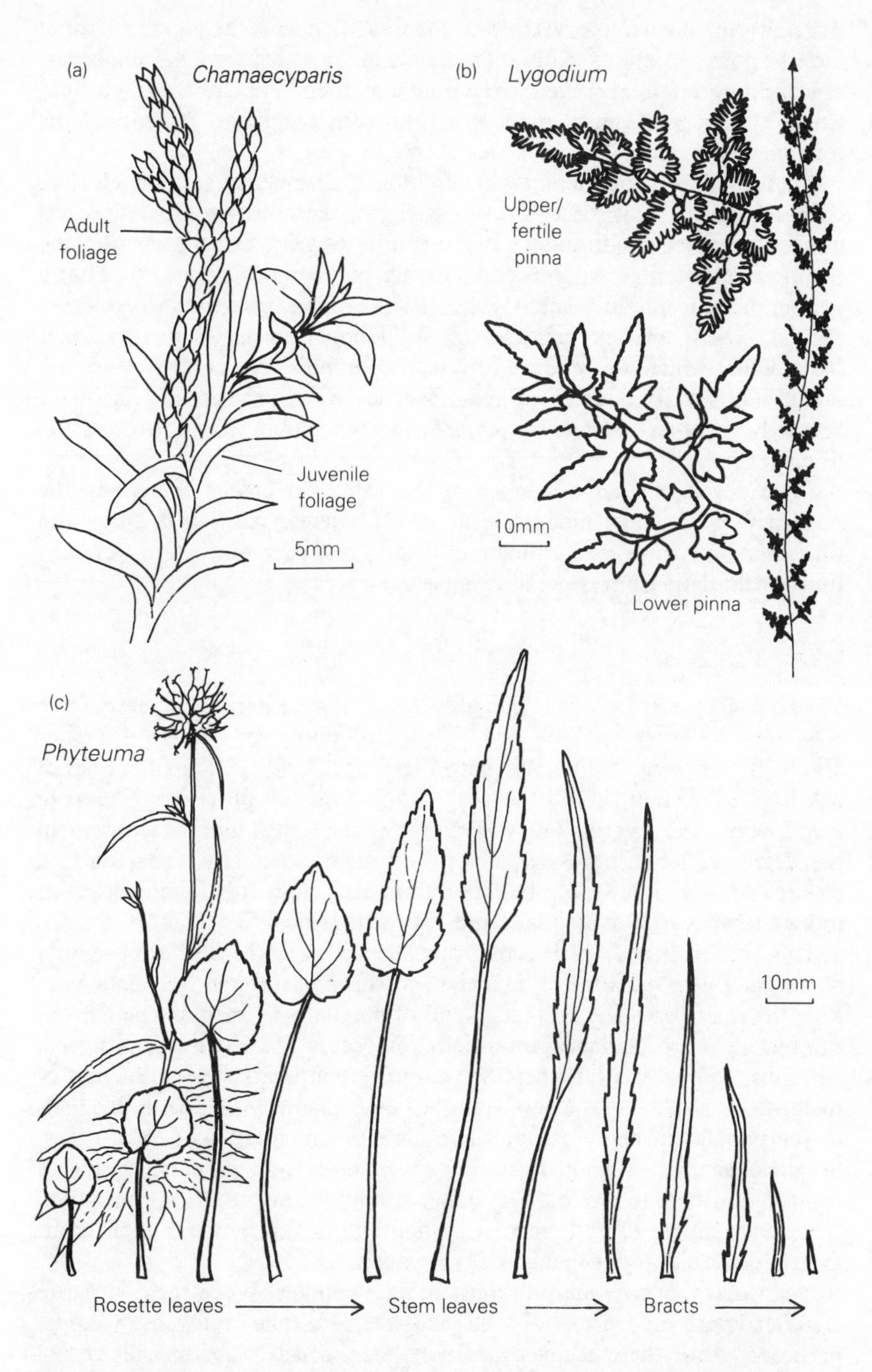

Fig. 1.4. Heteroblastic development: (a) *Chamaecyparis thyoides*, juvenile and adult foliage; (b) *Lygodium* sp., a climbing frond with fertile and sterile pinnae; (c) *Phyteuma scheuchzeri*, rosette leaves, stem leaves and bracts.

tive flowering plants, like *Nymphaea*, the transition from the vegetative shoot into the flower is gradual with intermediate modules between leaf and **bract**, bract and **perianth**, and even perianth and **stamen**. The most marked transition is between stamen and carpel but even the carpel (Chapter 5) is fundamentally a modified fertile leaf.

Another kind of developmental variation is that which accompanies the changing season. Not strictly tied to ageing or reproduction different kinds of shoots may be produced at different times of year. In oak, *Quercus*, the spring shoots stop growing in mid-summer but some restart growth after a resting phase to produce lammas shoots which have different shaped leaves (Fig. 1.5a). An extreme example of this kind of change is the budscales (**cataphylls**) which are produced by many perennial plants at the end of a season's growth, in place of normal foliage leaves, to protect the apical bud. Similarly there are seasonal differences in the anatomy of the wood tissue of many trees.

Some developmental changes can be related directly to increasing the efficiency of the shoot module in its own particular position in space and time, or to a change in the function of the shoot over time. However, this functional relationship is not always clear.

1.3.3 Variation

Variation is a feature of all living systems. For example the leaves of a single oak tree vary in shape (Fig. 1.5a). Different trees of a single species also have on average a different shaped leaf (Fig. 1.5b). Different species of oak have a different shaped leaf (Fig. 1.5c). This last difference is used to help identify the species. The variation is of the same kind in each case, in the degree of lobing, the size of leaf and many other characters, but the source of variation is said to be different in each case. It is found within an individual, between individuals, and between species.

The trees in Fig. 1.5b all come from the same wood. They are part of a single variable **population**. The mean leaf shape of different populations of a species may also differ. It is this kind of population variation which is the raw material of speciation. Natural selection occurs by selection for or against particular individuals but these are only the temporary holders of part of the genetic variation of the population which is shared with other individuals in the population because they share parents and breed with each other. Selection against individuals results in changes in population means of characters. Reproductive barriers between populations results in speciation. The measurement of patterns of variation within and between populations gives a picture of the first stages of speciation.

One variant, or race, may abruptly replace another. When these races are associated with ecologically distinct areas they are called ecotypes or ecodemes. Sometimes these ecological variants are formally taxonomically recognized. For example *Parnassia palustris*, Grass of Parnassus, has two varieties in Britain (Gornall, 1988) (Fig. 1.6b): var. *palustris*, which grows in the tall grass of wet meadows, has a long stem, small flowers and leaves and a weakly

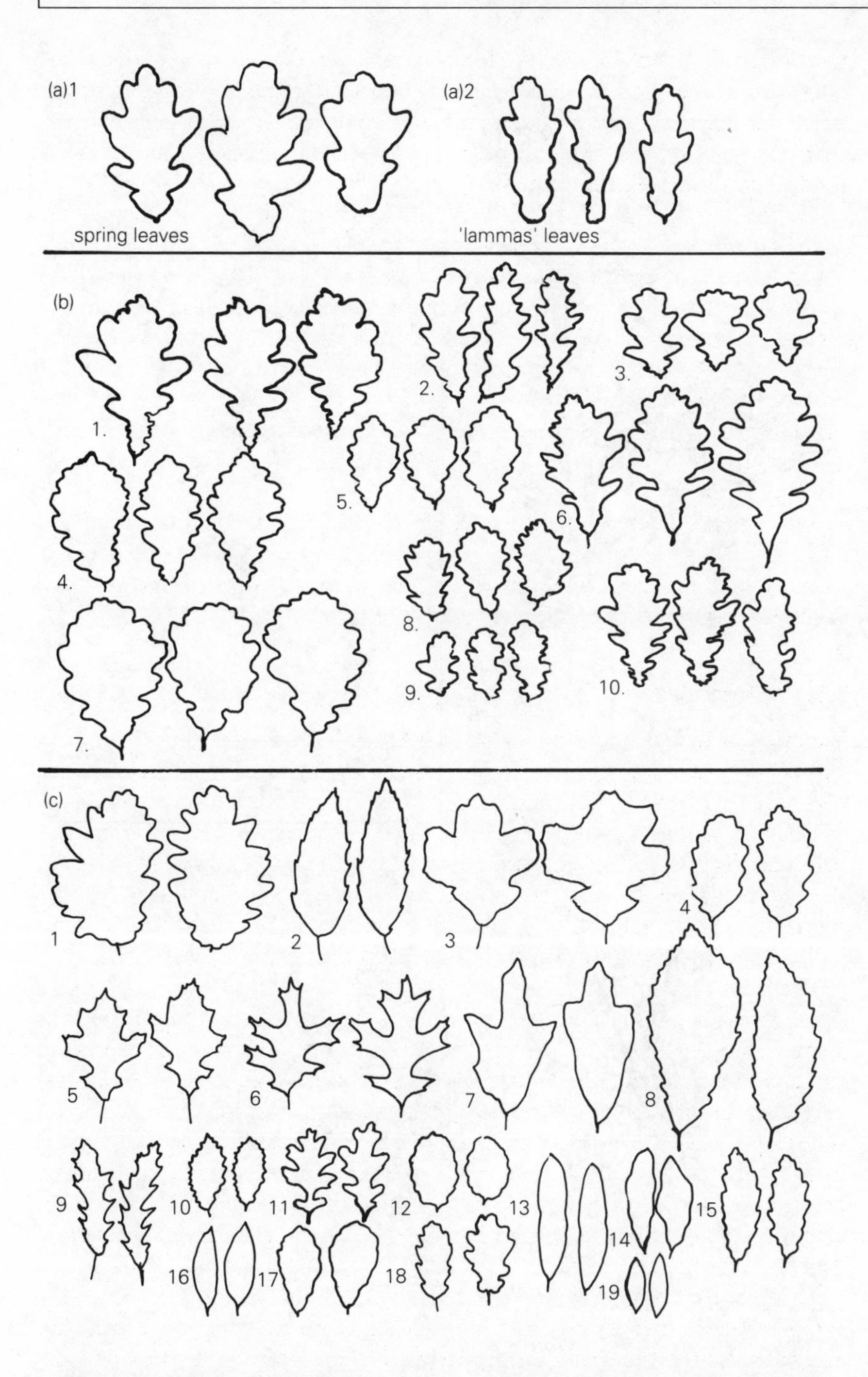

Fig. 1.5. Leaf shape, sources of variation: (a) *Quercus robur*, variation within an individual, 1. spring leaves, 2. late season 'lammas' leaves; (b) *Q.petraea* leaves (in 3s) from different trees in a single variable population; (c) paired leaves from different species.

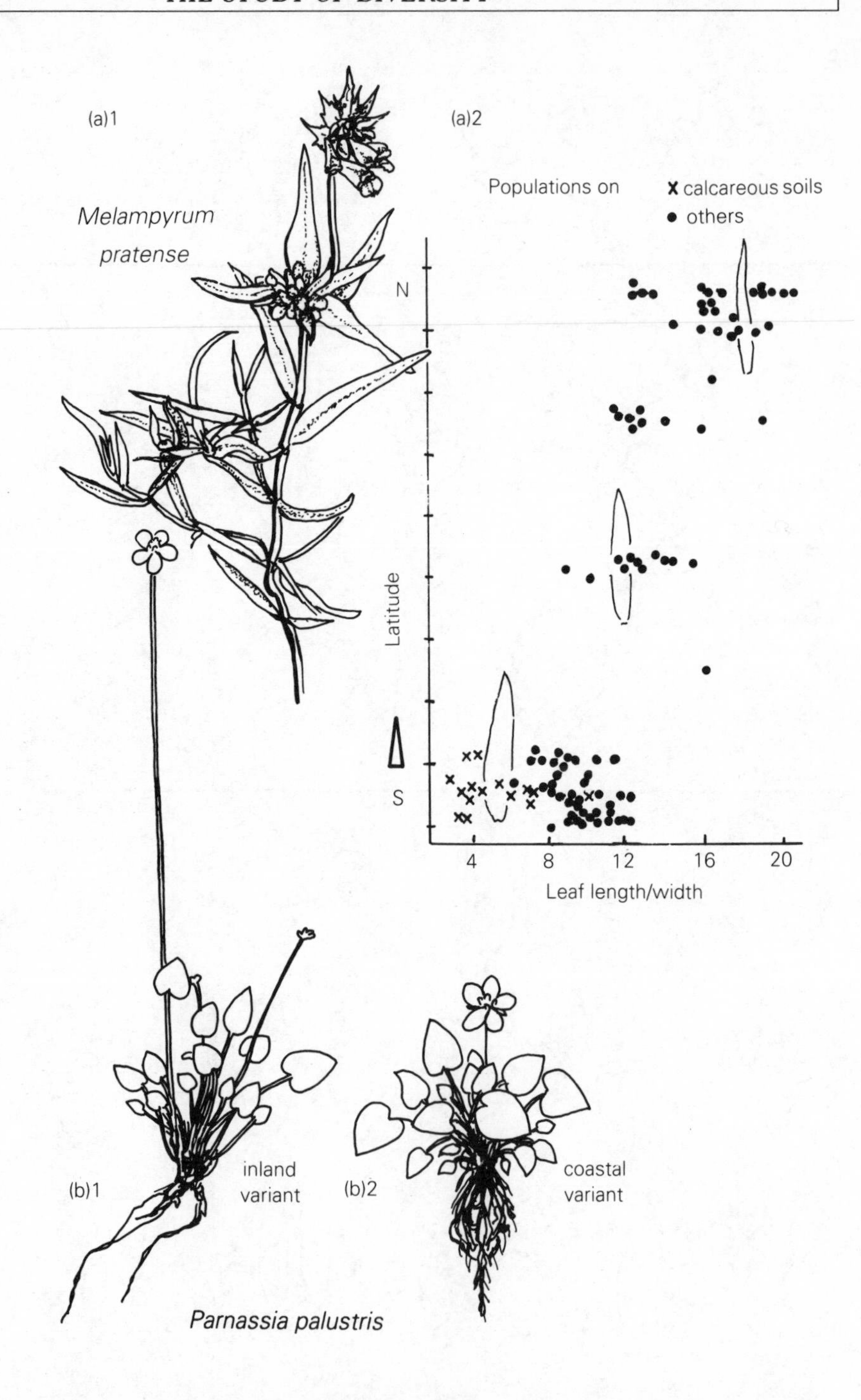

Fig. 1.6. Variation which has a geographical or ecological pattern: (a) *Melampyrum pratense*, a topocline in Britain (leaf shape with latitude) and a calcareous ecotype; (b) *Parnassia palustris*, coastal and inland variants (genecodemes).

branched rhizome; and var. *condensata*, which grows in the turf of coastal habitats, has a short stem, large flowers and leaves and an extensively branched rhizome.

For some continuous characters the mean value of the character changes regularly along a geographical distance, a pattern which is called a **cline**. When the cline is expressed over a large geographical distance it is called a **topocline**. In Fig. 1.6(a) a cline in leaf shape for the species *Melampyrum pratense* with latitude in the British isles is illustrated. Alternatively, where the differences between plants are qualitative, clinal variation can still occur as the proportion of plants of each kind changes. In some cases a clear correlation has been demonstrated between the changing character and an environmental variable. Commonly a climatic factor is identified as the correlating variable in a topocline. On a smaller scale the correlating environmental variable may be altitude, shade, aspect, slope or some other determinant of the **microclimate** or a character of the soil. These small scale clines are often called ecological clines, or **ecoclines**.

These clinal patterns and other environmental correlations suggest the first stages in evolution, an adaptive shift in a character due to natural selection. However it is an extra step from demonstrating a clinal relationship to demonstrating that the environment is actually the selective agent determining the plant characteristics. In the example of *Parnassia* the differences can be readily regarded as having an adaptive value, the result of natural selection, as a response to different levels of shading and exposure in their different environments but patterns can arise by chance. For example a cline can be created by migration of plants from two original areas, in which alternative characteristics were established by historical accident, into a third area. This may explain some of the pattern in *Melampyrum*. Spurious variation and clines may exist in characters which are not adaptive themselves but are genetically or developmentally linked to adaptive characters.

Sometimes a rather haphazard pattern of variation is observed with neighbouring populations containing different variants or where a population is polymorphic. In these cases the adaptive significance of the variation is even harder to understand. There are circumstances in which a range of variants is selected for at a single location. There may be frequency dependent selection where the most abundant variant is selected against. It is also possible that these patterns of diversity are the result of chance events; mutation, migration and changes in population size.

Some of the complexities of trying to understand plant variation are illustrated by the *Melampyrum* example. *Melampyrum* shows many different kinds of variation. Within *M.pratense* at least three kinds of variation have been detected; geographical, ecological and local (Smith, 1963). As well as leaf shape, there is geographical variation in range of characters. Within a geographical region there is some clearly ecological variation with, for example populations on calcareous soils having broader leaves, but there is also variation in flower colour which is more difficult to understand.

In Britain it has been possible to record some of these morphological patterns in a formal taxonomic hierarchy:

Melampyrum pratense
subsp. *pratense* (on acid soils)
var. *pratense* (white or pale-yellow flower)
var. *hians* (golden-yellow flower)
subsp. *commutatum* (on calcareous soils)

The two subspecies are distinguished by a range of characters. A purple-lipped variant is only given the lesser rank as form *purpurea* because it is of sporadic occurrence. Most of the clinal variation is not recognized because of the overlap in characters between regions and ecotypes.

Much of the pattern of variation in flower colour in *Melampyrum pratense* is likely to have arisen by chance. Certainly the haphazard distribution of form *purpurea* suggests this. The more constant varieties, var. *pratense* and var. *hians*, may record the accident of a pattern of colonization of Britain shortly after the Ice Age, rather than being adaptive now. If a population was founded by a golden-yellow individual all the plants in that population were golden yellow.

Species exist as series of isolated and semi-isolated populations. The degree of isolation between populations and the size of the populations are important factors determining the patterns of variation which arise. In the process of sexual reproduction there is a chance sampling of the total variation in each generation. Thousands of male and female gametes are produced but only a few take part in fertilization. Thousands of zygotes are produced but only a few grow into mature reproductive adults. As a result of this sampling process gene frequencies can change by accident. This process is called **genetic drift**.

In a small isolated population chance effects are magnified because they are not swamped. A similar process can occur in populations which become very small, for only a temporary period. Also marginal populations founded by a few individuals may be different from central ones. As a result of these accidental processes, populations which are very different from each other can arise without natural selection.

The pattern of variation is complicated because the patterns in Britain run counter to those in the rest of Europe. In some parts of Europe the golden-yellow var. *hians*, which is only taxonomically recognized in Britain because it occurs in pure populations, turns up sporadically, mixed with the normal pale-flowered plants. In addition the characters used to distinguish the British subspecies, subsp. *pratense* and subsp. *commutatum*, partly overlap with those used to distinguish other recurrent ecotypical variants called autumnal (found in woods and scrub), aestival (found in meadows), montane (found in mountain pastures) and segetal (found in cornfields). These ecological variants flower at different times in the year (De Soó and Webb, 1972) and recur in different species of *Melampyrum*; a parallelism which is a strong indication of natural selection acting in a similar way in different species.

The functional significance of character variation, even if there is a simple cline, is not easy to understand. It may not be possible to translate the observed differences into functional ones. Are the *Melampyrum* plants growing on calcareous soils adapted to the soils or are they there by accident? How does having a broad leaf adapt them to calcareous soils? Why is a golden-yellow flower found in pure populations in Britain but turns up only sporadically in Europe? Only a beginning has been made to try to understand character variation. For example, in *Melampyrum arvense* a difference in branching pattern results in a greater seed set for the more branched plants (Kwak, 1988).

1.3.4 Adaptation

Darwin's theory of natural selection provided a way in which biological diversity might have arisen by a process of **adaptations** becoming established. It was Darwin's insight to see that inherited differences between individuals could by natural selection lead to species which are better adapted to their environment. If different populations experienced different circumstances, different sets of adaptations would be selected. If this process of selection was combined with **reproductive isolation** between the differently selected **populations** then a new species would arise.

That is the theory, but it is still true that very few unequivocal examples of natural selection have been observed in nature. Physiological characters, such as the heavy metal tolerance which allows some plants to grow on contaminated soil, have been well studied. However, other characters, seemingly simple to understand, like the ability to produce cyanide to deter herbivory, have proved to be very complicated. The adaptive value of morphological or anatomical variation is even more complicated than that of simple physiological characters, and is much less well studied because the characteristics are usually quantitative in nature and blend into each other. The study of adaptation seems even more difficult, because the unit of selection is not a single character but the whole individual, the sum of all its adaptations.

There are problems in the use of the words 'adaptation' and 'adapted' anyway. They are sometimes used in a physiological or sensory context when they have a different meaning. In an evolutionary sense, as they are used throughout this book, many writers have noted their ambiguity. It makes no sense to use the word as if 'adapted' was a synonym for 'fitted'. To say that an organism is adapted or fitted to its environment is stating the obvious for if it was not fitted it would not be able to live within that environment (Lovtrup, 1987). This use of the words 'adaptation' and 'adapted' is said to be non-scientific because the advantage characters confer is not measurable.

Attempts to make the study of adaptation scientific, to measure relative adaptation, have centred on measuring another quality, 'fitness', i.e. reproductive success, as a kind of summation of the effectiveness of all the adaptations of an organism. However even this measurement is extraordinarily difficult (Lewontin, 1977). Other measurements of adaptation have centred on establishing 'optimality models', assuming adaptations maximize the efficiency of some function of the plant. However, in practice such models are unscientific, because they cannot be proved one way or the other, they are just modified to fit the available facts (Gould and Lewontin, 1979). Voltaire disparaged this attitude as a belief that 'everything is for the best in the best of all possible worlds'.

In spite of these theoretical difficulties, it is very useful to view the world of biological diversity through Darwinian spectacles. Without the Darwinian view one either has to suppose that the variation is haphazard, nice to look at, but actually not understandable, or one has to resort to some kind of mysticism. In the latter case it would be as if there are some mysterious interior forces which mould development and which are inexorably creating diversity. This harks back to an old-fashioned use of the word evolution, to describe the unrolling of inherent variations. This is certainly a valid way of looking at the world and may even be partly true but it is something of a cop-out. It is not open to test. Anything and everything could be 'explained' by reference to unknown inner forces.

It is true that most so called adaptations are merely speculations, untested hypotheses, and if it is suspected that a structure has a particular function, like the part of a machine, it is incumbent on us as scientists to test that hypothesis. In fact only a tiny proportion of variations, supposed adaptations, have actually been tested in this way. However, that does not mean that the concept is a faulty one. There are many examples of taxonomic groups which have radiated to occupy diverse environments. This radiation has then been accompanied by striking diversification in some of the most obvious, especially vegetative, probably adaptive, features of the plants. So-called 'convergent' evolution gives us more solid evidence for adaptation than radiations do. There are the many striking examples of convergence, where similar looking structures have arisen separately in unrelated groups, probably because they have been selected for under the same environment. Convergence provides dramatic evidence for functionality, the adaptive nature of some characters. The ways in which convergent structures adapt the plants to their particular circumstances may not be fully understood, but adaptive they surely are.

Structures and mechanisms are adaptive if they manifestly fit the organism to its environment, that is if they aid survival and reproduction. However, it is important not to slip into teleological thinking. Teleology is a theory that all things were designed to fulfil a purpose. Not all differences between species are necessarily adaptive and in many cases close examination of an 'adaptation' has also revealed its limited importance to the individual and the complexity of its relationships with other variations in structure. There are limits to what can be understood by the **reductionism** of studying an adaptation in isolation. Adaptive value is contingent on a particular set of circumstances; on all the other characters of the individual to which the adaptation is tied and on the place and time in which the individual exists. The circumstances change as the individual grows or its surroundings change. The surroundings include other individuals of the same or of other species, which are also changing. The state of 'being adapted' is in constant flux.

Likewise, the process of adaptation of a species is unending because the environment is constantly changing, both physically, and biologically as other species become more effective at competing for the same resources. Each species is positioned within a complex network of relationships which changes over time. Evolution is unending because of what has been called the 'Red Queen hypothesis'. The Red Queen says to Alice: 'it takes all the running you can do to keep in the same place'. And so an evolutionary lineage which ceases to run, ceases to evolve, and is always threatened with extinction if its circumstances change.

It is not clear how much of the variation observed is non-adaptive, with little or no effect on the survivability or reproductive success of a plant. It might be questioned, if two very different-looking plants can grow alongside each other successfully, how many of the differences between them are of adaptive significance? However, this is a false question because the environment of a plant does not just consist of its geographical location, with its particular climate and soil, but is in part defined by its own **characteristics**. A plant with small leaves and shallow roots may actually be growing in quite different conditions to one growing next to it with large leaves and deep roots. These two sets of characteristics may represent two different but equally successful strategies for growing in the one location.

Alternatively a single characteristic may be an adaptation to a range of evolutionary situations. An example is found in the 'hedgehog' plants of the Mediterranean steppe vegetation (Fig. 1.7). These plants are small, rounded, woody shrubs, cushion plants with many thorns or spines. There are several species with the same form which have evolved convergently in a number of quite different families. As is often the case with convergent evolution the similar looking structures have arisen in quite different ways. For example spines have evolved from branches in *Launaea cervicornis* (Asteraceae) and *Teucrium subspinosum* (Lamiaceae), from the leaf petiole and rachis in *Astragalus balearicus* (Papilionaceae) whereas *Smilax aspera* (Liliaceae) has prickles. The value of the adaptation seems obvious, namely as a protection from grazing. However, the thorns also protect the plant from the physical environment; from cold and/or dry winds and/or from the very high light levels of the exposed environments where the plants grow. The cushion growth form also provides a high moisture environment and shade for the leaves. It also maximizes nutrient storage in woody tissue in the highly seasonal habitat, and it aids litter retention for nutrient recycling.

Perhaps all of these adaptations are important, but not all of the time, or in every situation. The different species have slightly different **microhabitat** preferences and the characteristic 'growing as a hedgehog plant' may have a different value, a different purpose, for each of them. Adaptive values are also different for different individuals of the same species growing in different areas. The adaptation, 'growing as a hedgehog plant' is certainly not the only possible one in the Mediterranean steppe environment. Other quite different adaptive forms are exhibited by other species. There are large

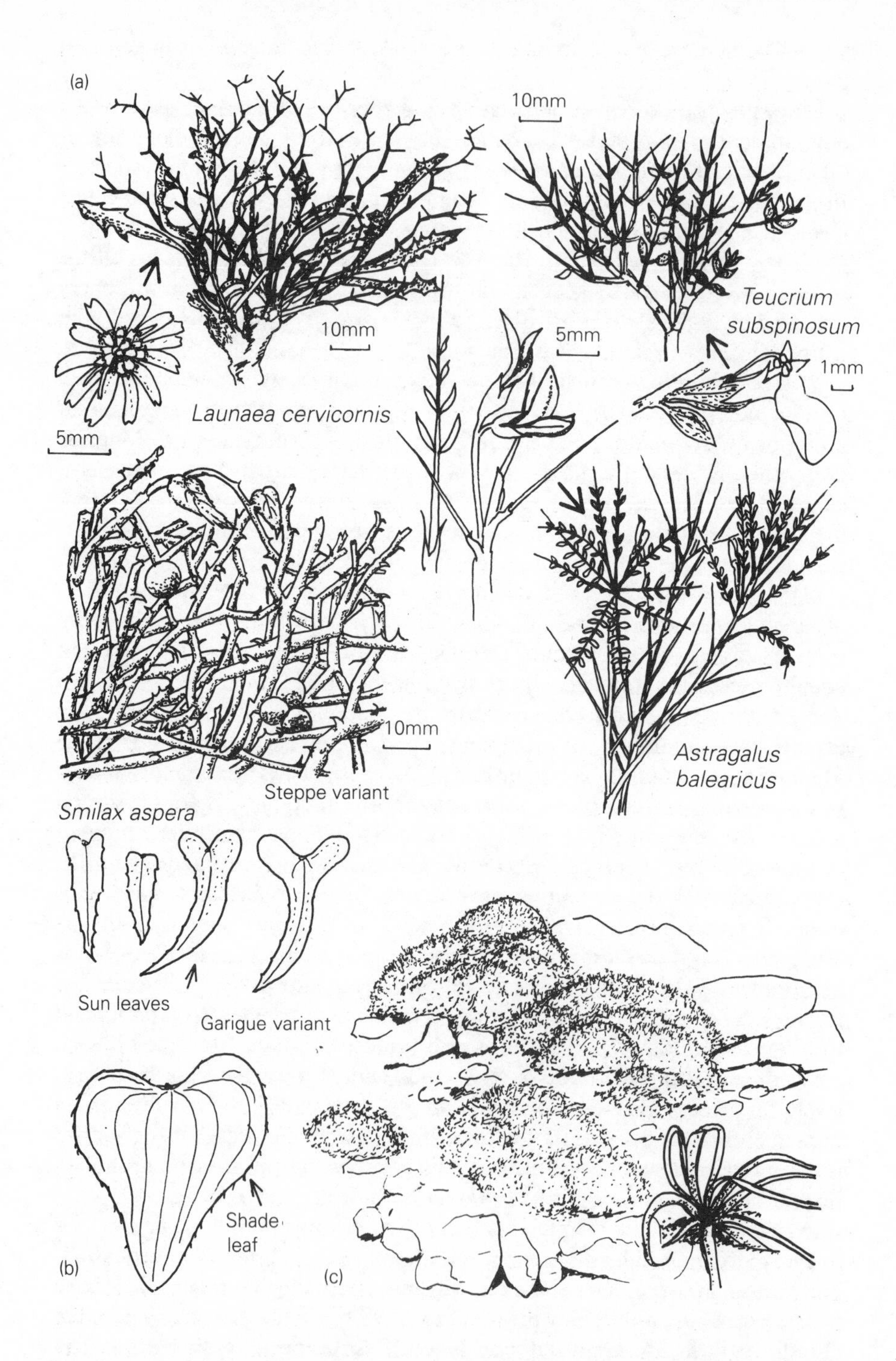

Fig. 1.7. Convergent evolution of spiny cushion plants of the mediterranean steppe: (a) four species from Mallorca; (b) sun and shade leaves from garigue variant of *Smilax aspera*; (c) general view of the vegetation.

sclerophyllous(hard-leaved) shrubs, delicate bulbous plants, herbaceous perennials and annuals.

In the 'hedgehog plants' growth form is more strongly genetically determined in some species than others. One species which grows as a hedgehog plant is called *Smilax aspera* (Fig. 1.7). In sheltered and shady circumstances it normally grows as a climber, and there is a great deal of variation in leaf size and shape depending upon the degree of exposure of the leaf. As a hedgehog plant, it almost lacks leaves. It is not clear how far the steppe variant and the woodland variant differ genetically but there is a large measure of plasticity in the growth of both forms though they may be genetically different and are recognized as different taxa. Another species of hedgehog plant, *Astragalus balearicus*, has the same form when it is growing by the coast, on the top of mountains or even in the shade of a mature oakwood. It is possible that many characteristics of species are not adaptations at the time at which they are observed though they may have had an adaptive value at an earlier stage in the evolution of the species. Later they might have become less important and were retained only because they were part of the development process of newer more important adaptations or just because they were not disadvantageous.

In the long run evolutionarily successful species are those which manage to preserve as many variations, potential adaptations, within the species as possible, whether they are presently adaptive or not. In many cases it seems that species which have been successful in a changing environment are those which in some way are pre-adapted to the new set of conditions. A good example is the moss flora of the polar regions. Many of the polar mosses are species also found in areas with less extreme environments (Longton, 1988). It seems likely that they evolved either before or during the Tertiary period at a time when their present-day tundra niche did not exist, when cycads were growing in Alaska and Antarctica. Coniferous and deciduous forests persisted well into the late Pliocene in high latitudes. When the trees were eliminated a few of the mosses which were already around survived to become the dominant plant form. They were preadapted to resist very low temperatures and long periods of darkness.

A fundamental criticism of adaptationist hypotheses is that they do not allow the possibility that many variations are not adaptive in a Darwinian sense. The adaptationist approach is essentially teleological, seeking to find the purpose of every biological structure. Characters may have been shaped by other forces, environmental modifications of a plastic form, by developmental constraints or by evolutionary accidents, such as genetic drift in small populations.

It is not surprising if patterns of variation appear complex because we do not understand all the processes at work. Attempts to understand patterns of variation are bound to be difficult. Even attempts simply to describe natural patterns of variation in the ordered hierarchy of a classification are bound to have their limitations.

2 The plant body: plant behaviour

2.1 PLANTS ARE DIFFERENT FROM ANIMALS
2.2 THE ENVIRONMENT OF PLANTS
2.3 PLANT GROWTH AND BEHAVIOUR

SUMMARY

This chapter is about the basic structure of land plants and the way plants behave. It is easy to forget that plants are living organisms. We associate movement with life. Plants seem static. Some plants look like pebbles but they are not stones. If plants do not have life surging through them like animals, it trickles through them in a constant stream. Plants are actually very strange living creatures indeed.

2.1 PLANTS ARE DIFFERENT FROM ANIMALS

Plants are autotrophic, photosynthesizing their own sugars from carbon dioxide and water, trapping and using light energy with the pigment chlorophyll. Land plants are sedentary, usually rooted to one spot for the whole of their life cycle. Plant structure is simple with relatively few cell types. Plants have no elaborate excretory system. Plants do not have a nervous system. Most plant cells are connected by protoplasmic connections, **plasmodesmata**, so that the plant body forms a **symplasm** or symplast.Plants cells are surrounded by a rigid cell wall made of cellulose and other materials. Plant structure is simple and growth occurs by the repetition of a few basic parts like leaves. Separation of somatic cell lines and germ cell lines occurs very late in development. Plants can reproduce very effectively **clonally**. Plants are developmentally **plastic**. Mature plants of the same species may have very different sizes and shapes. Plants may have indeterminate growth. Plants are nearly potentially immortal with for example, some bristle cone pines, *Pinus longaeva*, over 11 000 years old.

2.1.1 Plant cells and tissues

Different types of plant cells can be distinguished by the shape, thickness and the constitution of the cell wall, as well as by the contents of the cell. As human beings we take advantage of the diversity of cell types. There are the storage cells of food plants which contain starch and protein. There are the unicellular hairs (trichomes) of fibre plants, up to 6 cm long in cotton

(*Gossypium*), and stem fibres of hemp (*Cannabis*) and flax (*Linum*), which we spin and weave into cord and cloth. There are the wood (xylem) cells of trees which we use in construction and for making paper. Very hard wearing paper used to make banknotes is made from cotton fibres.

The protoplasm of some cells, especially **xylem** cells, disintegrates at maturity leaving the 'dead', but still functional, rigid skeleton of the cell wall. Large parts of the plant body may be composed of these dead cells. Plant cells are diverse but just a few basic types can be recognized in most land plants. In different combinations they form the tissues of the plant.

2.1.2 The cell wall

The main component of the cell wall, but still only 20–40%, is cellulose. This is a structural material. It is a fibrous polysaccharide which is 'spun' through a rosette of cellulase synthase enzymes embedded in the plasma membrane, **plasmalemma** (Giddings *et al.*, 1987). The cellulose is anchored to the inside of the pre-existing cell wall. Each molecule of cellulose is a ribbon-like polymer of between 1 000 and 15 000 ß-glucose units up to about 5.0 μm long. Each elementary microfibril of cellulose, which has a diameter of 3.5 nm, contains about 40–70 molecules spun off together in a highly H-bonded structure of great strength. The elementary microfibrils may be arranged in patterns of great complexity, the result of many 'spinning' rosettes acting together in an array. Larger fibrils, 20–25 nm in diameter, form chains or rods, with alternating dense crystalline areas, called micelles and linked by unordered fibrils. Fibrils may be further combined to form macrofibrils up to 0.5 μm in diameter.

It is the orientation of the various fibrils and the composition of the many other substances, such as the hemicelluloses, which fill the spaces between them, which confer the mechanical characteristics of each cell (Fry, 1989). These substances, which include proteins, act either as glues or lubricants. Important constituents are the pectic substances which glue cells together at the middle lamella. Some cell walls, especially in wood, become impregnated with a matrix of lignin, a complex polymeric resin which helps to bind the fibrils together, and prevents them shearing apart. Other cell walls are impregnated or encrusted with the fatty substances, cutin or suberin, making them impermeable to water.

The composition of the cell wall is not fixed. Growth-promoting plant hormones make the wall extensible. In the maturation of fruits, cells become less firmly glued to each other. In **abscission**, cell walls break down. As a response to fungal infection or incompatible pollinations materials are laid down in cell walls to prevent invasion. Components of the cell called oligosaccharides, which are derived from partial breakdown of cell wall polysaccharides, can act as plant hormones, mediating cell-to-cell signalling (Albersheim and Darvill, 1985).

2.1.3 Parenchyma, collenchyma and sclerenchyma

The primary cell wall, which is formed by all plant cells, tends to have a rather disorganized net of cellulose fibrils, with many fibrils oriented transversely to the main axis of the cell. The cell can stretch and expand as it grows. A secondary wall may be produced. Those cells with only a primary cell wall are called **parenchyma** or **collenchyma** cells (Fig. 2.1). In parenchyma cells, the cell wall is uniformly thin except in areas with dense **plasmodesmata**, the primary pit fields, where it is even thinner. Parenchyma cells can be many shapes, rounded, irregular or elongated.

Collenchyma cells have a thickened primary cell wall. They may vary from short and isodiametric ones to long and fibre-like. The cell wall may be thickened just at the angles of the cell or along one or more faces of the cell. The cell wall in collenchyma is rich in pectic substances. It is more organized than a parenchyma cell wall, with many layers of fibrils. The fibrils are arranged, in alternating layers, in parallel either transversely or longitudinally. The cell wall is plastic, allowing the cell to deform without splitting or snapping and yet give strength to the plant as it grows.

Parenchyma cells and collenchyma cells may make up tissues, called parenchyma or collenchyma respectively. Parenchyma cells are found in many other kinds of tissues and may be specialized in various ways to carry out different functions such as photosynthesis or storage. Two important parenchymatous cells are the sieve tube elements and companion cells of the phloem, which function in the translocation of nutrients. Collenchyma tends to be found only as a strengthening tissue in a peripheral location in the plant, but similar kinds of cells may be found elsewhere.

The other main cell type, **sclerenchyma,** has a secondary cell wall. The secondary cell wall is produced inside the primary wall, after the cell has elongated or enlarged. It makes the cell elastic, allowing it to deform, but returning to its original shape after the stress is removed. The fibrils are regularly arranged: in parallel to each other, mainly longitudinal to the main axis, and in a weave with alternating layers at different angles. The more acute the mean angle of the fibrils to the main axis of the cell the greater the stiffness of the cell. Hemp fibres, with a mean fibril angle of 3°, are four times as stiff as cotton **trichomes** with a mean angle of over 30° (Vincent, 1982). However, if the fibrils are orientated more transversely, the cell is less likely to break, because the helically wound fibres can buckle inwards rather than snap when put under strain. The secondary wall is often produced unevenly in bands, rings, helices or lamellae, conferring different characteristics to the cell; or more continuously, and then it is pitted. The secondary cell wall does not form over **pits** or primary pit fields but it sometimes arches over the pit as a dome with an aperture at its apex. Coming in pairs in adjacent cells these kinds of pits form a **bordered pit** pair. A tertiary non-cellulosic cell-wall may also be produced in some cells.

Many sclerenchyma cells lack a protoplast at maturity. They form a dead structural element in the plant and/or a conducting system in the xylem tissue. However even those sclerenchyma cells which are dead at maturity

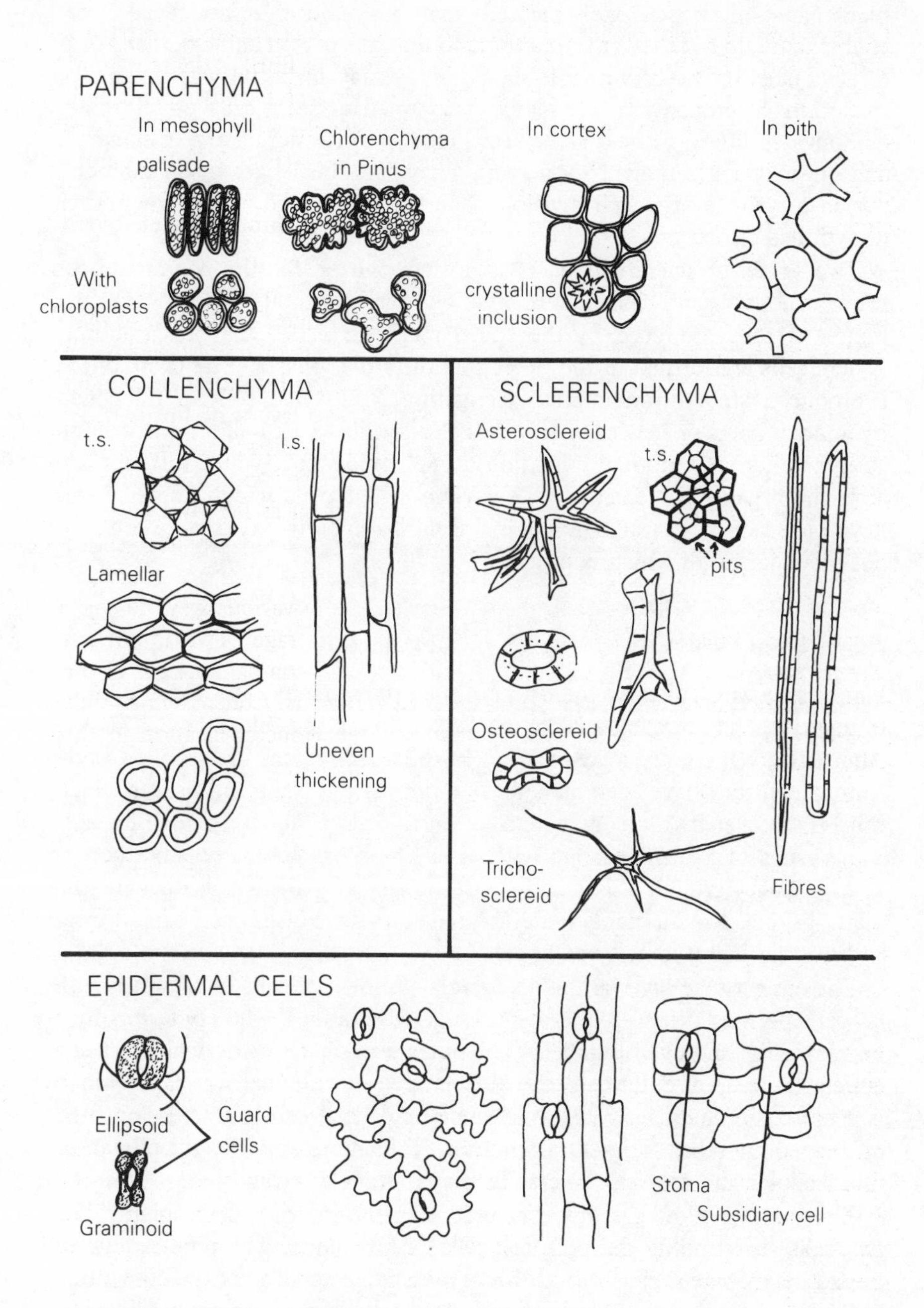

Fig. 2.1. Cell types.

may have a prolonged life before dying. They become the receptacles of plant metabolites. Sclerenchyma cells may be found in a specialized sclerenchyma tissue but they are commonly found interpersed among other kinds of cells in other tissues, as **idioblasts**. They may make up a large part of the xylem. Sclerenchyma cells may be isodiametric ('stone cells' or **sclereids**) or elongated fibres. Fibres can be very long, up to 10 cm in hemp (*Cannabis*), and up to 55 cm in ramie (*Boehmeria*). They gain this length by an extended period of growth after cell division. The secondary wall is laid down after growth has ceased.

Two kinds of sclerenchyma cells are specialized for the conduction of water; **tracheids** and **vessel elements**. The vascular tissue arises in a special region of cells produced by the apical meristem called the **procambium**. Xylem cells are formed in the procambium in two phases. The **protoxylem** is produced while the organ is still elongating. Protoxylem cells are thickened by annular rings or helices of secondary cell walls, which allows the cells to stretch. There is a gradual transition to **metaxylem** cells which differentiate after the organ has elongated. Metaxylem cells have a wider diameter and have wide secondary wall banding in a scalariform pattern or a more continuous secondary wall which is pitted.

2.1.4 Plant tissues

Plant tissues are called by the name of the predominant kind of cell present in them, and so there is parenchyma, collenchyma and sclerenchyma tissue. Alternatively there are mixed tissues in different regions of the plant body from which they have been named. The various cell types and tissues which can be discovered in a single stem of *Cucurbita* are illustrated in Fig. 2.2. The tissues of a plant change with age. The most notable change occurs when primary growth ceases and the secondary growth, which thickens the plant, begins. Not all plants undergo secondary growth. It occurs in all trees and is described in detail in Chapter 6.

The outer tissue layer is the **epidermis**. Epidermal cells fit together with no gaps between them except at pores called **stomata**. The epidermis may be very thick and even lignified. The outer wall of the epidermal cell has a cuticle made up of cutin and may also have an epicuticular wax layer which is sculpted in characteristic ways. Some epidermal cells may be elongated or shaped in other ways as trichomes. Trichomes can be unicellular or multicellular and are very diverse in shape. Some are glandular.

The stomata allow gaseous exchange with the interior of the plant. The two cells surrounding the pore are called **guard cells**. The pore is formed during development by a breakdown of the middle lamella between the guard cells. There are basically two kinds of guard cell: elliptic and graminoid. The cell walls are irregularly thickened and in addition often have a cuticularized ledge around the pore itself. The guard cells open and close the pore by expanding or contracting in volume (see section 7.3.2). The size and distribution of stomata vary both within plants and between species. The

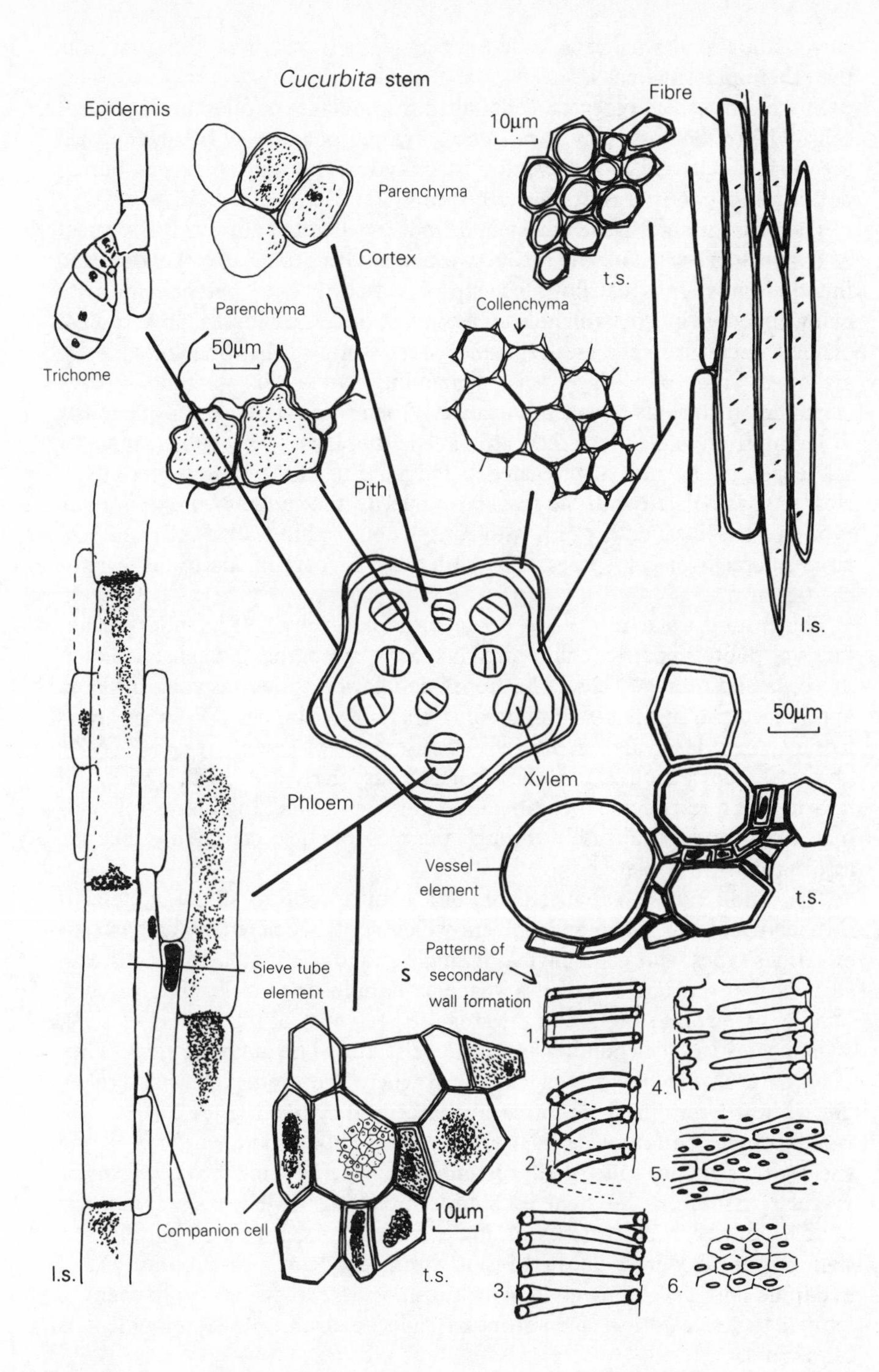

Fig. 2.2. Cells and tissues in a stem of *Cucurbita*. 1–6 illustrate different kinds of secondary wall observed in the xylem.

surrounding epidermal cells, called subsidiary cells, may look different from the other epidermal cells.

In the root some epidermal cells called **trichoblasts** develop into root hair cells. The trichoblasts may have a dense cytoplasm and may be smaller than the normal cells. They are evenly distributed in the root epidermis but do not necessarily form a **root hair** (see section 2.3.8).

The main ground tissue of a stem or root is called the **cortex**. It is present in both stems and roots and lies between **vascular bundles** or the **stele** and the epidermis. It is usually made up of loosely packed parenchyma with many airspaces but may include collenchyma or sclerenchyma. In succulent plants the cortical cells are large and water storing cells. Sometimes there are other kinds of storage cells, containing starch, oils, tannins or even crystalline inclusions of calcium oxalate. When these cells are conspicuously different from the others they are called **idioblasts**. The collenchyma or sclerenchyma is usually aggregated at the margins of the stem, especially in ribs, or is associated with the vascular tissue. In the centre of the stem there is often a tissue called the pith with conspicuously thin walled cells and very large intercellular air spaces. The pith may be continuous or have large cavities in it.

The ground tissue of the leaf is called the **mesophyll**. It includes various kinds of photosynthetic (chlorophyllous) cells, strengthening cells, conducting cells and storage cells. The upper and lower epidermis may differ in appearance and in the distribution of stomata or trichomes. Different layers like a spongy layer and an adaxial palisade layer of columnar cells may be observed (see section 2.3.4). The leaf has vascular traces in the veins and midrib which runs down the stalk of the leaf: the petiole. Idioblastic sclereids may be distributed diffusely through the mesophyll or commonly they are associated with veinlet endings.

The conducting or vascular tissue of a stem or root is called the **stele**. It is made up of the xylem and phloem which contain sclerenchymatous cells of various types, and parenchyma. A single strand of the stele in the primary stem of many plants is called a **vascular bundle** or fascicle. The vascular bundles are arranged in a ring towards the periphery of the stem (Fig. 2.3). In monocots vascular bundles appear to be scattered through the stem. They often have an obvious protoxylem canal where the protoxylem has broken down. In the root the stele forms a solid central cylinder (Fig. 2.4).

Meristems are found particularly in buds. They produce differentiating tissues. One tissue, called the **procambium**, differentiates into the primary vascular tissue. In the stem each procambial strand becomes a vascular bundle. In the root there is a single central procambial strand. In the stem the protoxylem is located inside the metaxylem, a distribution called **endarch**, and development is centrifugal. In the root the protoxylem is located to the exterior, a distribution called **exarch**, and development is centripetal.

The **vascular cambium** is a special meristem which produces secondary vascular tissues (section 6.1.1). Vascular bundles are of two sorts; open or closed. Closed bundles lack a cambium and are unable to produce secondary

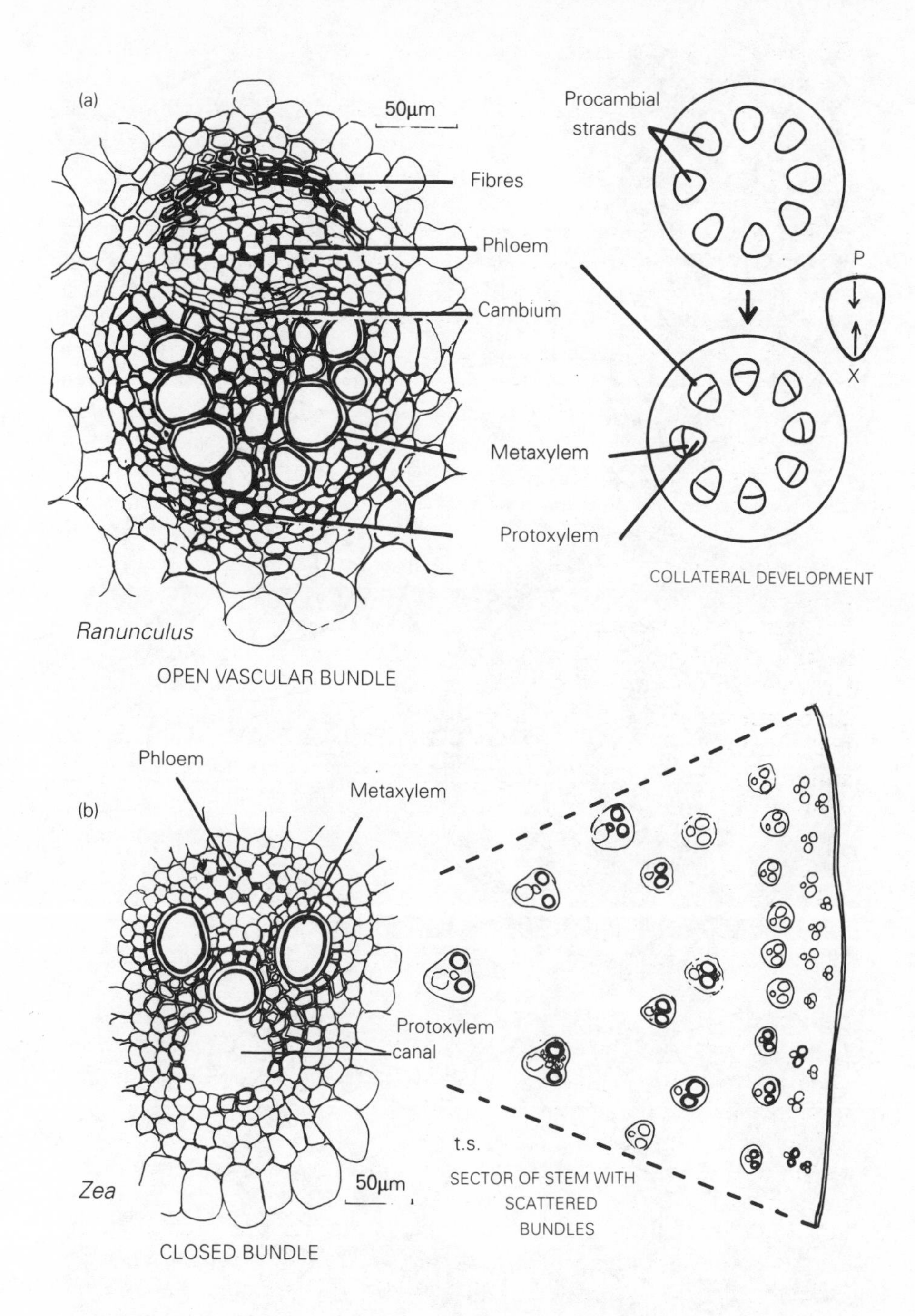

Fig. 2.3. The stele in the primary stem. (a) Shows the typical pattern of endarch development in a ring of 'open' vascular bundles in a dicotyledon, e.g. *Ranunculus*. (b) Shows the typical scattered 'closed' bundles of a monocotyledon, e.g. *Zea*.

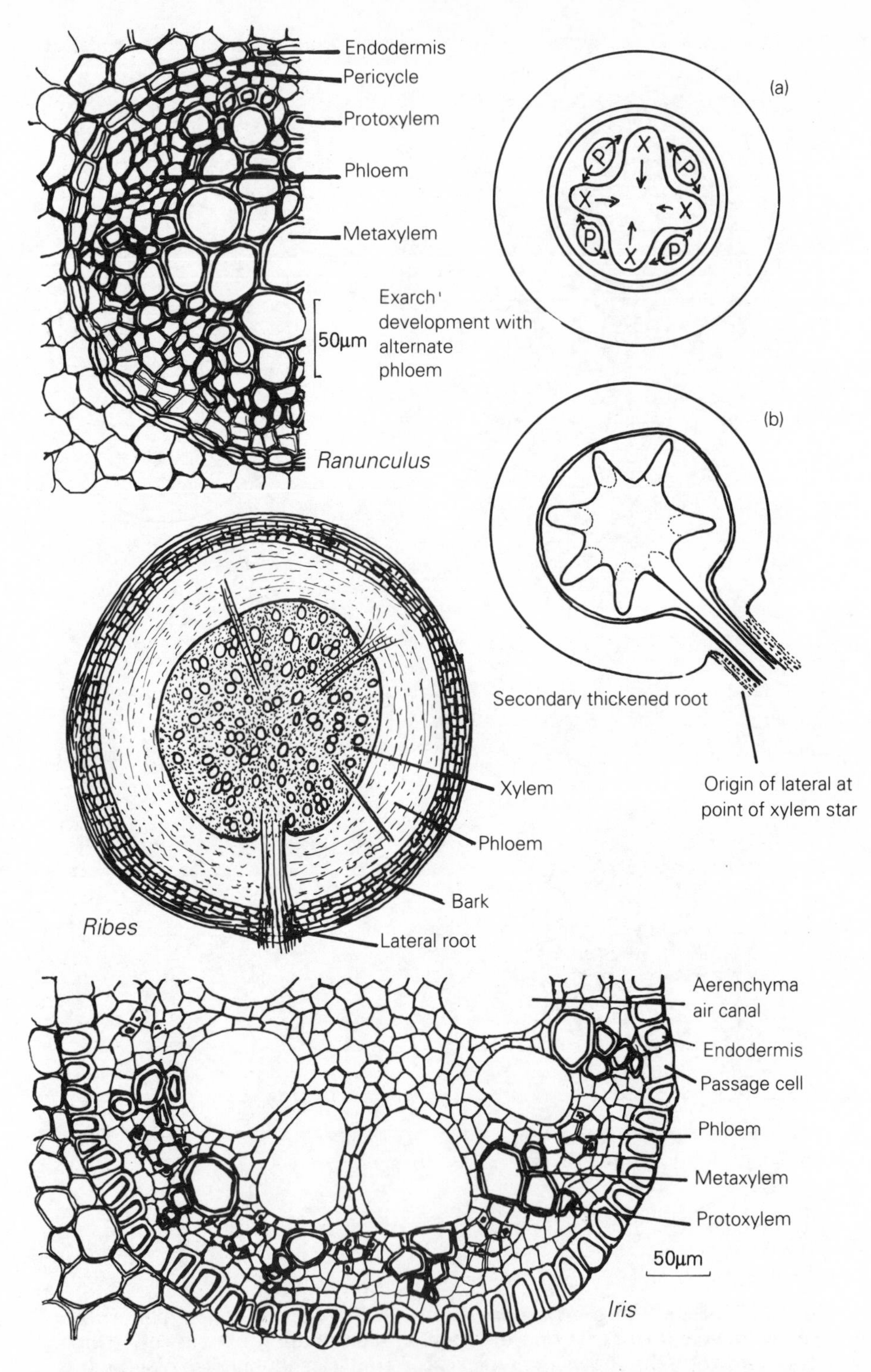

Fig. 2.4. The stele in the root. (a) Alternate and exarch development of the stele; (b) endogenous origin of a lateral root in a heptarch root. *Ranunculus* has a tetrach stele (half drawn). *Ribes* is a secondarily thickened root. *Iris* has a polyarch stele (half drawn).

vascular tissue. Open bundles can grow by the cambium dividing to produce secondary xylem and phloem.

In some plants there is an obvious layer of different cells surrounding the vascular tissue called the **endodermis**. The cells in the endodermis are irregularly thickened. There walls may be thickened, lignified and suberized, just on the lateral (anticlinal) walls in lines called casparian strips, or the two lateral walls and the inner wall are thickened. Thin walled **passage cells** are sometimes retained in positions opposite the protoxylem. The endodermis is less obvious in the stem and less obvious in aerial stems than rhizomes. It is sometimes called the starch sheath because in the developing stem it has abundant starch. The endodermis controls passage of solutes in and out of the vascular tissue. A layer of parenchymatous cells called the pericycle separating the endodermis from the stele may be recognizable.

In some plants like mosses and liverworts the structure of the plant is very simple. Leaves may only be one cell thick (**unistratose**). There may be few obviously different tissues. Even a clearly differentiated vascular system is not present, hence they are sometimes called non-vascular plants. They have also been called 'lower plants' because of their apparent simplicity.

2.2 THE ENVIRONMENT OF PLANTS

Each land plant lives in two distinct physical zones: the aerial zone and the subaerial zone. The subaerial zone is the soil in terrestrial plants, water in aquatic plants, or, in some epiphytic and all parasitic plants, within the tissue of other plants. The subaerial environment varies in water content, chemical composition, nutrient availability, availability of oxygen, presence of other organisms, and in the structure of the physical matrix. The aerial environment can be measured in terms of light, temperature, humidity and the presence of other organisms. Both physical zones vary over time as well as spatially.

The relationships of a plant to its environment are extremely complex, changing over a single day, over the seasons and over the lifetime of the plant. They are further complicated because they include not just the physical parameters of climate and soils but biotic factors, the complex net of relationships to other living organisms. There is competition with other plants of the same or different species for resources. There are interactions with micro-organisms in the soil, on the plant surface or within its tissues. There are relationships with animals as herbivores, dispersal agents or cultivators.

Unlike animals, plants have a very limited ability to choose their environment. They cannot move to a more favourable area. However it is important to remember, that plants are not passive acceptors of the physical environment. They can react to it, for example by changing the orientation of their leaves, or by growing. In many ways plants create their own environment. They absorb carbon dioxide and release oxygen. They create shade and modify air currents. By opening their stomata and increasing transpiration, plants can markedly reduce the temperature of their leaves. The vegetation is an important determinant of climate, especially the microclimate of the aerial zone. With the destruction of rainforests we are beginning to realize the importance of vegetation for global climates.

The environment of the soil itself is formed by a complex interaction in which the living plant plays an important part. Transpiration by the plant transports water from the soil to the air. The organic material from decomposing vegetation is an essential component of the soil matrix. It stabilizes soil aggregates and modifies the chemical behaviour and water retention characteristics of the soil. The rate of colonization of purely mineral soils like dune sands, volcanic ash or glacial sediments by plants can be studied by recording the degree of incorporation of organic material.

Two different axial systems have evolved in land plants to exploit the aerial and subaerial zones; the shoot and root or rhizoidial systems respectively.

2.2.1 Roots and shoots

Almost all land plants, apart from terrestrial algae, have a shoot system though it may be very simple, small or modified. The thallus of thalloid liverworts (section 7.2.1) seems to be closer to the structures of some algae than a shoot system but its relationship to a more normal shoot system can clearly be seen in a group like the Metzgeriales which includes both thalloid and leafy liverworts. Most land plants have a root system, though roots were not present in early land plants and never evolved in bryophytes or whiskferns. They have rhizoids which are unicellular or multicellular filaments. Rooted plants may have rhizoids at one stage of the life cycle. Roots may be absent in some plants after having been lost by evolutionary adaptation, or may be present only early in development and replaced later functionally by modified stems. The root and shoot systems of plants adapted to different modes of life are often highly modified. This can be seen in the aquatic plants such as the Podostemaceae, or parasitic plants like the Balanophoraceae.

Developmentally, roots and shoots arise in different ways. A plant has a primary shoot and root system which are connected by a transitional region the **hypocotyl**. The shoot and root develop from the counterpart embryonic systems, called the **plumule** and **radicle** respectively. It is common in later development for roots also to be produced directly from a shoot. These are called **adventitious** roots. The grasses are one group in which the primary root system is replaced almost entirely by adventitious roots. It is also possible for roots to give rise to shoots. In some plants the primary shoot aborts early and is replaced by secondary shoots arising from the roots or hypocotyl.

The stem combines the function of mechanical support and arrangement of the photosynthetic system and reproductive organs, with the conduction of water, mineral ions and photosynthates. The shoot system is not always photosynthetic. The aerial shoot system of some parasitic plants such as *Orobanche*, and saprophytes like *Monotropa*, does not photosynthesize. In many plants the shoot system can grow horizontally either in the soil, as a

rhizome, or at the soil surface, as a **stolon**, spreading the plant laterally. The shoot system of some plants can be modified as a storage organ; either a tuber, bulb or corm.

The root system combines the functions of searching for and exploiting the mineral nutrients and water of the soil, conducting these to the shoot system, with the anchoring of the shoot system. Some roots are contractile; they shorten, concertina fashion, to pull the base of the plant under the soil. Contractile roots are especially common in plants with bulbs and corms. Roots usually arise in the soil but aerial roots are also quite widespread. The roots may act as storage organs forming either a swollen, fleshy tap root or root tubers. In many species, perhaps most species, the functioning root system should not be viewed in isolation from its fungal associate, the **mycorrhiza** (section 7.5.2). Together they exploit the soil.

2.2.2 Transport

Intercellular communication in plants can occur through the **symplasm** via plasmodesmatal connections between cells or by the movement of water outside the plasmalemma in the **apoplasm** (apoplast). Plant cell walls provide a continuous apoplasmic network reaching throughout the plant restricted only where the cell walls are suberized, as they are in the casparian strips of the endodermis. The resistance to water movement is 10–100 times greater in the symplasm than the apoplasm. In the **stele** the plant has two specialized transport systems, one apoplasmic and one symplasmic; the **xylem** (apoplasmic) for the transport of water and the **phloem** (symplasmic) for the transport of dissolved solutes, sugars and, in woody plants, mineral ions.

The phloem contains sieve elements, either **sieve cells** or **sieve tube members** and **companion cells**, as well as other parenchyma cells and fibres. The sieve elements have cell walls which are not lignified. In some cases strength is conveyed by the presence of phloem fibres. Dissolved sugars are transported in the phloem protoplasm. The sugars are transferred from cell to cell across areas in the cell wall called sieve areas or sieve plates. These are modified primary pit fields where the connecting strands differ from normal plasmodesmata by being much thicker and having the pore lined with a cylinder of a carbohydrate called callose. Companion cells arise from the same meristematic cell as the sieve element to which they are closely attached by many plasmodesmata. Companion cells are not present in all plants but their place is taken by other parenchyma cells, which are called albuminous cells in gymnosperms.

Water is conducted in the xylem in cells called **tracheids** or **xylem vessel elements**. There is a **transpiration** stream: as water is lost from the aerial parts of the plant by transpiration, it is drawn into the roots of the plant because of the water potential gradient which has been created. A continuous column of water is drawn through the plant like a rope of many threads. Mineral nutrients are transported in the flow of water.

Water can also be forced up the xylem by root pressure. The accumulation of solutes in the xylem sap creates a negative osmotic potential drawing in water from the root cortex. The increased volume of water in the xylem is forced up the plant. This process is important at night when transpiration is not occurring and in Spring in deciduous trees.

In plants of very humid environments and in aquatic plants where transpiration is impossible, mineral nutrients are transported in an active, energy-requiring way. In parts of the plant especially in the leaves, there are secretory areas called **hydathodes**. A water potential gradient is created by the activity of **transfer cells** in the hydathodes and mineral nutrients are carried in the flow. Water is actively pumped out of the stele so that it escapes through modified stomata to the exterior. In terrestrial plants this process, called guttation, may be observed as the appearance of small drops of pure water on the leaves of the plant.

2.3 PLANT GROWTH AND BEHAVIOUR

Movement is a feature of animals but it is not quite true to say that only in science fiction do plants move. When touched a wave of movement passes over the sensitive plant (*Mimosa pudica*), as it closes its leaves (Fig. 2.5). This is very un-plant like. The bladderwort, *Utricularia*, responds very rapidly to the touch of its *Daphnia* prey. The lid of its bladder flips open and its prey is sucked inside to be digested. The suddenness of the response in these cases is allowed by an electrical signal, an action potential like that in animal nerve cells, passing along parenchymatous cells linked by plasmodesmata. The signal causes sudden changes in the permeability of the plasmalemma of the target cells. Water floods out and hinge cells change shape. In *Mimosa pudica*, special pulvinus cells at the base of the leaflets rapidly lose turgor so making the leaf close. In *Utricularia* the catch on the trap door is sprung. Sudden explosive movements are fairly common in seed and spore dispersal though these are not triggered electrically. Changes in turgor or pressure within a closed plant organ mediate most of them. In *Utricularia* active excretion of water from the bladder resets the trap. There are also a broad range of slow plant movements, tropisms, which can be revealed by time-lapse photography. Some movements are more or less reversible. These are mediated by changes in turgor though less explosively than the ones described above. There are heliotropic responses, orientating flowers or leaves to the light or opening and closing of flowers, especially at dawn and dusk.

These reversible changes are not the main way in which plants behave. From our zoocentric point of view it is easy to forget that the way plants behave is by irreversibly changing the pattern of growth. A single stimulus can bring about the change. In an experimental situation short periods of wind and fine rain or even a gentle touch, brought about irreversible changes to the growth of *Arabidopsis* plants (Braam and Davis, 1989). Receptors in the cell wall or plasmalemma trigger a response by activating ion channels. Plants show some similarities with animals in this; they both have long distance signalling, both electrical and hormonal and in the *Arabidopsis* experiments, calmodulin genes were switched on. Calmodulin is well known

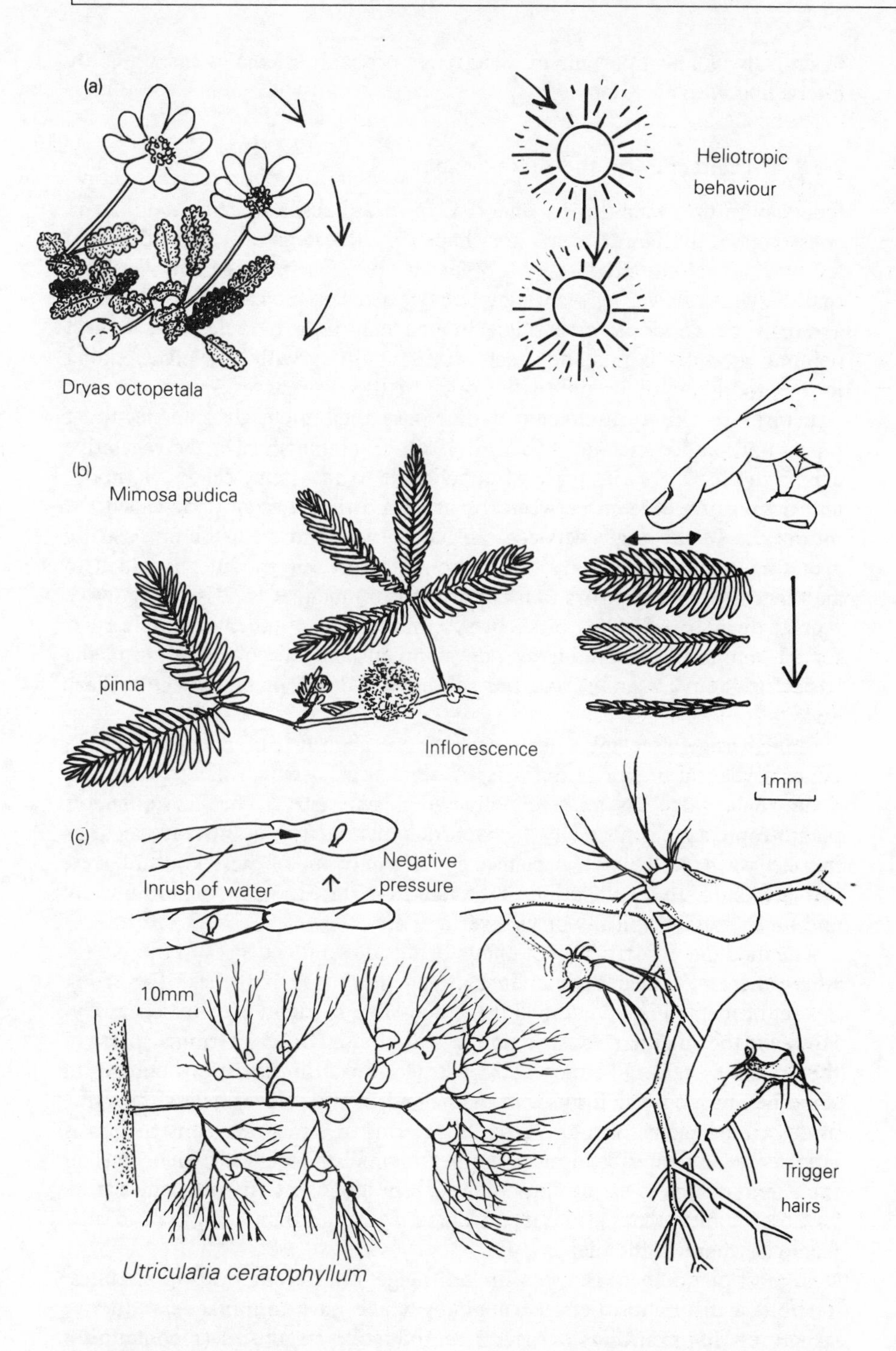

Fig. 2.5. Plant tropisms. (a) *Dryas octopetala*, heliotropic behaviour; (b) *Mimosa pudica*, thigmotropic response; (c) *Utricularia*, the carnivorous bladderwort, thigmotropic response.

in animals as an important messenger. It probably mediates responses by interacting with Ca^{2+} ions.

2.3.1 Architecture of the shoot system

One way to understand plant shoots is to regard them as being modular in construction. Different forms are shaped by the degree of repetition and arrangement of the modules. **Modules** can be identified at many levels so that the whole flowering plant may be regarded as a module, or a branching axis may be a module or a single branch may be a module. One way of defining modules is that they each represent the growth of a plant ending in the production of a reproductive organ, either an inflorescence or another structure. In this terminology modules are then themselves made up of repeat units called metamers (Barlow 1989). Each metamer of the vegetative axis is the leaf, plus its place of attachment to the stem, called the **node**, and the portion of stem between the node and the adjacent node called the **internode**. In the angle between the base of leaf and stem there is usually a bud called the an **axillary** bud. One metamer, one module, is added to another, as the plant grows to form a stem or branch. It is possible to derive a great diversity of forms of shoot systems from a number of simple rules for the timing of the production, development and rate of extension of the basic leaf–stem metamer, and the subsequent development of the axillary buds.

Similar vegetative metamers may behave slightly differently becoming either a vertical or horizontal axis. A vertical axis, which is also called an **orthotropic** axis, has an essentially radial symmetry. The horizontal, or **plagiotropic** axis, is normally dorsiventrally flattened (Fig. 2.6) but can arise in other ways (see below). A plant may have only one of these kinds of axes, or both kinds. In a few species, a branch may start growing orthotropically and finish plagiotropically or vice versa.

The modular construction of plants breaks down to a degree in those trees where there is a clear distinction between trunk and branches. The trunk is a set of metamers so modified it is impossible to identify them separately. However the modular construction is still obvious in the terminal parts of the branch system. The modular construction is also difficult to observe in some highly modified forms and in thalloid plants. Nevertheless the value of describing plant form and growth in terms of a modular construction is emphasized by physiological experiments which have shown that most of the energy which goes into producing a reproductive structure comes from its close vicinity, that is from its own module, rather than being translocated from elsewhere in the plant.

In most plants there is a very limited range of different kinds of modules. There is a distinction between modules which have terminal reproductive structures (**hapaxanthic**) or lateral reproductive structures (**pleonanthic**). The former are determined, having **determinate** growth. They cease to grow after a while, often after producing a reproductive structure. The others have **indeterminate** growth. The axis is potentially immortal. A whole plant

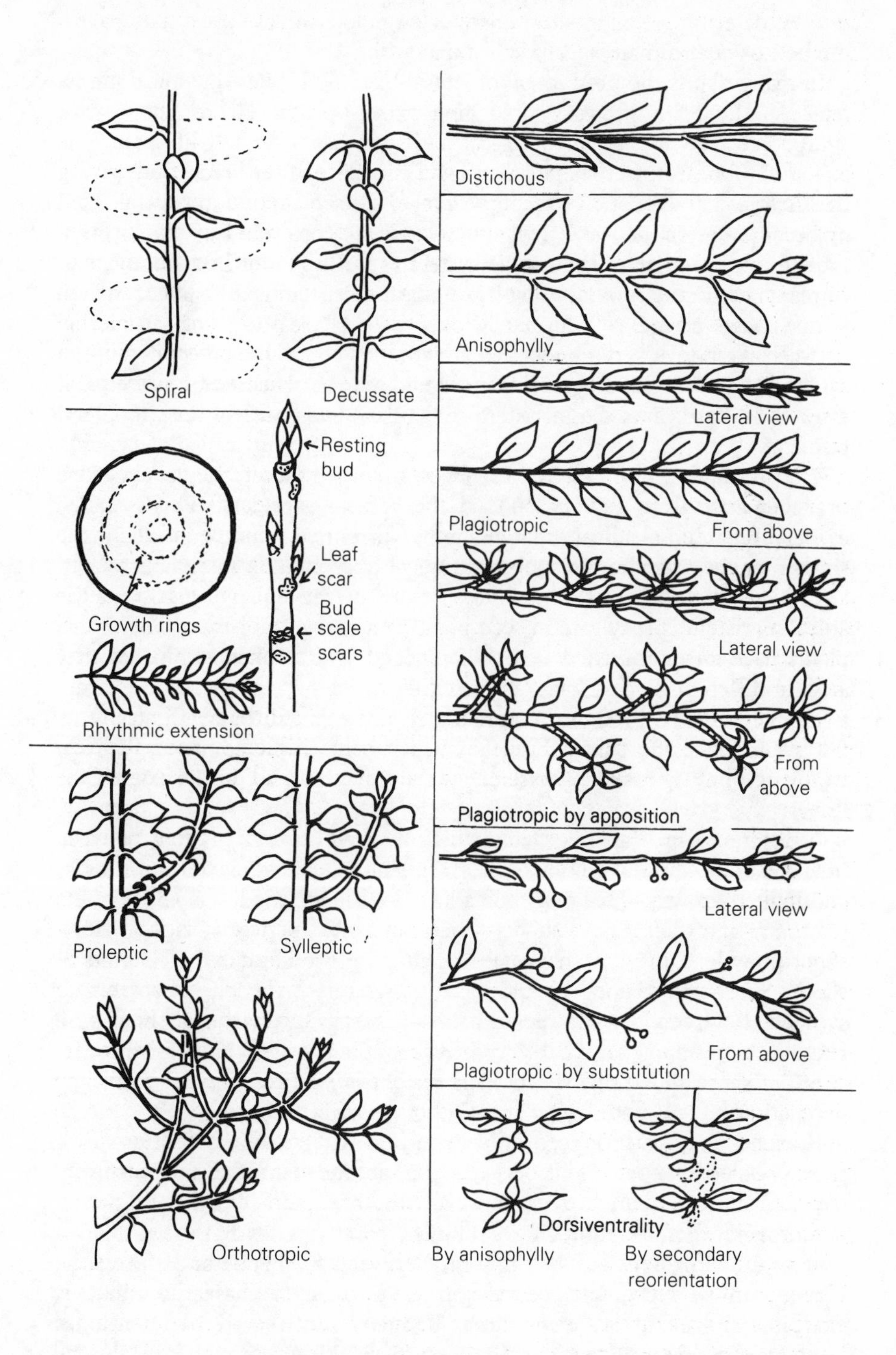

Fig. 2.6. The architecture of shoot systems (following Hallé *et al.*, 1978).

may be determinate or indeterminate or an indeterminate plant may have a mixture of determinate and indeterminate axes.

In many plants the production of lateral buds multiplies the apical meristem. As the plant grows a branching axis is produced. In plants with **sympodial** growth each shoot apical meristem has a limited lifespan. The axillary bud nearest to the apex then takes the place of the apical bud, giving the shoot a characteristic zigzag appearance. The apical bud may be aborted or become determined as a reproductive organ. This kind of **substitution** growth can occur in a plagiotropic system as well as an orthotropic system. In plants with **monopodial** growth the apical meristem is long lived, though it may have periods of dormancy over a harsh season. The distinction between sympodial and monopodial growth may be hard to detect because in sympodial growth the superseded apical bud may be obscured by subsequent growth. In both cases the meristems of axillary buds can give rise to lateral branches.

The arrangement of lateral branches follows that of the leaves since branches arise from axillary buds at the base of each leaf. This is called **axillary** branching. An alternative, rather primitive pattern, is where the shoot branches dichotomously by the apical bud dividing more or less equally. True **dichotomous** branching, involving the equal division of the apical meristem is very rare in seed plants but is seen in some palms. Some plants have an unbranched axis. In branched species lateral branches may arise in different places, for example either at the base or distally. Alternatively they may be produced continuously along an axis or rhythmically. A few plants have short-lived branches which behave like compound leaves. In the bryophytes branches arise below the leaves and not in the axils of the leaves.

A difference in branch systems which is often observed is that between **long** and **short shoots** (section 6.2.1). The short shoots have short internodes and limited growth. The long shoots have longer internodes. Some conifers like *Larix* and *Cedrus* have shoots which can switch between long- or short-shoot growth. Different shaped leaves may be produced by each kind of shoot. Short shoots may be produced as laterals on a long shoot or, in sympodial systems, by the replacement of forward growth of a shoot by a lateral. The terminal part of the original shoot then has very short internodes: it is the short shoot and the branch system is said to be plagiotropic by **apposition** (cf. plagiotropy by substitution described above).

Branches may be produced immediately so that growth is continuous, a process called **syllepsis**. This occurs in herbaceous plants and many tropical trees. In most temperate trees, however, the bud meristems have a period of dormancy before producing a branch, so that growth is rhythmic, a process called **prolepsis**. The difference between a sylleptic and a proleptic branch can be recognized because in the former the basal internode is elongated and in the latter there may be many very short basal internodes sometimes marked by bud scales or scars. Rhythmic growth can also be recognized by regular changes in leaf size and by the formation of growth rings in the wood.

In some plants the axis is very short so that the leaves are crowded together in a rosette. The leaves are arranged in a complex mosaic filling light space. Orthotropic shoots combine the functions of gaining height with catching light. Plagiotropic shoots act primarily to occupy light space. The leaf arrangement, or **phyllotaxis**, in orthotropic shoots is usually either spiral or in pairs opposite each other but with each successive pair at 90° to the former, called a **decussate** arrangement. In plagiotropic shoots the leaves may either be arranged distichously, strictly on two sides of the shoot, or have a spiral phyllotaxis, but in both cases they are arranged in one plane. If the phyllotaxis is spiral this involves reorientation of the leaves or the twisting of the shoot further back from its apex. **Anisophylly** may be present with the leaves held in a lateral position larger than those which are either above or below the shoot.

2.3.2 Diverse leaves

One of the most important areas of diversity in land plants is in their leaves and yet the first land plants did not have leaves. Leaves are specialized appendages for the reception of light. Almost all living land plants either have leaves or if leafless have evolved from plants with leaves by their secondary loss. The latter have a photosynthetic stem or, in the case of some epiphytic orchids, photosynthetic roots. Leaves have originated in many different ways. In some species there are structures called **cladodes** which look and function like leaves but are actually modified stems or branch systems (Fig. 2.7). The degree of resemblance to true leaves varies. In some cases as in *Phyllonoma* and *Ruscus*, it is difficult to be sure about the nature of the 'leaf' until it produces a flower or fruit on its margin or surface, something a true leaf never does. Fern fronds are leaves which bear the reproductive structures **sporangia**.

Each leaf arises from a leaf primordium produced from a meristem. In higher plants leaves have a bud in the **axil**, in the angle between the lower part of the leaf and the stem. Leaves are very diverse in structure and development. Commonly a stalk, the **petiole**, a midrib and a blade (the **lamina**) can be recognized. Some leaves are sessile, lacking a petiole. The function of the petiole is to hold the leaf away from the branch to maximize reception of light. Petioles are often flexible so that the leaf shakes in the breeze. This may aid cooling and deter insects from landing. The petiole is the site of **abscission**. The midrib is a continuation of the petiole into the blade. It is usually thickened and carries strengthening and vascular cells.

Some leaves have a clasping or sheathing leaf base or petiole. In the family Polygonaceae the leaf sheath is extended into a kind of collar called an **ochrea**. Grass leaves have two **auricles** and a **ligule** at the junction between the leaf sheath and lamina (Fig. 8.2). Many leaves have a **stipule** or pair of stipules arising below the leaf. Stipules are flaps or projections, sometimes leaf-like, which may serve to protect the developing bud and leaf **primordium**. Many aspects of a leaf vary, including the margin, the texture, thickness, hairiness, glandulosity, colour. Leaves may be modified to form

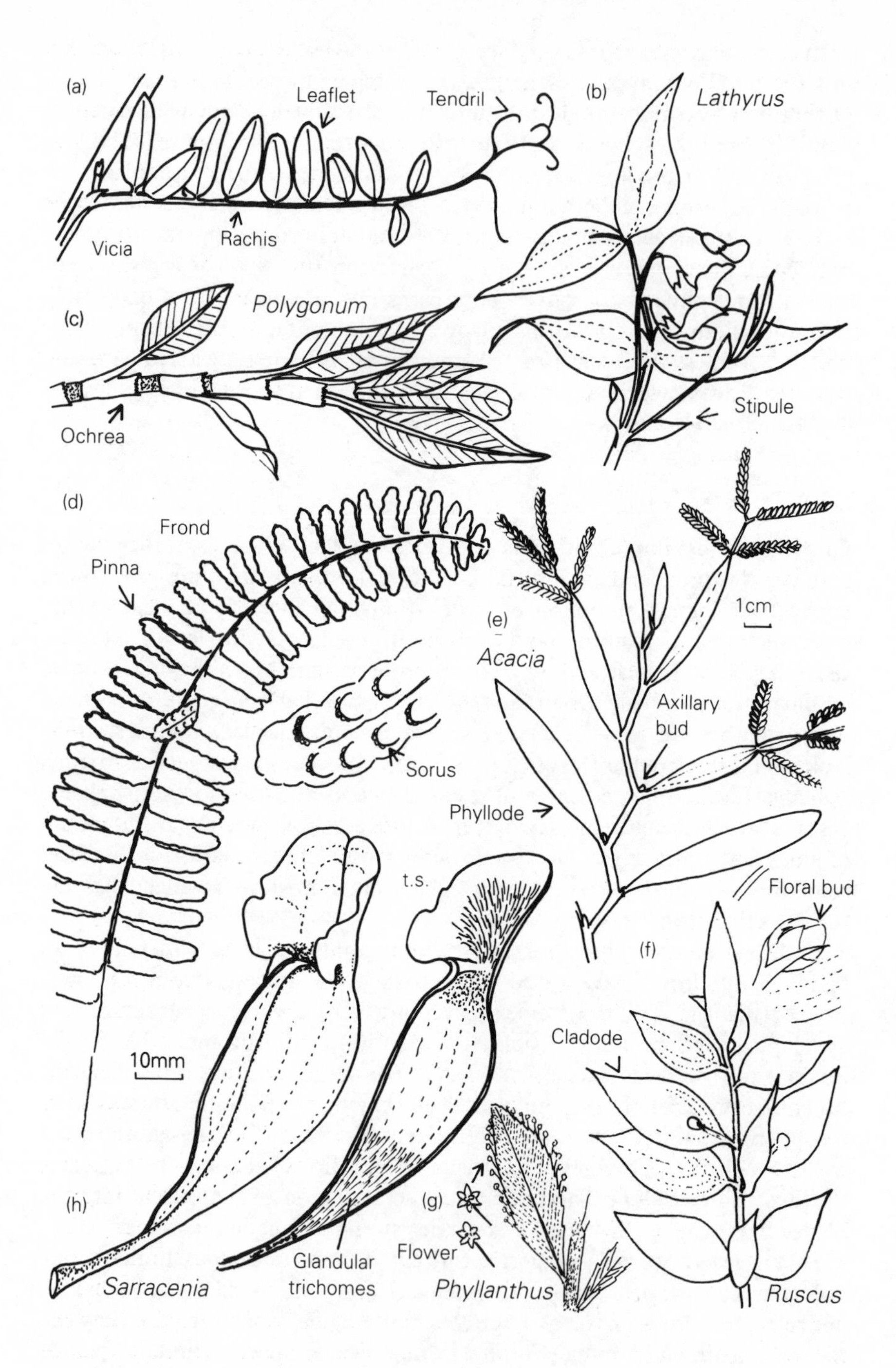

Fig. 2.7. Photosynthetic organs. Leaves in the broadest sense come in many forms. Leaves have been modified to form many other structures like the pitcher of *Sarracenia*.

scales, bracts, sepals, petals, thorns, tendrils, flasks, bladders (*Utricularia*, Fig. 2.5), storage organs, and root like structures. The last occur in the water fern *Salvinia* (Fig. 7.8). The leaves of carnivorous plants are particularly diverse (Fig. 7.12). Leaves vary in size from tiny scales a few millimetres in size to the huge compound leaves of some palms and ferns.

Compound leaves are divided **pinnately** or **palmately** into leaflets or segments. The segments of a pinnate leaf (**pinnae**) may be further divided (into pinnules) and subdivided up to five times. The midrib of a compound leaf is called the **rachis**. Each segment of a compound leaf arises from a separate primordium. This can be seen very clearly in the unfolding fern frond with its tightly packed primordia, in the fern crozier. Some plants have pseudo-compound leaves where leaf segments arise by the splitting along predetermined lines of weakness of a single lamina. Pseudo-compound leaves are found in banana, palms and the ubiquitous house plant *Monstera* for example. Compound leaves maximize the photosynthetic area while releasing the potential mechanical strains a single surface would suffer.

One thing almost all leaves, share is that they are **determinate** structures. With a few exceptions, leaves, even in evergreen species, have a limited life span, which is only one season in deciduous species. In the very long-lived bristle-cone pine (*Pinus longaeva*), leaves can live for up to 33 years. Unusually these leaves, and those of some of other evergreen species, have a capacity for secondary growth, though it is limited to the production of some secondary phloem. There are a few exceptional species with **indeterminate** leaves. For example the leaves of *Welwitschia*, a peculiar gymnosperm, grow continuously from a basal intercalary meristem while the apex erodes away. The basal intercalary meristem of many monocot leaves is different, in that it does not provide indeterminate growth. Another example of indeterminate growth is found in the fronds of some climbing ferns like *Lygodium*. These behave like shoots by growing continuously from an apical meristem, producing pinnae to each side as they grow forward.

2.3.3 The venation of leaves

Leaves may lack any vascularization even if they have a thickened midrib or costa. The vascularization if present may be simple and unbranched. For example the needle like leaves of conifers have one or two central unbranched vascular bundles. Leaves with a broad lamina have complex patterns of venation. A primitive pattern, which is seen in many groups including the ferns and gymnosperms, is a dichotomously branching, open venation. The veins do not interconnect and have blind endings. Occasionally adjacent veins may anastomose, as seen in *Ginkgo* (Fig. 3.8). In some ferns there is a more regular pattern of cross-bridges between the veins giving a kind of reticulate pattern, though the underlying **dichotomous venation** is still obvious.

The leaves of flowering plants, especially dicots but including some monocots, and of the gymnosperm *Gnetum*, have a **reticulate** pattern of venation (Fig. 3.10). First there is the vascular bundle or group of vascular bundles in the midrib. These give rise to secondary veins which in turn give rise to tertiary veins and so on. The order or degree of lateral venation varies between different species. The finest veins surround areas called **areoles**. All cells of the mesophyll in an areole are very close to a vein. The system of venation is called closed because either the veins anastomose or, if they end blindly, the veins which give rise to them anastomose. In a reticulate closed venation each part of the leaf can be served by many different routes so that if any part of the venation is damaged, perhaps by herbivores, no part of the leaf is isolated.

The finest veins make up by far the greatest part of total vein length, 95% in *Amaranthus* (Fisher and Evert, 1982). Veinlets may be very simple consisting of only a few tracheids. The xylem is primary xylem because the finest veins are produced while the lamina is expanding so the tracheids have to be able to stretch. Phloem may also be present but sometimes it is only present in the previous order of veins closer to the midrib. The vein may be surrounded by an obvious sheath of sclerenchyma or parenchyma cells. **Transfer cells** are found as a sheath surrounding the phloem. They are **companion cells** or specialized parenchyma cells.

With dichotomous venation the water diffusion pathway to the mesophyll and palisade cells is long because the veins are often quite far apart. In the cycads and conifers, however, this is compensated by the presence of a transfusion tissue surrounding the vascular bundles. Short, wide, nearly isodiametric tracheids are arranged radially around the bundle. In the cycads there are two transfusion tissues, one between the vascular bundle and the endodermis and another between the spongy mesophyll and the palisade parenchyma.

Most monocots and some dicots have parallel venation. The veins run longitudinally down the leaf connected by thin **commissural bundles**. Alternatively there is a midrib from which veins run out laterally in parallel. The arrangement is related to the way the leaf expands, either longitudinally or laterally. In those with longitudinal venation there is a basal intercalary meristem. The potential for unlimited (i.e. indeterminate) growth by the activity of this meristem is limited because of the problem of conducting water into the lamina across this relatively immature region. In plants like *Yucca*, the leaves are so slow growing that cells start to differentiate within parts of the meristematic region, thereby maintaining a vascular connection. Nevertheless even in these species the potential for leaf extension from the base is limited to very young leaves.

The differences between the parallel venation of monocots and the reticulate venation of dicot leaves has led some workers to suggest that monocot leaves are derived from the petiole, or petiole plus midrib, of the dicot leaf (Kaplan, 1975). Leaves of this sort are known from other groups, where they are called **phyllodes**.

2.3.4 The leaf lamina

Many of the most important evolutionary developments in leaf structure have been in leaf anatomy (Chapters 6 and 7). The tissue between the upper and lower epidermis is the **mesophyll** (Fig. 2.8). The evolution of a spongy mesophyll tissue with a network of intercellular air spaces connecting with stomata was an important evolutionary step because it improved the uptake of carbon dioxide. The upper part of the mesophyll, the palisade layer, became specialized as a photosynthetic layer. Columnar cells maximize the interception of light. It is the combination of the mesophyll and palisade tissues which creates a more efficient photosynthetic unit than either tissue working alone (Parkhurst, 1986).

Leaves are supported by the turgor of their cells. In many cases extra support is provided by sclerenchyma in a **bundle sheath** around the vascular bundles. The bundle sheath may extend as a girder to reach either one or both epidermises. Non vascular fibre bundles may also be present. Often the margins of the leaf are also strengthened by the presence of sclerenchyma and a thick epidermis. Such **sclerophyllous** leaves are common in areas of seasonal drought. Sclerenchyma, sclereids as well as fibres help to make the leaf unpalatable to herbivores.

Usually the epidermis is different on the upper or **adaxial** surface from the lower or **abaxial** surface. The lower surface usually has stomata. Sometimes these are protected by being sunken in chambers, furrows or by the presence of hairs. The adaxial surface is usually simpler, presenting a smooth mosaic of epidermal cells. Stomata are few or absent. It sometimes has a thick waxy surface giving it a **glaucous** appearance. Some plants like *Eucalyptus*, hold their leaves vertically, in which case the distinction between abaxial and adaxial epidermises is not present. The epidermis may contain specialized cells many of which are modified from unicellular or multicellular trichomes. They are hair shaped, scale-like or globular. For example there are the glands of insectivorous plants, the salt glands of halophytes, stinging hairs and nectaries. Some trichomes seem to function as antiherbivore devices making the leaf unpalatable. Alternatively, trichomes may protect the leaf from too much sun. Silica cells in the epidermis are also protective. Trichomes are often associated with stomata, helping to prevent too much water vapour loss from the open pore.

2.3.5 The leaves of mosses and liverworts

The leaves of mosses and liverworts, are also often called microphylls, but they are not homologous either to the microphylls of pteridophytes or even to each other. They share the characteristic of typically arising in three ranks from a tetrahedral apical meristem cell. The existence of a group of thalloid liverworts (Family Fossombroniaceae) with the thallus lobed and folded into leaf like structures has suggested a mode of origin of liverwort leaves from a thallus or vice versa. The leaves of liverworts lack a central thickened midrib. A midrib (costa) is observable in many moss leaves but only in some

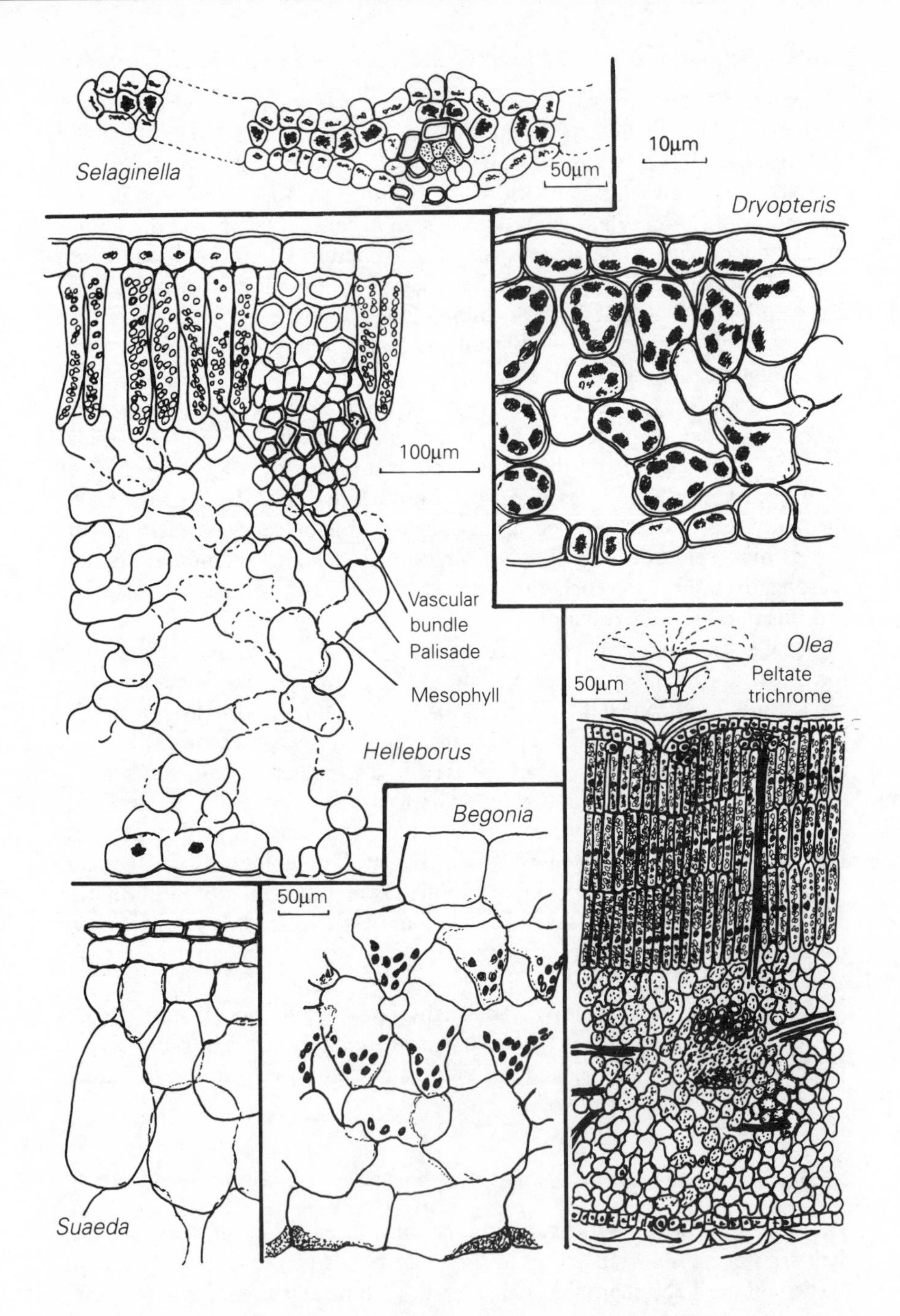

Fig. 2.8. Mesophyll variation. Most leaves have a palisade layer of columnar cells. In the sclerophyll of *Olea* there are three layers of palisade cells. In the shade-adapted *Begonia* the palisade cells are obconical. A palisade layer is absent in *Selaginella* and poorly developed in the fern *Dryopteris* and the succulent *Suaeda*.

does it appear to have a vascular function. The **leaf trace** arising from the midrib of the leaf may only just penetrate the cortex of the stem. Even in *Polytrichum*, a moss with a relatively well developed water conducting system, some of the leaf traces do not connect with the conducting tissue in the centre of the stem. The leaf trace may act as a kind of wick or be simply for anchorage and support of the leaf.

The lamina of moss and liverwort leaves is typically very thin, usually only one cell layer, but several mosses show considerable complexity in leaf structure (section 7.2).

2.3.6 Meristems

Plants grow by cell division and cell extension. Meristems are regions in the plant where cells divide. Meristem cells produce two kinds of daughter cells; those which extend and differentiate into specialized cell types and others which remain meristematic. In some cases there is a small population of cells which is relatively inactive but by dividing occasionally provides the actively dividing meristematic cells (Steeves and Sussex, 1989).

Meristems are found in many parts of the plant, though usually they are restricted to identifiable areas. Nevertheless they may arise almost anywhere from any living cell, but especially from those which are relatively undifferentiated. Differentiated cells, destined to become meristematic, at first go through a process of redifferentiation. They become less vacuolate and more cytoplasmic before the first divisions occur.

Apical meristems are found at the tips of the shoots and roots and are carried forward by the mass of their daughter cells differentiating and elongating behind them. Basal meristems, which push a differentiating organ out ahead of them, are found in some plants. Lateral meristems allow the plant to undergo what is called secondary growth; to increase its girth. The **vascular cambium** is a lateral meristem in the vascular tissue and the **phellogen**, or cork cambium, is one in the bark. These are meristems surrounded on both sides by differentiated tissues. A general term for a meristem which is located between its daughter cells is an intercalary meristem, of which the cambium and phellogen are special examples. Intercalary meristems may arise in many other parts of the plant. Finally there are marginal and plate meristems in the leaves and the reproductive organs which shape these structures.

All the cells of a meristem are genetically identical and totipotent, except in a **chimera**. However, it seems that the development of cells becomes channelled as a result of their positions relative to each other so that, for example, some become cortical cells and others become vascular cells. Some workers have described four different kinds of primary derivative tissues of the apical meristem. These in turn give rise to different types of mature cells. Thus there is the **protoderm** giving rise to the epidermis, the **procambium** giving the primary vascular tissue, the **ground meristem** giving the cortex and pith, and **promeristems** giving rise to other meristems.

An important feature of meristems is that they are functionally integrated.

If an apical meristem is experimentally divided into small isolated units, each unit can regenerate an entire apex, whether it comes from the centre or the margin of the meristem. Somehow marginal cells sense that they are now central cells of the new unit and start behaving as such. Similarly central cells which are placed experimentally in a marginal position behave as marginal cells.

Apical meristems are radial or elliptical and range in size over several orders of magnitude; from less than 50μm to over 3 000 μm. The largest are found in single stemmed trees, called pachycauls because the single stem is so broad, such as the Cycads and Palms. In plants such as whiskferns (*Psilotum*), clubmosses (*Lycopodium*), horsetails (*Equisetum*), and some ferns (e.g. *Dryopteris*), a single apical meristem cell, shaped like an inverted tetrahedron or pyramid, can be recognized (Fig. 2.9). It divides on each of its lower sides and then the daughter cells divide again several times to produce packets of cells. Something similiar seems to happen in mosses (e.g. *Fontinalis*, Fig. 2.9).

In seed plants and some ferns the apical meristem cannot be traced back to a single cell but there is a dome of dividing cells. Different layers or regions of cells can be distinguished by the orientation of the cell divisions taking place and by the size, stainability and the internal architecture of the cells. In ferns the outer layer is very obvious because the cells are elongated anticlinally, perpendicular to the surface. Commonly in the angiosperms (e.g. *Forsythia*, Fig. 2.9), an outer layer or layers of cells of the apical meristem, called the tunica or mantle, and an inner mass of cells, called the corpus or core, are distinguished. In the outer layer(s) usually only anticlinal divisions occur though in the monocots and gymnosperms some periclinal divisions have been observed. In the inner layers cell divisions are in many different directions.

In Gymnosperms like *Pinus* (Fig. 2.9) an apical zonation has been described. There is a group of apical initials which are slightly larger and have a low affinity for cell stains. They give rise to a group of central mother cells from which regular files of small vacuolated cells arise, called rib meristems (Steeves and Sussex, 1989). A similar apical zonation has been noted in some flowering plants.

The shoot apical meristem is surrounded by leaf and bud **primordia**; these are very obvious in those species with an elongated meristem like *Elodea* (Fig. 2.9). The primordia are small bumps enlarging at the margin of the dome of the meristem. They are left behind as the meristem grows forward. As the leaf primordium differentiates, a small package of cells in the axil of the leaf base remain meristematic and gives rise to an axillary bud apical meristem. The time between the production of each successive leaf/stem metamer in the meristem has been termed a **plastochron**.

Root and shoot apical meristems are different in some ways. Root meristems are clearly separated from the rest of the root by a well-defined elongation zone where the cells derived from the meristem elongate. There is a root cap, ahead of the root apex, with its own meristem (Fig. 2.11). The root cap protects the apical meristem proper as the root pushes through the

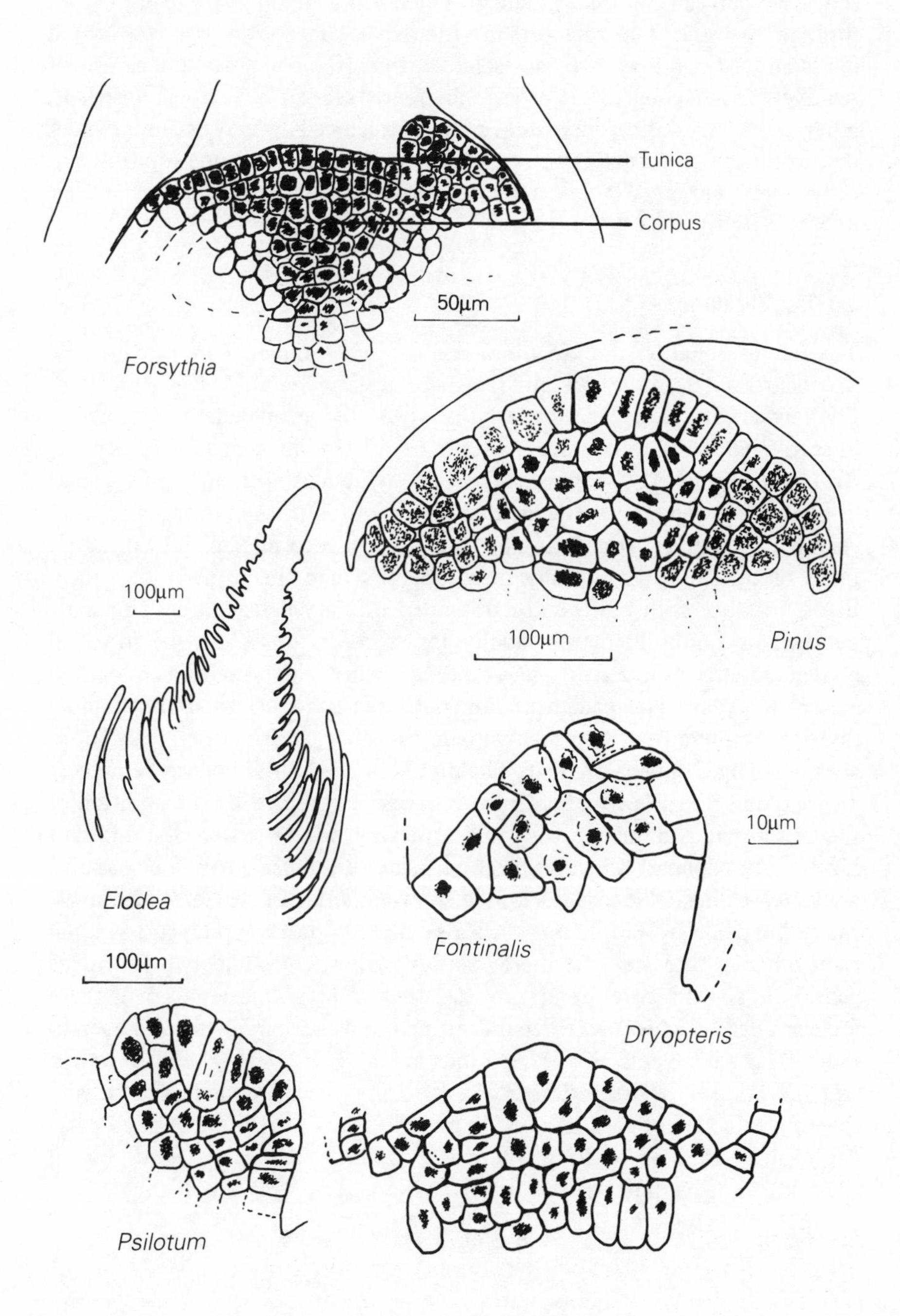

Fig. 2.9. Shoot apical meristems in longitudinal section.

soil. Root cap cells produce mucilage which lubricate the passage of the root through the soil. The root cap meristem is derived from the root apical meristem, but it may become isolated from it by a quiescent centre of relatively inactive cells. In this case the meristems are said to be closed. In other cases the root cap meristem may produce cells which become part of the epidermis or cortex of the root proper. It seems that one kind of root apical meristem may be converted into the other by small variations in their relative activity.

2.3.7 Phyllotaxis

The rate of initiation and growth of the leaf primordia or more properly the plastochrons determines the phyllotaxis or arrangement of leaves on a stem. Phyllotaxis can be seen particularly clearly in the arrangement of tubercles in cacti (Fig. 2.10). Successive tubercles have been numbered for *Ferocactus*, *Acanthocalycium* and *Gymnocalycium*. The phyllotaxis is often a spiral (helix) or more technically, a genetic (generative) spiral, illustrated in the *Ferocactus* example. Secondary connecting parastichies between metamers in different turns of the spiral are usually more obvious than the primary sequence. Plants produce both right- and left-handed primary parastichies. The angle between primordia in the parastichy approximates to 137.5°, the so called golden section or Fibonacci angle, especially where the primordia are packed closely together. The golden section produces the most even and gradual division of the apical dome maximizing packing of primordia at different sizes and stages of development. The numbers from the Fibonacci sequence, 1,1,2,3,5,8,13,21,34...., where each number is obtained by the addition of the two previous numbers, turn up in the way we use to describe different kinds of phyllotaxis. Lines connecting primordia lying on the same radius are called orthostichies, as illustrated in *Mamillaria*. In a plant with leaves, the orthostichies would be lines of leaves directly overlappping and shading each other on the stem. If there are five orthostichies and two full turns round the spiral before the first overlap then the phyllotaxis is 2/5. If there are eight orthostichies and three full turns round the spiral before an overlap then the phyllotaxis is 3/8. Other kinds are 5/13, 8/21, 13/34 and so on. In fact the number of orthostichies relates to the size of the apical dome and rate of production of plastochrons. In a single stem, therefore, the phyllotaxis can change from 2/5 to 3/8 phyllotaxis as growth speeds up. Note how all these kinds of phyllotaxis give approximations of the Fibonacci angle.

$$\frac{2 \times 360°}{5} = 144° \qquad \frac{3 \times 360°}{8} = 135° \qquad \frac{5 \times 360°}{13} = 138.5°$$

Phyllotaxis is one of the few examples where differences between mature plants can be described in terms of a developmental formula. The phyllotaxis of a plant may be regarded as the result of an interaction between two different space requirements; the requirements for arrangement of leaves to

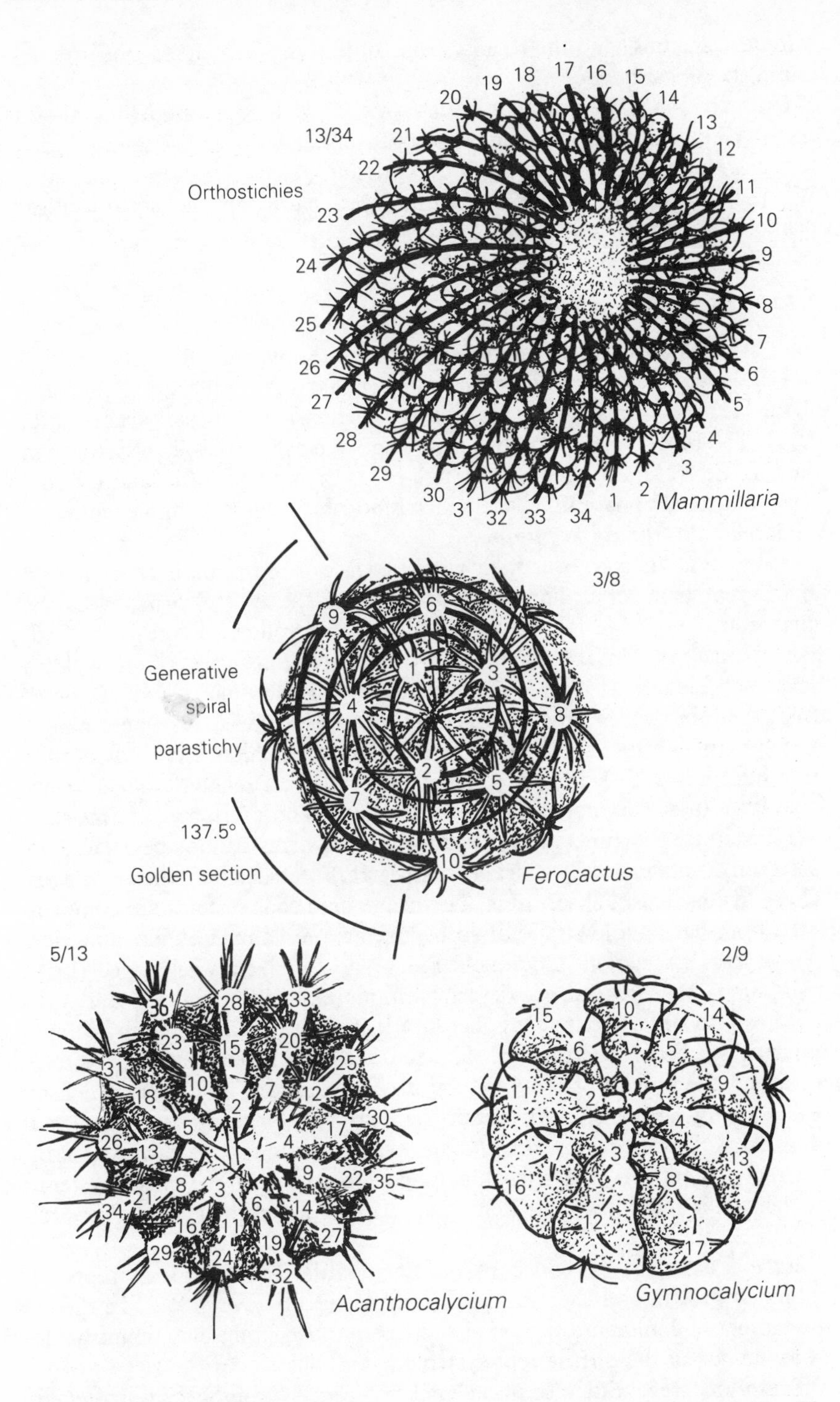

Fig. 2.10. Phyllotaxis, illustrated by patterns of tubercle arrangement in different genera of cactus.

minimize shading, and the requirements of packing developing leaf primordia in the apical bud.

How are the patterns actually determined? Computer modelling shows that two controlling compounds, one diffusing down from the apical meristem and the other up from the differentiating vascular tissues could do it. Where both the hypothetical compounds reach a critical concentration a leaf primordium is initiated (Chapman, 1988).

2.3.8 The root system

Root systems are very different from shoot systems. Initially roots are fine and delicate. The root apex is much smaller than that of the stem, generally about 0.2 mm in diameter, allowing it to penetrate most soil spaces (Fig. 2.11). The root cap protects the apical meristem as it forces its way through the soil. Cells from the root cap slough off as the root advances and the root cap produces abundant mucilage which lubricates its passage and encourages a microflora in the **rhizosphere**.

Roots do not absorb over their whole surface. Absorption is concentrated in the root hair zone. The small diameter of the root also increases the surface area available for uptake relative to the volume of the root. Root hairs increase the surface area even further. Three categories of root systems have been identified (Fig. 2.11) based on the distribution of root diameter classes and the degree of development of root hairs (Baylis, 1975): graminoid root systems are fine and delicate with profuse root hairs and as a result a very high absorptive area to root volume ratio; magnolioid root systems have large diameter roots and fewer root hairs; and there are intermediate root systems. The functioning of these root systems cannot be viewed in isolation. Roots generally exist in a mycorrhizal association with a fungus whose hyphae aid in absorption. The magnolioid root systems are compensated from having a low ratio of absorptive area to volume by having a rich mycorrhizal association. Graminoid roots tend to have a weak mycorrhizal association. **Mycorrhizae** are described in more detail in section 7.5.2.

Roots also differ from shoots because it is usually impossible to see any regularity in their construction. The soil is a very heterogeneous environment and the growth of roots is very plastic to exploit it. Different orders of branching can be recognized but these are very different from the kind of modules described from shoots. However, in some species each region of the root system serves a particular region (module) of the shoot system, though water and mineral nutrients can mix laterally in the xylem (Klepper, 1987).

Some species may produce roots of two different sorts; long roots of unlimited growth and short determinate roots (Kubíkovà, 1967). In dicots the taproot is dominant, at least at first, before becoming indistinguishable from the rest of the diffuse root system.

There are great difficulties in classifying root systems because of their plastic development (but see Fig. 7.10). Nodes and internodes cannot be identified though adventitious roots usually arise at stem nodes. The very

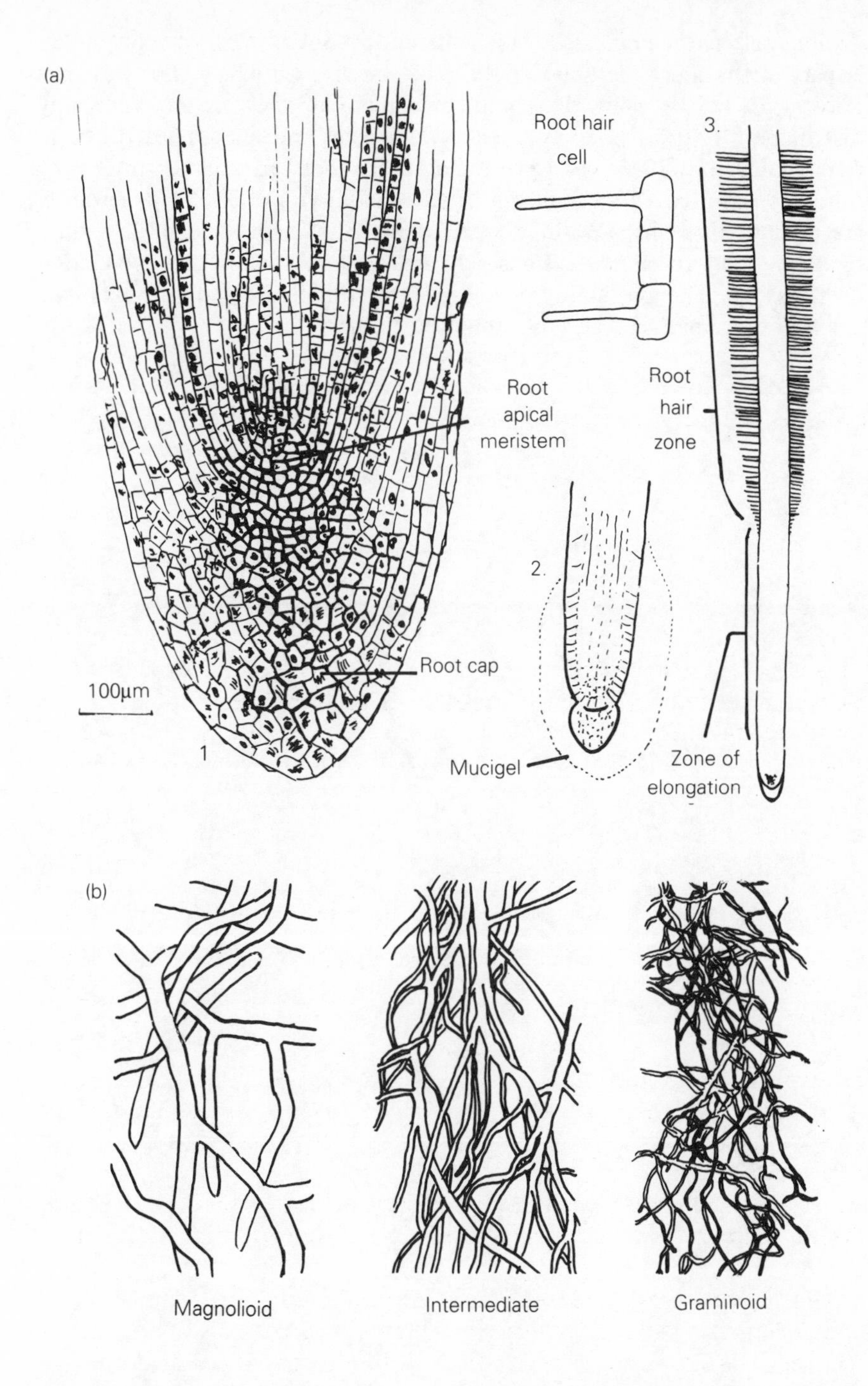

Fig. 2.11. Root characteristics. (a) 1. root apical meristem, 2. root mucigel (after Rayner (1927) (Fig. 33)), 3. root zones; (b) root system types (after Baylis, 1975).

flexible pattern of branching which the **endogenous** origin of roots allows is part of the great plasticity of the root system as a whole. Laterals may arise at any place along the length of a root but they are not randomly distributed. They arise opposite the xylem poles and may appear to be in ranks (Charlton, 1983). However, the lateral roots arise either thickly or thinly as required. Water and mineral availability and ease of penetration are patchy. Even in what may seem to be a homogeneous soil a regular branching pattern rarely develops. There is also variation in root production over seasons. The availability of soil resources, moisture and nutrients, can change very rapidly because of drought or rain.

The first land plants: patterns of diversity

3

3.1 PLANT FOSSILS
3.2 ALGAL ANCESTORS
3.3 AN EARLY COMMUNITY OF LAND PLANTS
3.4 DIVERSIFYING ON LAND
3.5 THE AXIS AND ITS APPENDAGES
3.6 THE AGE OF GYMNOSPERMS: EXTINCT SEED PLANT GROUPS

SUMMARY

Land plants diversified rapidly once established on the land. The adaptive radiation of land plants can be studied in two ways; the evolution of reproductive structures, and the increasing diversity of vegetative structures. The main stages of the vegetative evolution of land plants are described in this chapter. This includes, the evolution of the stele, and of the roots and leaves. Reproductive characters are mentioned but not described in detail until the next chapters.

3.1 PLANT FOSSILS

Fossils may be large and visible to the naked eye (macrofossils) or only visible under a microscope (microfossils). The most important microfossils in land plants are spores and pollen grains. Fossils are preserved in a number of different ways.

Permineralized macrofossils are those in which the plantmatter has been infiltrated by minerals like silicates. The minerals have then precipitated out to provide a rock matrix which supports the organic tissues. Permineralized fossils are particularly useful for discovering the internal structure of plants. A classic example of permineralized fossils are those of the Rhynie plants described below. Because of the vicinity of an active volcano the Rhynie marsh was periodically flooded by boiling silica rich water. The water killed and sterilized the plants and the subsequent infiltration with silica preserved beautiful fossils in the chert rock. Some of the fossils have a vertical orientation which suggests that they were preserved *in situ*. Almost miraculously the details of cell structure may be preserved.

By cutting and polishing the petrified fossil the palaeontologist can obtain a thin section. Observed under a microscope it shows cellular detail like any section of a living plant. The peel technique, which is more commonly used these days, relies on the fact that the organic material of the plant is preserved within the rock matrix. A cut and polished surface is dipped in acid to dissolve the rock but not the organic material so that the organic material stands up in relief. When the surface is flooded with acetone and a cellulose acetate sheet is placed over it, the sheet partially dissolves and then hardens around the fossil. The cellulose acetate sheet plus the fossil can then be peeled off and mounted on a slide for microscopic examination.

Other fossils include the coalified compressions of plants. They result from the collapse of the plant material and the chemical altering of the residues under pressure. Sometimes the compressions have a thin film on them of the remains of the crushed plant or its waxy cuticle. The cuticles of plants are particularly resistant to decay. They can be removed and examined on a glass slide. Pollen and spores have a resistant outer coat, the exine, which increases the chances of their preservation in compression fossils. Maceration of the fossiliferous rock by applications of a strong oxidizing reagent and then a strong basic solution releases the spores or fragments of cuticle for examination. A third type of fossil is the cast. A mould or impression is formed around the soft tissues of a plant. When the plant tissues rot away the space becomes filled with sediment providing the cast. A fourth kind of fossil is the preservation of the silicified or calcareous coats of some organisms, notably the corals and diatoms.

A major problem in the interpretation of the evolution of plants from fossils is the patchiness of the fossil record. It is a very patchy ecological and geographical record. Plant fossils are found in sedimentary rocks, like shales and mudstones. As a result, the types of fossils preserved are likely to be biassed towards those kinds of plants which grew in the close vicinity of lakes and swamps where these sedimentary rocks were forming. Marsh plants are over-represented. Plants growing in dry, upland ecosystems where the rocks were eroding are likely to be under- or unrepresented in the fossil flora. Only fragments which were small enough and tough enough to survive being washed in streams and rivers down to the sedimentary basin are likely to have been preserved. The fossil record is also patchy in time. The sequence of geological ages has been named (Table 3.1).

However, the fossil record does not record a continuous sequence of the vegetation in any one place. Sedimentary basins appear and fill up. In addition it may be difficult to place fossils in an absolute time sequence to enable the cross comparison of floras from different parts of the world growing at the same time.

Another difficulty in interpreting fossil floras is the different levels of preservability of different parts of a plant. Soft tissues rot too quickly for fossils to be formed. The fossil plant is rarely preserved as a complete specimen but only as a number of separate fragments mixed with those of other species. Each kind of fragment is given a different taxonomic name. The separate parts of a coal swamp tree were given the following names (Thomas and Spicer, 1987): *Stigmaria*, the basal horizontal axis; *Knorria*, the bark impression; *Lepidophloios*, the impression of leaf scars; *Lepidophylloides*, the leaves; *Lepidostrobus*, the male cone; *Lycospora*, the microspore;

Table 3.1 Geological ages (m = million years)

Period	*Epoch*	*Time began (years)*	*Significant events*
QUATERNARY	Holocene	10 000	Civilization
	Pleistocene	2.5 m	Ice Ages
	Pliocene	7 m	Establishment of regional
	Miocene	26 m	differences including
TERTIARY	Oligocene	38 m	temperate floras and grasslands
	Ecocene	54 m	Large mammalian herbivores
	Palaeocene	65 m	Bees and butterflies diversify
			Extinction of dinosaurs
CRETACEOUS		136 m	Origin and diversification of angiosperms
			Ornithischian dinosaurs
JURASSIC		190 m	Rise of modern ferns
			Diversification of conifers
			Sauropod dinosaurs
TRIASSIC		225 m	Cycads and ginkgos
			Reptiles diversify
			Large herbivores evolve
PERMIAN		280 m	Greater aridity, spread of upland floras into lowlands and demise of lowland swamps
PENNSYLVANIAN		320 m	Rise of reptiles
			Diversification of insects
CARBONIFEROUS			Coal swamps, warm and humid
			Amphibians diversify
MISSISSIPPIAN		345 m	Arborescent lycopods and calamites
			First seed plants
DEVONIAN		395 m	Diversification of plants
			Terrestrial arthropods, collembolans and mites
SILURIAN		430 m	Origin of land plants

Lepidostrobophyllum and *Lepidocarpon*, female cone scale; and *Cytosporites*, the dispersed megaspore. Only as information accumulates, with the discovery of specimens showing a physical connection of diverse parts, can a whole picture of the plant be reconstructed. Even then three-dimensional shapes and orientations have to be estimated.

Another difficulty is that we know from living plants that a single individual changes in size and even in gross morphology as it grows and matures. Even within a plant, at any one time there may be different shaped leaves on different parts of the plant. Each kind of leaf may be given a name as if it is a different species. Living species contain many varying, genetically different, individuals within each population. Populations may vary geographically. Only because we can see the full range of variation of a species is that variation seen as part of a coherent pattern. In the fossil record only isolated representatives of part of a spectrum of variation may have been preserved and these variants may be given different names. Only as more and more fossils are discovered is a true picture of ancient ecosystems emerging. Important in this scientific process of piecing together the past is taphonomy, the study of the processes of fossil formation.

3.2 ALGAL ANCESTORS

The oldest fossilized cells are probably of bacteria and blue–green algae from the Gun-flint Formation rocks of western Ontario in Canada dating back 2000 million years ago in the Precambrian. The first eukaryotic cells may have been red algae, the Rhodophyta which arose in the late Precambrian, 1400 million years ago. It is widely accepted that land plants arose from the green algae, the Chlorophyta with which they share their photosynthetic pigments, cell wall chemistry, type of storage compound and some aspects of their reproduction. Fossil green algae are present in the Cambrian but they may have arisen much earlier. These fossils have been preserved because they are of quite complex multicellular thalloid organisms which secrete lime around the thallus.

All algal groups are primarily aquatic. There are between 7 and 11 divisions of algae in different classification schemes. They are separated on the basis of their physiology, chemistry and reproductive behaviour. Algae are particularly diverse in their reproductive characteristics. Some undergo sexual reproduction by the production and fusion of identical gametes a condition called isogamy. Others produces gametes which are similar but differ in size, a condition called anisogamy. A third kind, which are oogamous, produce non-motile female eggs, and small motile male sperm. Asexual reproduction is by fragmentation or the production of spores of several different sorts.

In several divisions some species have adaptations for the intertidal zone. A smaller number have forms which are truly terrestrial. Aquatic algae may be floating or fixed to the substrate. Many aquatic algae are unicells or form small floating colonies. The free floating brown alga (division Phaeophyta) *Sargassum* is unusual in being a large complex and free floating plant. In form it is like an intertidal plant which has escaped to the open sea.

Intertidal algae are complex multicellular plants. They show a range of adaptations for life in the inter-tidal zone. A strong holdfast grips the substrate. A strong and flexible stipe and midrib absorbs the stresses of waves. The frond is divided to allow water to flow around it easily. Air vesicles buoy up the alga and help to prevent tangling. The plant body is made up of a pseudoparenchyma of tightly packed filaments. In the stipe there is a specialized conducting tissue analogous to the phloem of land plants. It has sieve-tube elements, 'trumpet hyphae', which lack a nucleus at maturity.

Terrestrial algae are reduced members of the Chlorophyta and Chrysophyta. One common one is *Pleurococcus*, a green alga which only reproduces asexually. It is the green slime, small aggregations of cells, which coats tree trunks and the soil surface. The possibly related *Trentepohlia* grows in humid terrestrial conditions. It has prostrate and erect filaments and reproduces sexually by isogamy.

3.3 AN EARLY COMMUNITY OF LAND PLANTS

There were two major phases of diversification of land plants, the first after their origin 400 million years ago and then again after 200 million years ago. Land plants have diversified to occupy almost every kind of terrestrial habitat. Over the whole long period plant communities, vegetations have been evolving. Diversification has occurred as a result of competition between plants. There are fragmentary land plant fossils from the Silurian, *Cooksonia* from the Upper Silurian, 406 million years ago, is one. *Zosterophyllum* from 395 million years was more advanced.

Perhaps our earliest real picture of a terrestrial plant community comes from a remarkable fossil site in Scotland. In a small area near the village of Rhynie in Aberdeenshire, there is a small patch of volcanic chert which preserves the vegetation of a Devonian reed-like marsh from 370–380 million years ago (Kidston and Lang, 1917,1920a,b,c). At least six different species have been identified: *Rhynia*, *Aglaophyton*, *Horneophyton*, *Asteroxylon*, *Nothia* and *Lyonophyton* (Fig. 3.1). These plants already represent several different lineages of land plants. Some show many close similarities to earlier fossils. *Rhynia* is rather like *Cooksonia* and *Nothia* is like *Zosterophyllum*.

3.3.1 The adaptations of the first land plants

The land which land plants colonized was hostile. The thalloid algal life forms which lay in damper indentations and at the margins of pools and streams would not have had a profound effect on the microenvironment. The sun must have baked the land and the winds and rain must have scoured it. Soil development was minimal.

Land plants grow in an unsupporting environment. They require several adaptations: mechanical strength to support them in the air, to expose light catching surfaces; an anchoring system to prevent them being blown over; a conducting system to supply water to all parts of the plant; a system for obtaining mineral nutrients; a means of restricting water loss in the desiccating environment of the air; a means of reproducing and dispersing on land. They require all these things if they are to be successful.

It is a tall order but if the plant is small, and enters the aerial zone in only a limited way, many of the requirements for life on land are made less strict. Close to the soil surface the environment can be almost aquatic, even if only intermittently when it rains. Here the prime adaptation required for life on land is the ability to withstand dry periods in between the wet. The living bryophytes, mosses and liverworts, have diversified in this zone. There is some evidence that the very first land plants evolved in this niche. *Protosalvinia* from the Upper Devonian combines the characteristics of a

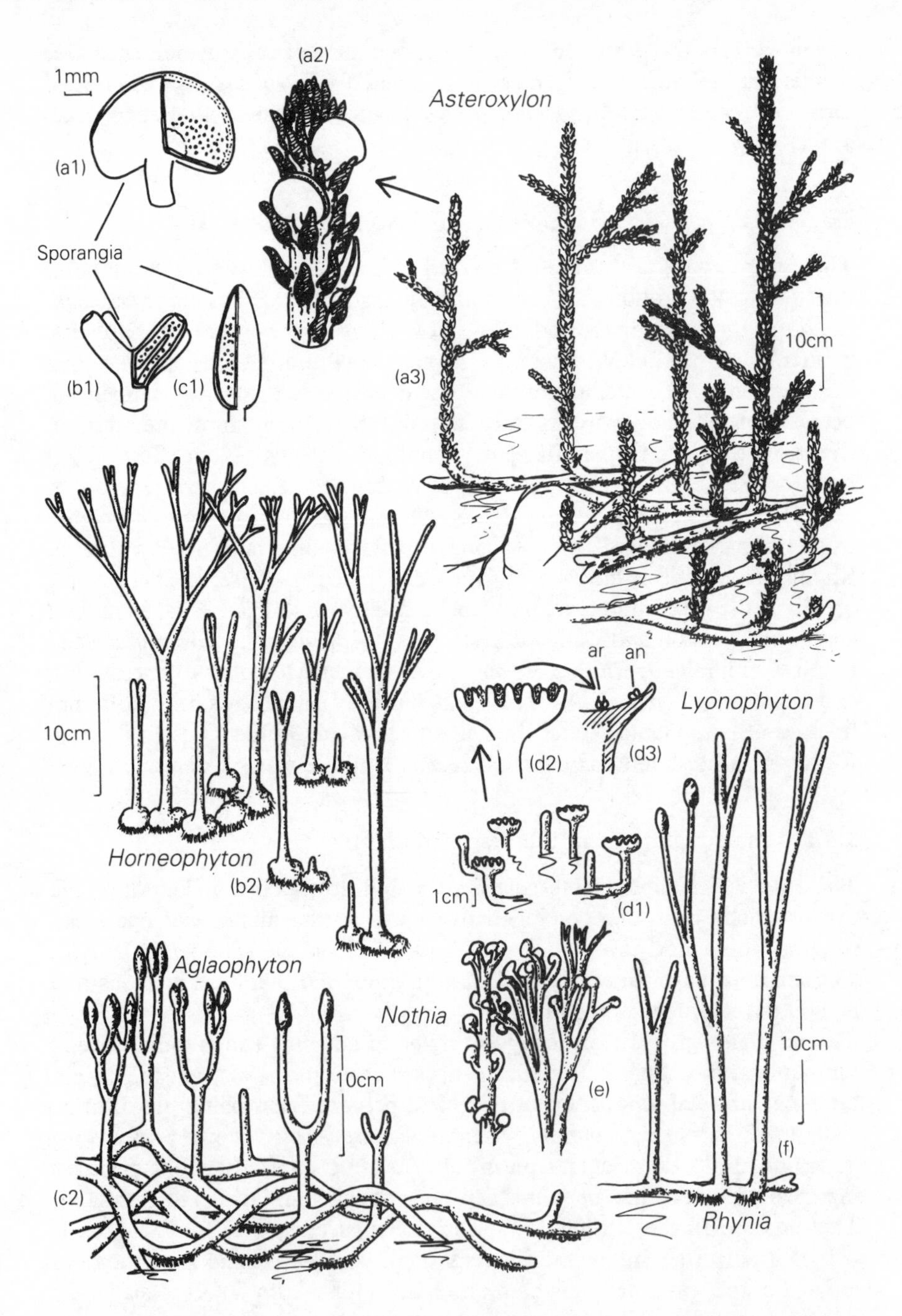

Fig. 3.1. The Devonian vegetation at Rhynie, an early marsh community. Sporangia: a1 = *Asteroxylon*; b1 = *Horneophyton*; c2 = *Aglaophyton*. Key: ar = archegonia, an = antheridia (Reconstructions after Chaloner and MacDonald, 1980 (a1,2); Kidston and Lang, 1920c (a3, f); Eggert, 1974 (b1,2); Remy and Remy, 1980 (d1-d3); Edwards, 1986 (c1,2); El Saadaway and Lacey, 1979a (e)).

littoral alga with that of a terrestrial thalloid liverwort. Other early possibly terrestrial plants had a pseudo-parenchymatous filamentous tissue like some algae, with large filaments enclosed in a matrix of small interweaving filaments (Burgess and Edwards, 1988).

3.3.2 Growing on land

The plants of the Rhynie chert however have many adaptations which show that they occupied the aerial environment, albeit one only a few centimetres above the soil surface. Most of the rhyniophytes have a simple dichotomously branched axis between 15 and 30 cm tall. In *Rhynia* some axes were less developed than the others having the appearance of side branches. *Asteroxylon* was a taller plant, growing up to 50 cm. It also, in some reconstructions, has a single main axis with definite lateral branches.

Roots had not yet evolved. *Horneophyton* had swollen corm like structures at the base of the stem with many rhizoids on the lower surface. *Aglaophyton* arose from a dichotomously branching horizontal axis which rose and fell to give it knee like joints as if it was a tiny species of mangrove. In the lower parts there were **rhizoids**. *Asteroxylon* arose from a branched horizontal axis. Fungi are associated with the terrestrial part of the plant, perhaps an early example of a mycorrhizal association.

Most rhyniophytes showed little adaptation towards catching light but there were stomata with guard cells which allowed the entry of carbon dioxide. This was an important advance, seen elsewhere in *Zosterophyllum*. It allowed greater areas to be exposed for gaseous exchange within the plant tissue while restricting water loss. A cuticle was present on the external surfaces. *Asteroxylon* was the only rhyniophyte with leaves but they lacked a vascular connection to the vascular system of the stem.

Most of these plants have a central vascular strand (section 3.5.1) though this was apparently lacking in *Aglaophyton* (Edwards, 1986). Large surface areas exposed to the air for gaseous exchange would dry out without a ready supply of water. The lignified xylem system of vascular land plants provides a conductive system while resisting the collapsing stresses which arise. Various kinds of tubes, some 45 μm long and 20 μm in diameter, others 200 μm long with internal annular thickenings, have been identified from Silurian rocks. Since these are associated with spores and cuticles they may represent the early kinds of vascular system of land plants. *Orestovia*, a plant from the Lower Devonian, had a vascular cylinder of long tracheid like tubes, with spiral, annular or reticulate wall thickenings. However the first plants where we can see a vascular system which compares directly with those of modern plants is in the plants of the Rhynie chert.

Two living kinds of plants give some idea of what the earliest land plants were like though both have advanced features (Fig. 3.6). The whisk fern *Psilotum* is a plant with a simple dichotomously branching stem which sometimes grows as a humus epiphyte. It lacks roots. It differs from *Rhynia* in having lateral sporangia and microphylls (see below). The club mosses, *Lycopodium*, share many features with *Asteroxylon*, but they have roots.

3.3.3 Reproduction on land

There are two problems which land plants had to overcome in reproducing on land. First, they had to have sexual structures which would allow the production, liberation and fusion of gametes in an essentially dry environment. Secondly they had to have some means of dispersal of **propagules** in the air. We know very little of the sexual adaptations of the earliest plants, but quite alot about the dispersal of propagules. In all living land plants these two aspects of reproduction are carried out by different phases of the life cycle, alternating generation after generation. The two phases of reproduction require different adaptations and in all living land plants the two phases are carried out by different looking plants, i.e. the **alternation of generations** is heteromorphic. The sexual phase is carried out by the gametophyte. It bears the sex organs, the male **antheridia** and female **archegonia**. The **sporophyte**, the dispersal phase, produces spores in organs called sporangia. It seems likely that a similar alternation of generations was present in the earliest land plants.

It is interesting that the alternation of generations also occurs in algae but here there are some species which have an homomorphic alternation of generations, with identical plants at each phase of the cycle, except one produces gametes and the other spores. The evolution of a heteromorphic alternation of generations was required for life on land.

Spores, with a desiccation resistant coat and sometimes with the characteristic trilete scar from being produced in tetrads, are possessed by all land plants. The appearance of air dispersed spores in the fossil record is evidence for the presence of land plants. Rhyniophytes are pteridophytes, spore producing plants without seeds.

The dominant plants of the Rhynie chert are sporophytes. They have sporangia with different shapes from the lobed columns of *Horneophyton* (El-Saadawy and Lacey, 1979b) to the large (7 mm) flattened type of *Asteroxylon*. In *Rhynia*, *Horneophyton*, and *Aglaophyton* the sporangia were in a terminal position like those in the earlier *Cooksonia*. In *Asteroxylon* and *Nothia* they were in a lateral position as in *Zosterophyllum*.

Fertilization occurs after the liberation of motile sperm into water so that they can swim to the egg in the archegonium. The antheridia and archegonia must be located near the soil or where there is standing water or moisture or where rain splash temporarily creates a wet environment. In living plants, the gametophyte is a small and inconspicuous plant occupying a niche close to the soil surface. Therefore it is not surprising that there is little fossil evidence of early gametophytes. One fossil from the Rhynie chert, called *Lyonophyton*, has been described as a gametophyte. Unusually it is not very small and has stems several centimetres tall terminating in a bowl shaped structure in which there are archegonia and antheridia. This bowl is analogous to the splash cups of mosses. Self-fertilization was possible and cross-fertilization may have occurred if sperm were splashed from the antheridia on one axis to the archegonia on another. Presumably other Rhynie gametophytes are small, difficult to recognize or preserved poorly.

3.4 DIVERSIFYING ON LAND

In the 30 million years after *Cooksonia* first occupied parts of the land, a great diversity of kinds of land plants evolved and invaded the land. They were so successful that they modified the earth, changing its environment. In the middle of the Silurian period 420 million years ago there was perhaps only one major group of land plants. Diversification was rapid. By the beginning of the Middle Devonian seven major groups of land plants can be recognized. New forms were adapted to colonize different, perhaps more purely terrestrial environments. There were many short herbaceous species, relatives of *Asteroxylon*. There were small shrubby zosterophylls with well-developed rhizome systems. There were the tall *Psilophyton* and *Pertica* with well developed lateral branch systems. There were trimerophytes with a clear distinction between sterile and fertile branches. Much of this early diversification was plants adapting to new niches, colonization of new environments by the evolution of new adaptations.

New more effective competitors evolved. A primitive fern-like plant called *Hyenia* in the order Cladoxylales may have been an aggressive colonizer, the bracken (*Pteridium*) of its day. It sent out 5 cm thick branching horizontal rhizomes from which arose densely spaced upright branches. One of the most important competitive adaptations was an increase in height. By the end of the Devonian there were many trees (Fig. 3.2).

Pseudosporochnus was another member of the Cladoxylales. Its 'fronds' consisted of dichotomously branching axes. The progymnosperm *Eospermatopteris* looked superficially like a tree fern growing up to 12 m high. It had a profuse pattern of branching with the final divisions in a decussate arrangement. The related but later *Archaeopteris* was up to 18 m tall. It had frond like branch systems and small wedge-shaped leaves. The progymnosperms stand somewhere between the pteridophytes and seed plants. *Pseudobornia* had the characteristic nodal stem with whorls of secondary branches of the group which was much later to give rise to horsetails. Diversification continued apace. Though *Asteroxylon* was soon superseded its group, the lycopods, became the most diverse group with a large number of herbaceous forms and also trees. *Cyclostigma* was a dichotomously branching tree up to 8 m high and with a trunk 30 cm in diameter. It had grass-like leaves up to 15 cm long and 2–3 mm wide.

Not all plants were large. The bryophytes were evolving by the Middle Devonian to occupy the new niches created by the trees; shaded and moist habitats in which a premium was placed on being small.

Old groups, like those containing *Rhynia* and *Psilophyton*, were superseded, either limited by changes in climate or out-competed by newer groups. By the end of the Carboniferous 280 million years ago, 15 different major groups of plants can be recognized with certainty. Much of this later diversification was in different kinds of trees and plants adapted to growing as an understorey in woods and forests. Competition between plants for space and light was encouraging diverse new forms.

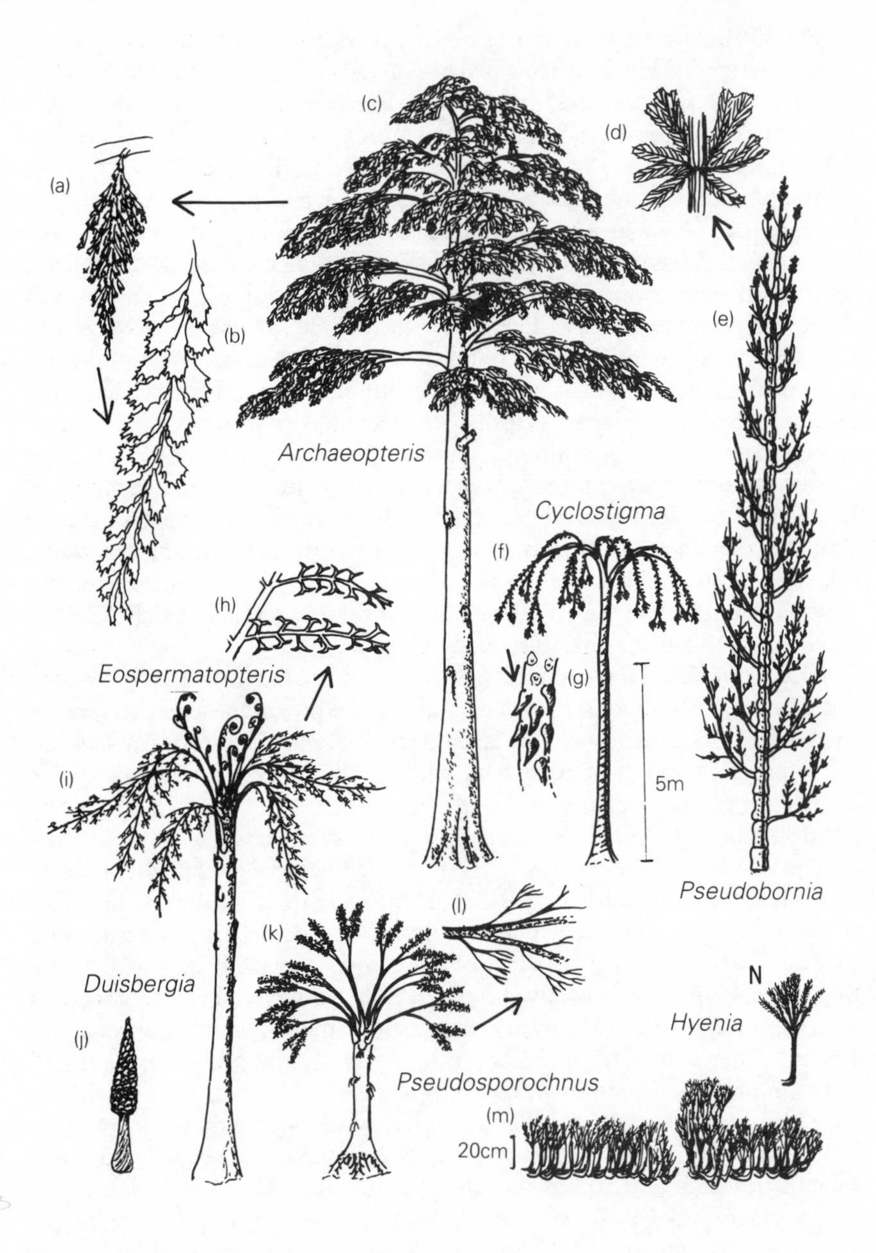

Fig. 3.2. Devonian trees and shrubs. (Reconstructions after Beck, 1962 (c), 1971 (a,b); Schweitzer, 1967 (d,e), 1972 (m,n); Chaloner and MacDonald, 1980 (f,g,j); Goldring, 1924 (h,i); Leclercq and Banks, 1962 (k,l)). All trees to the same scale.

3.4.1 Carboniferous swamps

The early great adaptive radiation reached its fullest expression in the late Carboniferous swamps. The North American and European parts of Pangaea (Eurameria, Laurussia) were in an equatorial position. The erosion of mountains and the infilling of basins produced a vast flat terrain. Rivers and streams meandered through swamps and mudflats. Peat deposits built up which were later converted into coal. After the collision of Laurussia and Gondwana warmer and wetter conditions and the loss of pronounced latitudinal gradients allowed a common equatorial vegetation to reach further north (Raymond, 1985).

The swamps had trees of great heights rising to over 50 m (Fig. 3.3). Two kinds, *Lepidodendron* and *Sigillaria* have been well studied. Branching dichotomously at each stage the meristem was divided equally down to the smallest division. Leaves which abscised from the lower part of the plant, were concentrated in spirals on the branching part of the axis and were awl-shaped borne on prominent leaf cushions. *Sigillaria* looked like a very peculiar palm with great tussocks of linear leaves at the apex of the trunk, arranged not in spirals but in parallel lines. The trunk was unbranched or had a few massive dichotomies near the apex, giving tussock-like apices.

Perhaps the strangest part of these Carboniferous trees was their massive platform like base. It supported them in the swamp muds. Huge stigmarian axes, rhizophores up to a metre in diameter, extended 12 m or more from the base of the trunk. Like the upright axes, they branched dichotomously but these subterranean axes had aerating tissue. Older parts bear the spiral scars of roots which have abscised as the rhizophore expanded.

Forming a kind of first understory in the swamp were *Calamites*, looking a little like living horsetails but massively expanded. Their jointed stems rose up to 10 m in height. One subgenus had just a few branches scattered over the nodes. Another had a rich bushy pattern of branching at every node, with secondary and tertiary branches (*Crucicalamites*). Leaves were linear and in whorls (verticels) of up to 40 at each node. The upright axes grew from massive horizontal axes which bore adventitious roots. The roots were adventitious in verticels at the nodes. At the base of the trunk they grew out acting like the prop roots of living mangroves. The growth of these horizontal axes must have made *Calamites* a very effective competitor, spreading it vegetatively, rather like a living bamboo.

Another kind of tree, which had a flared base and prop roots like a mangrove, was the seed plant, *Cordaites*. It grew to about 30 m with a trunk diameter of 1 m. The tree had a few stout branches bearing long, possibly leathery, strap like leaves up to 1 m in length and 15 cm across. Looking very similar to living tree ferns there were ferns like *Psaronius* and 'seed ferns' (gymnosperms) like *Medullosa*. *Psaronius* had its trunk supported by being clothed in a dense mantle of adventitious roots. These ferns may have grown as an understorey like tree ferns in temperate rain forests of today, taking advantage of breaks in the canopy.

The primitive pattern of branching and the rather unspecialized arrange-

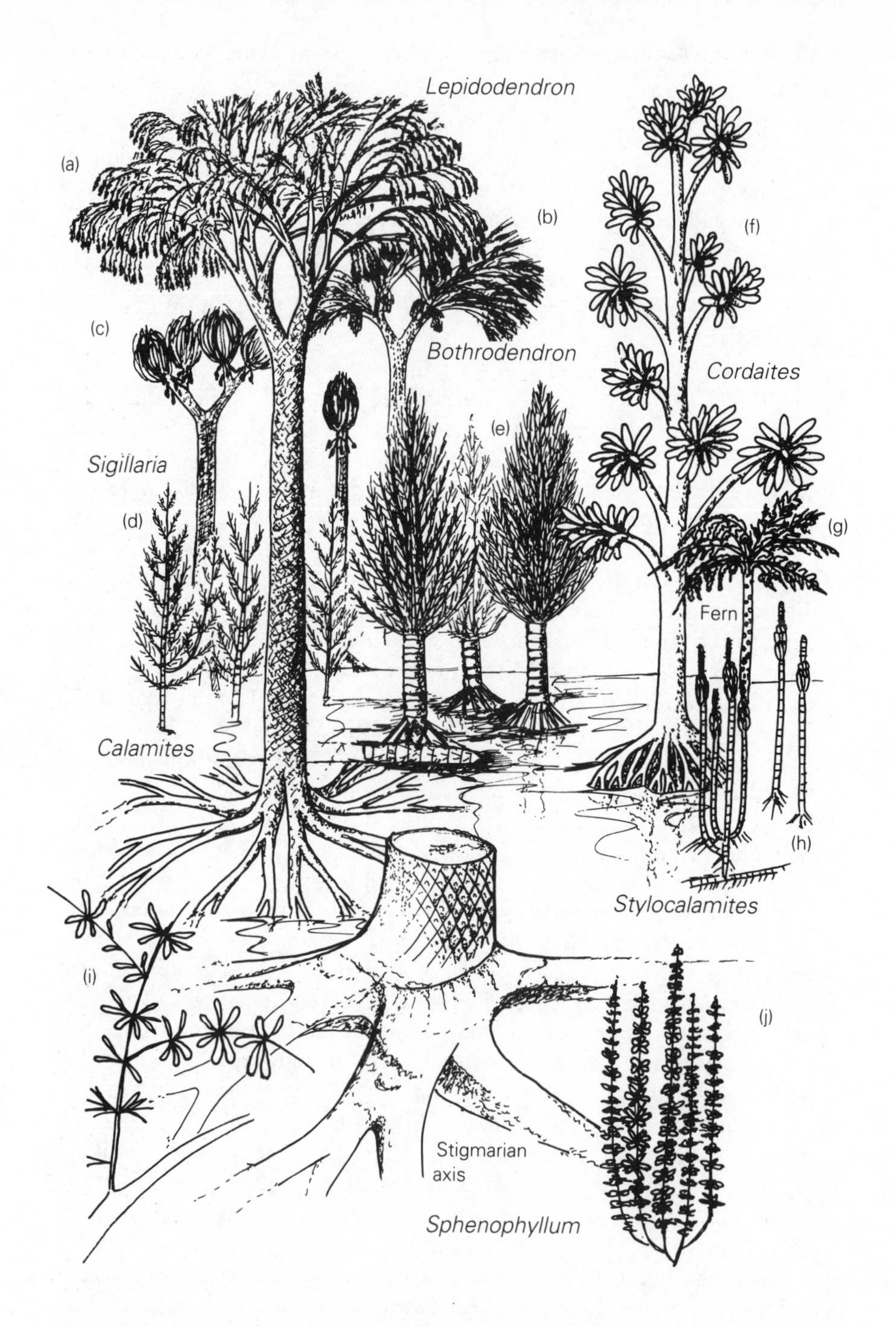

Fig. 3.3. The Carboniferous swamp. (Reconstructions after Andrews and Murphy, 1958, Hirmer, 1927 (a,b,c,d,e,h,i,j); Magdefrau, 1956 (d,h); Cridland, 1964 (f); Rothwell and Warner, 1984 (f); Boureau, 1964 (d); Morgan, 1959 (g)).

ment and morphology of leaves may not have created much shade but plants may have grown very close together, supporting each other physically. The great height suggests that they were competing for light. One very interesting feature of these Carboniferous trees is that many of them had determinate growth, with each upright stem dying after it had reproduced and not undergoing stem thickening. This was certainly true of *Calamites* but may have been equally true of dichotomously branching *Lepidodendron*. Each vertical axis probably grew very rapidly to a mature size, taking advantage of a gap in the canopy, and then there was a period of stasis while it produced sporangia before it died to create a new gap.

The swamps did not just have trees. Other species were like vines or lianes reaching up through and resting on the trees. *Lyginopteris* was a frond-bearing vine with a branched stem. *Sphenophyllum* was a scrambler with slender stems, no more than 7 mm in diameter but several metres long. The stems branched dichotomously and also had lateral branches. The leaves were arranged in verticels like *Calamites*, and sometimes wedge shaped. There was the small shrubby scrambler *Callistophyton*. There were herbaceous ferns and lycopods which may have grown as epiphytes. The fallen trunks of trees provided a home for the scrambling *Selaginellites* but probably the swamp trees were not burdened by the rich clothing of epiphytes we see in the swamps of today. There were non-swamp trees like the Voltziales, with its whorls of branches and short pointed leaves, resembling the living conifer *Araucaria*. It was adapted for growth on dry land and survived into the Permian after the disappearance of the coal swamps.

In the high latitudes of Siberia (Angaraland) a distinct flora had diversified in the late Early Carboniferous. In the cooler higher latitudes plants were smaller and patterns of competition were different from those in the coal swamps. Diverse primitive herbaceous lycopods were able to survive.

The Gondwanan flora of the southern hemisphere had no Upper Carboniferous coal swamp vegetation (White, 1986). As in Angaraland there was a low-growing hardy vegetation. As Gondwana drifted into cool southern latitudes adaptations to cold seasonal climates appeared. Some fossils show a marked seasonal banding in their stems, with periods of growth alternating with dormancy. *Brasilodendron* was probably a low shrubby lycopod growing in swampy areas which iced over in winter. *Cyclostigma australe* was another lycopod which had a thick protective cutinized surface. The flora has been named for one of its seed ferns called *Rhacopteris*, which occurs through the whole Late Carboniferous. The seed ferns diversified. *Botrychiopsis plantiana* had complex leaves with foliose pinnae. The leaves of *Dactylophyllum* were finely dissected. Both had large aphlebiae, a spathe like bract which may have protected the delicate pinnate fronds in development. In the late Carboniferous there was an ice age in Australia and many species disappeared though some like *Botrychiopsis* survived into the Permian.

3.4.2 Pteridophyte survivors and derivatives

Living pteridophytes are generally small in comparison to their Carboniferous relatives. There was a hiatus in the diversification of land plants at the beginning of the Permian (280 million years ago). Global climates became drier. Vast areas of the great super continent of Pangaea became deserts and the coal measure swamps were eliminated from the northern hemisphere. Pteridophyte groups which had included the swamp trees *Lepidodendron*, *Sigillaria* and *Calamites* (and the lianes which depended on them like *Sphenophyllum*) were eliminated in some cases as a result of the inability of their horizontal underground axes, the rhizophores with their massive apices, to grow in drier soil. Smaller forms were favoured.

Evolutionary trends for reduced size and the production of a subterranean basal storage organ, which allowed the plant to survive over a dry season are illustrated by three living genera: *Isoetes*, *Stylites* and *Phylloglossum*. The quillworts, *Isoetes* and *Stylites* (Fig. 3.4) are very closely related.

Isoetes has about 70 species, widely distributed around the world. *Stylites*, with two species, was not discovered until the 1950s growing beside a lake in the Andes of Peru. They have been regarded as representing the evolutionary end of a series of smaller and smaller variants which started with Carboniferous trees like *Sigillaria* which grew over 30m tall. The quillworts are small tufted plants, in which in the above ground axis has disappeared. There is a short unbranched corm in *Isoetes* and dichotomously branched corm a few centimetres long in *Stylites* which represents the modified stigmarian axes of their ancestors. The only leaves present are fertile ones, **sporophylls** a few centimetres long. Both the genera grow in wet soil at the margins of lakes and the like. Between them and *Sigillaria* there is an eloquent sequence of fossils which illustrate the progressive reduction of the vegetative part of the above ground stem: *Pleuromeia* a few metres tall from the Triassic, and *Nathorstiana* about 12 cm tall from the Cretaceous. *Stylites* is so reduced it even lacks stomata.

Phylloglosum has a similar tufted habit to the quillworts but it has a tuft of vegetative leaves and short sporophylls in a cone at the end of an elongated stem. *Phylloglosum*, grows in parts of Australia and New Zealand. Its leaves appear in the winter and in the dry summer months the plant survives as a tuber. It is related to the clubmosses *Lycopodium*.

All the Carboniferous diversity of Sphenophytes has given rise to only one living genus, *Equisetum*, the horsetails (Fig. 3.5). There are less than 30 species of *Equisetum* but it is not a rare plant. The largest is the tropical *E.giganteum* with stems up 13 m long. Since its stems are only 2 cm thick it cannot support itself and grows by sprawling over other plants. Other species of *Equisetum* are much shorter. They have determinate aerial stems arising from rhizomes. The rhizomes grow and branch very vigorously so that some species are strongly invasive weeds. Most grow in damp soil or even in shallow water but *E.arvense* is tolerant of a wide range of drier conditions. The stem has a characteristic ribbed and jointed appearance and the leaves form a collar which sheaths the stem at each node. The stem and

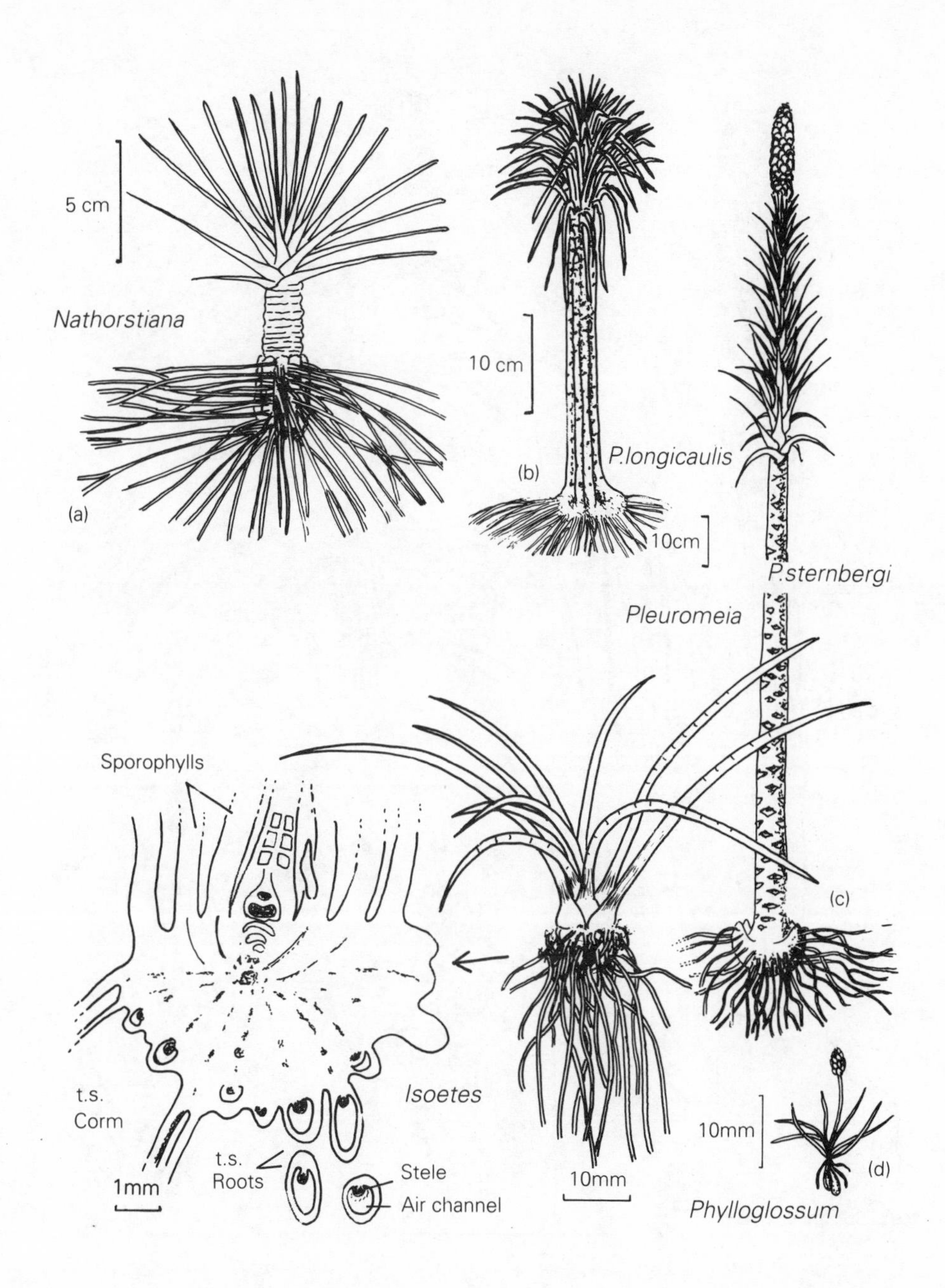

Fig. 3.4. Evolutionary reduction in size. *Pleuromeia* and *Nathorstiana* are extinct intermediates between tall Carboniferous trees with massive corm like bases and the small living descendant *Isoetes*. *Phylloglossum* is an even tinier living pteridophyte. (Reconstructions after Magdefrau, 1956 (a); Retallack, 1975 (b); Hirmer, 1933 (c); Bower, 1935 (d)).

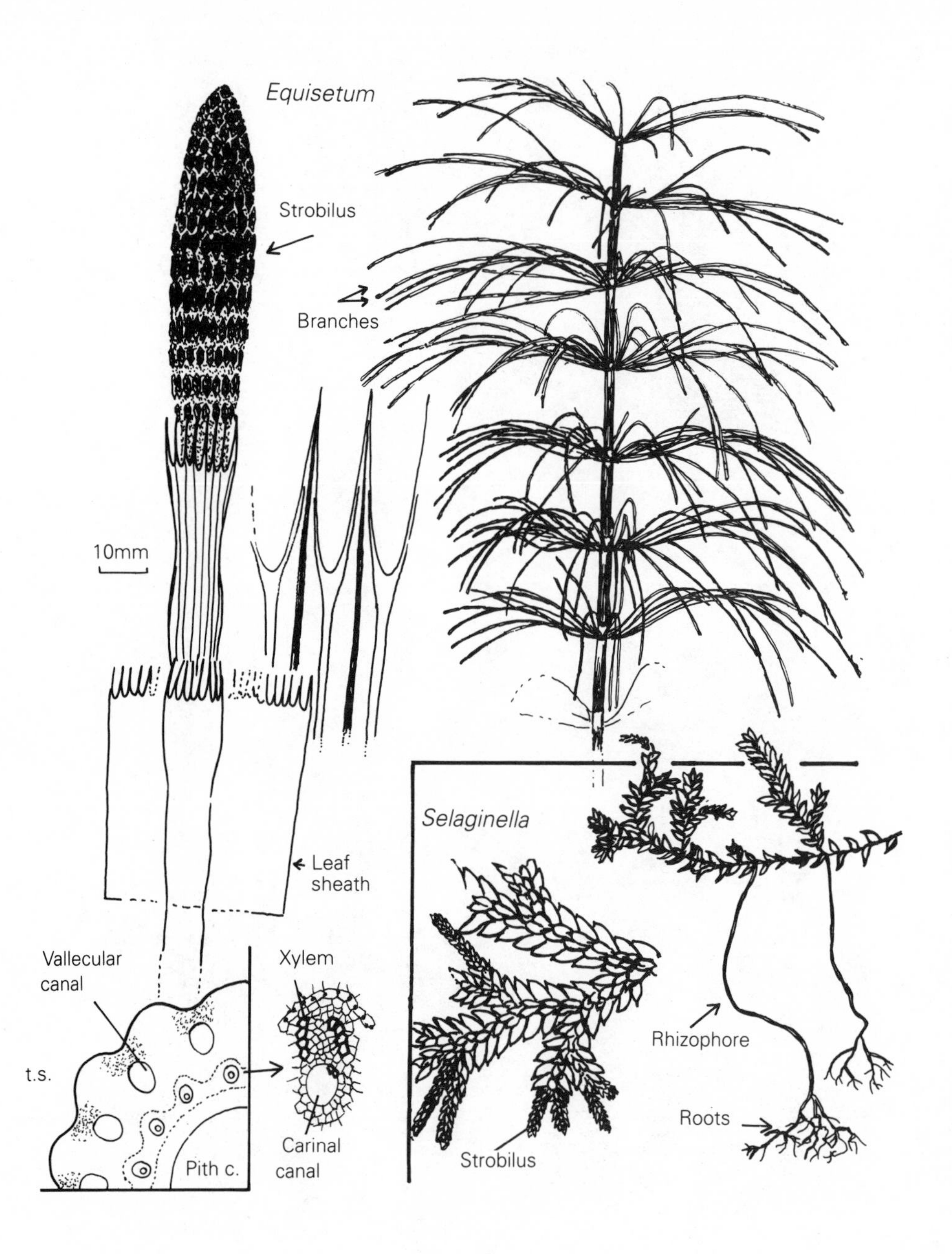

Fig. 3.5. Living pteridophytes I. *Equisetum* and *Selaginella* are two pteridophytes whose relationships can be traced back to the Carboniferous.

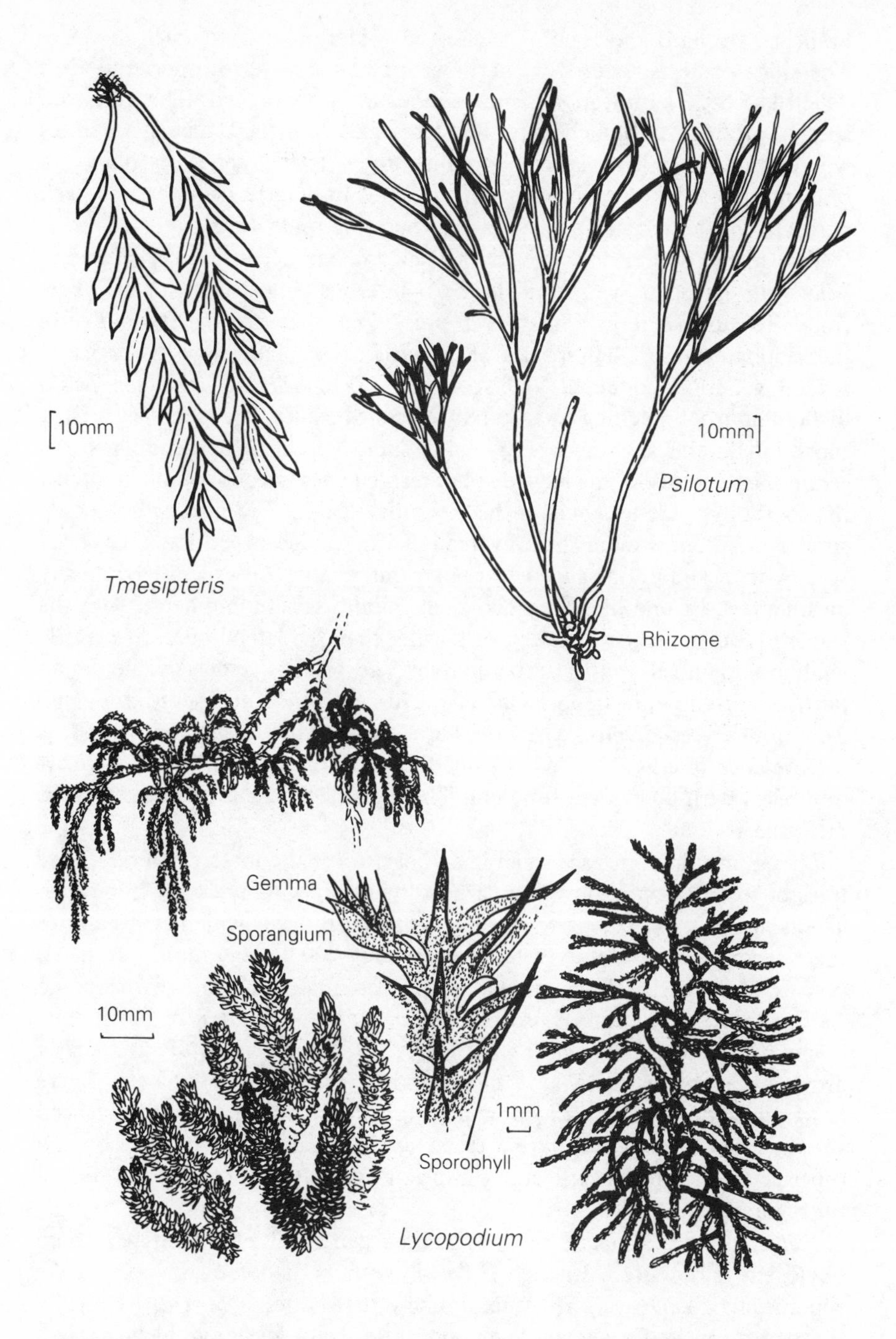

Fig. 3.6. Living pteridophytes II. *Psilotum* and *Tmesipteris* are taxonomically isolated from other plants. They usually grow epiphytically and lack roots. The main genus of clubmosses *Lycopodium sensu lato* has diverse pendulous, and upright forms. It has been divided into many different genera.

branches are hard and rough because the epidermal cells are rich in silica. The silica protects them from herbivores and may even support the stem which has little lignification. Some species are unbranched. The advanced species produce unbranched fertile stems and branched vegetative stems whereas the primitive species produce their fertile cones (strobili) on branched axes. The branches arise from buds in whorls at the base of each node. In the subterranean axes this is where the roots arise.

Two genera of pteridophytes, both small and herbaceous, *Lycopodium* and *Selaginella* have survived from the Carboniferous almost unchanged. Apart from the modern ferns, *Selaginella* and *Lycopodium* are the most diverse pteridophytes: over 400 species of *Lycopodium*, usually now divided into several genera, and about 700 species of *Selaginella*. *Lycopodium* has a dichotomously branching axis, spirally arranged leaves, endogenously arising roots and lateral sporangia (Fig. 3.6). There are upright, pendulous and scrambling species. Some have distinct fertile cones (strobili) while in others the sporophylls are identical to the vegetative leaves. *Lycopodium* has a very similar fossil ancestor in the Devonian called *Lycopodites*. Some *Selaginella* species with spirally arranged leaves are rather like *Lycopodium* but many have flattened frond-like branched stems, and a scrambling habit, with the leaves in four ranks, the upper pair smaller than the lateral ones. *Selaginella* is almost identical to the Carboniferous *Selaginellites*, indeed some would put them in the same genus. One characteristic which distinguishes *Selaginella* from *Lycopodium*, a characteristic it shares with *Isoetes* and *Stylites*, is the presence of a ligule. This is a small tooth at the base of the adaxial part of the leaf with no evident function. The ligule lacks a cuticle and produces mucilage.

Living pteridophytes are small in comparison to the massive pteridophyte trees of the Carboniferous. The largest living pteridophytes are tree ferns. The tallest of these belong to a group of 'modern' ferns which evolved after the Carboniferous. However there are about 200 living, mostly tropical, species of primitive ferns in the order Marattiales which are related to Carboniferous ferns like *Psaronius*. *Psaronius* grew about 8 m tall but its living relatives like *Angiopteris* and *Marattia*, which are quite frequently grown in botanic gardens, have short stems rarely longer than 60 cm. There has been a progressive reduction in height and some relatively advanced species like *Christensenia*, have a creeping rhizome. The stem is fleshy and tuberous and supports very large complex leaves, up to 6 m long and several times pinnate.

Psilotum and *Tmesipteris* are two related genera of pteridophytes whose phylogeny is obscure. Although *Psilotum*, with its dichotomizing axis, looks superficially like *Rhynia* it seems unlikely that either it or *Tmesipteris* is a survivor from the Devonian. They both differ from *Rhynia* in having leaves and sporangia borne laterally, but they are primitive in lacking roots. They have a branching rhizome with rhizoids and a rich mycorrhizal association which grows in humus. The aerial stem is erect or pendulous. *Tmesipteris* is usually unbranched. It has basal scale leaves like *Psilotum* but distally broad leaves which rather unusually are inserted sideways on. Both *Psilotum* and *Tmesipteris* are frequently found growing epiphytically.

3.5 THE AXIS AND ITS APPENDAGES

The **axis** is the stem and root system of the plant. The first land plants had simple dichotomizing axes where there were only small differences between the subterranean system and aerial system. The stele or vascular conducting system ran as a solid column in the centre of the axis. Early evolutionary changes can be observed in the axis, in its branching pattern, in the development of the stele, and in the development of roots. An early development was the evolution of lateral appendages, which increased the photosynthetic area of the plant, called leaves.

3.5.1 Evolution of the stele

The primitive **stele** was a central column, a protostele (Fig. 3.7). The phloem surrounds the xylem or is present as strands around the outside of the xylem. A similar kind of stele can be observed in the roots of many living plants. Differences are in the pattern of differentiation of the xylem either endarch, centrifugally, or exarch, centripetally, and in the number of points of development of the differentiated vascular tissue. A protostele is observed in the aerial axis of many fossil species and in a few living plants. Protosteles are of three kinds: a haplostele, an actinostele and a plectostele. A haplostele, which is circular in transverse section, was present in *Rhynia*, and is present in some species of *Selaginella*. In *Selaginella*, however, the stele is isolated from the rest of the axis in air channels connected only by columns of cells, the trabeculae, which are suberized like an endodermis.

In the fossil *Asteroxylon*, the living *Psilotum*, and some species of *Lycopodium* an actinostele is present. It is like a haplostele but has a wavy margin. This may have had a mechanical advantage. The stiff lignified xylem tissue internally buttressed the stem, but this cannot be useful in roots where it is the normal pattern for living plants. Possibly the shape of the actinostele has more to do with the developmental relationship between phloem and xylem, simply the result of xylem and phloem proliferating at different rates, and the necessity of accommodating both within an endodermis with a small a circumference as possible. It also allows a greater area of the metabolically rather inactive xylem to be in contact with the more active parenchymatous cells to its exterior. Similar reasons may explain the appearance of a plectostele in some species of *Lycopodium* where the protostele is divided into a set of parallel plates and small cylinders with phloem in between.

Some ferns have a protostele but as this expands it may be converted into a siphonostele. A siphonostele has a cylinder of vascular tissue surrounding a parenchymatous pith. Siphonosteles of various sorts are present in the aerial axes of some seed plants, and in the roots of monocots and other plants whose roots have a broad stele. In ferns the siphonostele has phloem both inside and on the outer surface of the xylem tube (amphiphloic) and

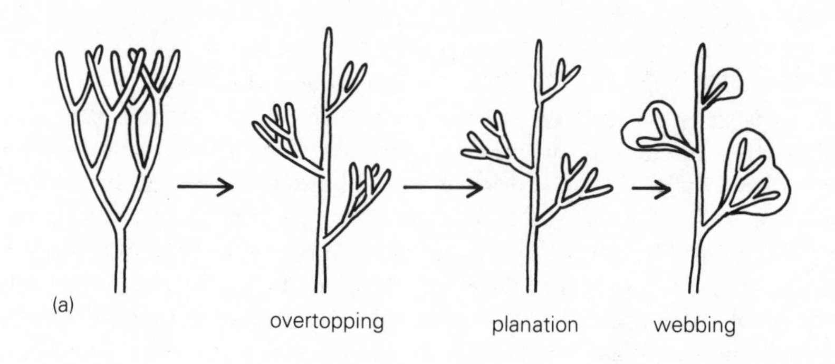

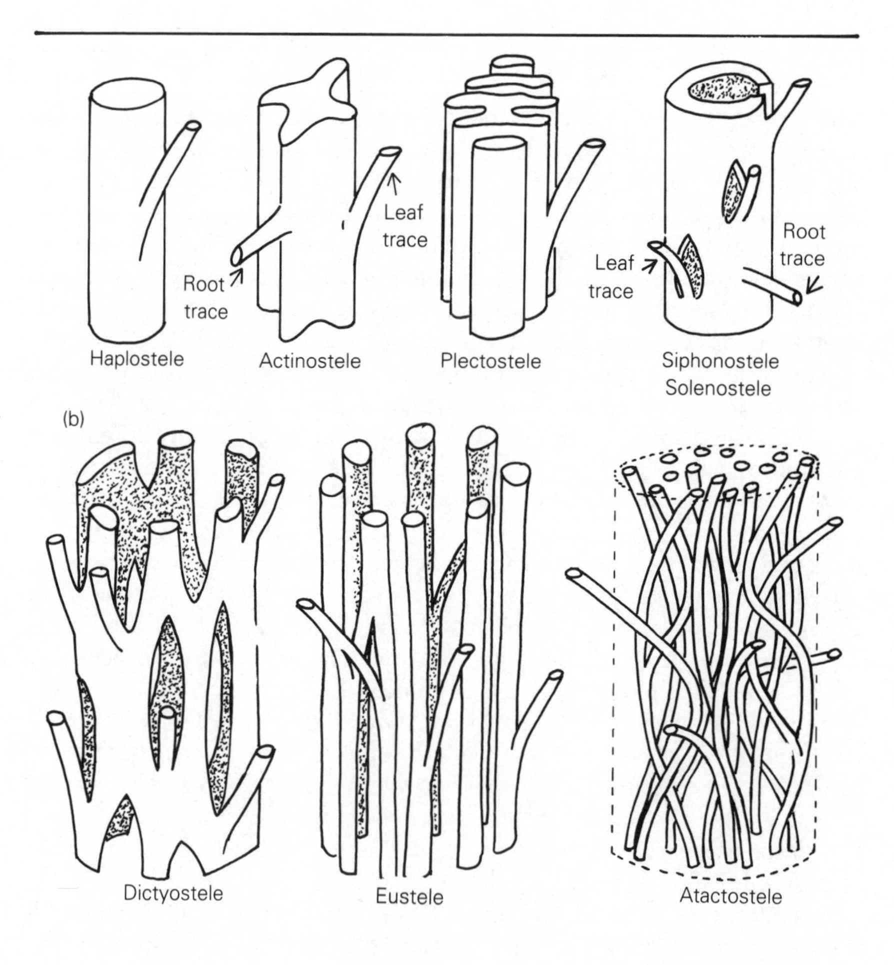

Fig. 3.7. The evolution of the plant axis and stele: (a) the telome theory after Zimmerman (1952); (b) stelar evolution after Stewart (1983).

is more properly called a solenostele. Most seed plants are ectophloic, with the phloem just on the outside. There are two endodermises in a solenostele, one inside and one outside.

In both seed plants and ferns the siphonostele/solenostele does not form a continuous cylinder but is a latticework. Vascular traces fork and anastomose as they pass up the stem. Traces vascularize leaves or lateral branches. A transverse section at any one level reveals a ring of separate vascular strands separated by gaps. In ferns they are leaf gaps left where part of the solenostele branches off to enter a leaf. Above the leaf gap parts of the remaining stele anastomose to fill the gap. Where leaf gaps are so close together that there is never a continuous cylinder a dictyostele is produced.

Seed plants and *Equisetum* have a eustele. This has some similarities to a dictyostele but has arisen from relatively discrete vascular strands which run the length of the stem and branch to produce leaf traces (Slade, 1971). The bundles are either collateral, with phloem on the outside of the xylem, or bicollateral, with phloem both internally to the xylem and externally. The existence of these axial vascular strands is another example of the modular construction of plants. The strands may remain separate (open) but more commonly they are connected at intervals by bridges. The advantage of this is that damage in any one part of the stele can be bypassed through another part of the stele. The stele acts as an integrated whole.

The transition between the protostele of the roots and the eustele of the stem in seed plants occurs in the region of stem below the seed leaves, the cotyledons, in a region called the **hypocotyl**.

The close developmental interrelationship of different kinds of steles is well illustrated in *Equisetum*. The stem of the young plant has a protostele. In a mature plant there is a eustele at the internodes which becomes a siphonostele at the nodes. The stem structure of *Equisetum* is peculiar in a number of respects. There is a large pith cavity, the central canal, and vallecular canals in the cortex, in each internode but not at the nodes. At the inner margin of each vascular bundle there are carinal canals which are closely associated with the vascular tissue and seem to function in conduction.

Monocots have a particular kind of eustele called an atactostele (Beck *et al.*, 1982). A transverse section through a monocot stem reveals a scatter of vascular bundles unlike the cylinder of bundles found in the eustele of dicots. However the pattern of the bundles is actually not very dissimilar to dicots except that the bundles are smaller, more numerous, and change their position relative to the surface of the stem as they pass up it. A major bundle moves inward gradually, then where it produces a leaf trace outwards rapidly, and then in again gradually. The pattern is more confused by the presence of leaf traces which may arise far below the actual origin of the leaf.

Several separate steles may be produced in a stem, either concentric, a polycyclic stele, as in some ferns like bracken, *Pteridium*, or running in parallel, a polystele, as in some dicotyledons.

Patterns of stelar morphology are greatly disturbed when secondary growth increasing the girth of the stem takes place (section 6.1.1).

3.5.2 Microphylls, megaphylls and the telome theory

The first land plants did not have leaves. Two distinct modes of origin for leaves have been proposed giving rise to **microphylls** and **megaphylls** respectively, but it is likely that there is a great deal of convergent evolution of leaves. Microphylls are simple appendages, called **enations**, of the stem which in some cases have become vascularized. Megaphylls are modified shoot systems. One important characteristic of microphylls is that the vascular leaf traces arise as an appendage of the central stele of the stem without disturbing it. In comparison, the leaf trace which reaches into megaphylls is a trace of the stele and normally leaves a gap, called the leaf gap, where it arises.

A complete lack of vascularization is found in the enations, of some early fossils, like *Asteroxylon*, which are spine and scale-like. *Psilotum* is a living plant with unvascularized enations. In both *Asteroxylon* and in *Psilotum complanatum* leaf traces arise from the central stele but do not enter the leaves. In *P. nudum* there is no leaf trace.

The living *Tmesipteris*, *Lycopodium*, *Selaginella*, and *Isoetes* all have microphylls. Those of *Isoetes* are several centimetres long. Since these living species are probably related to some fossil tree species from the Carboniferous with very large leaves, it is likely that, at that time, 'microphylls' were many centimetres long.

Megaphylls have arisen by the lateralization of a vegetative part of the branching axis: a process called overtopping (Fig. 3.7). Flattening of the lateral axis, called planation and webbing, between the branches produced the leaves. This sequence forms part of a very influential theory of evolutionary development called the telome theory, which was proposed by a botanist called Zimmerman (1952). The plant body is seen to be made up of lateral **telome** appendages, either fertile or sterile, and axial mesome segments. The complex plant body of higher plants can be interpreted in terms of overtopping, planation, webbing and two other processes, recurvation and coalescence.

There is said to be unequivocal evidence for the theory, of the origin of the fern megaphyll, from a sequence of fossil taxa: starting from the three-dimensional, sterile, lateral branch (telome) truss which was present in the fossil Trimerophytes (Stewart, 1983). The compound leaf of ferns evolved from only slightly webbed three-dimensional axes. Several variations occur in other groups. In the fossil progymnosperms only the terminal portion of the axis became webbed. An evolutionary trend from a three-dimensional branching system with helically arranged leaves to a flattened branch system with a distichous leaf arrangement can be observed in *Archaeopteris*.

The leaves of *Tmesipteris* illustrate the difficulties of determining the homology of leaves. Their peculiar lateral insertion has led some workers to suggest that they are in fact pinnae of a frond, a megaphyll (White *et al.*,

1977). This interpretation applies equally to *Psilotum* so that its dichotomously branching axis would then be considered a rachis. If this is so *Psilotum* and *Tmesipteris* may then be regarded as primitive ferns.

3.5.3 Roots

A problem in our understanding of the evolution of roots is that they were rarely fossilized. The earliest land plants did not have roots. They had branching horizontal or vertical subterranean stems (**rhizomes**) with **rhizoids**. Even a large tree like *Archaeopteris* had subterranean axes which were developmentally like stems. Trees of the Carboniferous had roots but the main dichotomously branching underground axis of trees like *Lepidodendron* (*Stigmaria*) have been called **rhizophores** because they were thought to be neither roots nor stems. These stigmarian axes bore deciduous spirally arranged lateral roots which were shed as the rhizophore matured, leaving characteristic round scars. The roots had a similar anatomy to those of the living *Isoetes*; the vascular cylinder lies to one side of a large central air canal. The presence of these kinds of air channels in the root is a common adaptation of plants growing in waterlogged soil.

The evolution of roots with their narrow diameter and plastic growth form allowed the more efficient exploitation of drier soils and the occupation of new areas. A root is an indeterminate axis with an apical meristem and a root cap. Lateral and **adventitious** roots arise **endogenously**. Some roots, like those associated with **mycorrhiza** and **nodules**, show **determinate** growth (section 7.5). Even in perennial plants the root system is not necessarily constant. Root biomass varies considerably in different communities. In temperate forests it is about 10% of the total biomass. In the tundra, prairie and steppe it approaches 90%. There may be considerable turnover of roots. Between 50 and 80% of all production goes into the roots in a range of vegetation types (Caldwell, 1987). Deciduous ephemeral short branch roots have been described in some gymnosperms, cut off at the base by an abscission layer (Tippett, 1982).

The embryo produces a primary root called the radicle. From this there arises the primary, or in seed plants, seminal root system. Adventitious roots arise directly from the stem or from the hypocotyl. They are sometimes called crown roots in monocots. Both primary roots and adventitious roots may be present in the adult plant. In many species, and especially in the pteridophytes and monocots only the adventitious system is present in the adult.

It may be a mistake to regard all roots as being homologous structures. The roots of *Lycopodium* are very different from those in higher plants. They arise endogenously in the stem, but they do not bear laterals. Instead they branch dichotomously. They arise in the aerial part of the stem and grow down within the stem before emerging at its base. They have root-hairs in pairs. *Selaginella* and *Isoetes*, but not *Stylites* also have dichotomously branching roots. *Selaginella* has aerial roots once called rhizophores because they were thought to be part stem and part root (Fig. 3.5). Where the aerial root enters the soil its rather blunt apex is converted into a more tapered form. In some species the aerial root has a root cap but in others this is only gained on entry into the soil (Barlow, 1987).

3.6 THE AGE OF GYMNOSPERMS: EXTINCT SEED PLANT GROUPS

Following the evolutionary hiccup of the Permian the rate of diversification of land plants, especially seed plants, but only gymnosperms at this stage, gathered apace again in the Triassic from 225 million years ago. Gymnosperms like *Archaeosperma* and *Moresnetia* had existed in the Upper Devonian vegetation, (Andrews, 1963; Rothwell and Scheckler, 1988) and in the Carboniferous vegetation they were a diverse and important element. Some, such as *Lyginopteris*, *Callistophyton* and *Medullosa*, have traditionally been grouped together as the seed ferns, the **pteridosperms**, though they represent different evolutionary groups (clades). In the Carboniferous vegetation gymnosperms were mostly smallish trees or scrambling members of the underflora. However after the Carboniferous they rose to full prominence. The possession of seeds was an important adaptation allowing dispersal in drier environments.

One group, the Glossopteridales, became abundant in the Permian and Triassic of Gondwana. They were trees up to 6m high or small shrubs. Their broad lanceolate entire leaves were probably carried in tight spirals or whorls on short shoots off the main branches. Different species had different patterns of venation: with an obvious midrib from which many lateral veins arose; with veins connected in a reticulate pattern; with parallel venation. In parts of Australia, cool temperate swamps developed, giving rise to a different kind of coal to that of the northern hemisphere Carboniferous coal swamps. Glossopterids with aerated roots grew in the swamp with many more on surrounding higher ground (White, 1986). There was a marked seasonality (deciduous leaves and growth rings in the wood). Other plants were present. The swamps were thick with horsetails, herbaceous lycopods and ferns. Even in the high latitudes there was a rich vegetation. Permian and Triassic peat deposits from Antarctica have fossil cycads with a strikingly similar internal anatomy to those of today except with slender stems 4 cm in diameter (Taylor and Taylor, 1989). Low light levels were apparently not a problem to growth.

Other pteridosperms were the Corystospermales, the Peltaspermales and the Caytoniales (Taylor, 1988). The corystosperms grew in parts of Gondwanaland. They had pinnate leaves, seeds in cupules with projecting micropyles and pollen-containing organs clustered on a rounded head. The peltasperm *Lepidopteris*, had pinnate fronds. The ovules were situated on a peltate sporangiophore. The Caytoniales had palmate leaves. Their reticulate venation and flat guard cells are angiosperm-like (Doyle and Donoghue, 1987). Not regarded as pteridosperms but also important as precursors, if not ancestors, of the angiosperms were the Cycadeoideales or Bennettitales. These were very similar vegetatively to the cycads but had a bisexual flower-like cone. There were short trees with thick unbranched trunks (Cycadeoidaceae) and others with slender branches with tufts of leaves at the base (Williamsoniaceaeae). They remained important throughout the Jurassic and became extinct in the Upper Cretaceous. Another possibly related group were the shrubby Pentoxylales from Gondwana.

3.6.1 The rise of living gymnosperms: cycads, *Ginkgo* and conifers

The living gymnosperms, the true cycads, Cycadales (Fig. 3.8), probably evolved from the medullosan seed ferns of the Lower Permian. Most are rosette trees with a short stout trunk. The leaves are pinnate. They became abundant in the Triassic and remained important throughout the Jurassic. In the Cretaceous they declined in importance. Three families with a little over a hundred species in total survive today. The Cycadaceae have pinnae with a single midvein. In the Stangeriaceae there is a midrib and dichotomously branching lateral veins. In the Zamiaceae there is no midrib and the simple or dichotomizing veins reach up from the rachis in wavy parallel arrangement. The most obvious innovation of the cycads over previous seed plants was the specialized fertile leaves, microsporophylls and megasporophylls, tightly arranged in separate pollen and ovule cones.

Perhaps the most famous example of a living fossil is *Ginkgo biloba* the sole representative of its order the Ginkgoales (Fig. 3.8). It is now restricted naturally to a small part of China. Close relatives such as *Ginkgoites* and *Baiera* but even *Ginkgo* itself can be identified as far back as the Permian from the leaf shape and venation. Some species had multilobed leaves. In the Jurassic ginkgos were abundant throughout the northern hemisphere.

In the Carboniferous an important tree was *Cordaites*, a mangrove with broad long strap like leaves (section 3.4.1). The Cordaitales are closely related and possibly even ancestral to the conifers (Rothwell, 1988). The Voltziales, which had first appeared in the Late Carboniferous, may have given rise to modern conifers. Transitional genera between the Voltziales and living conifer families have been discovered. The Voltziales diversified in the Permian, and remained important in the Triassic but underwent a relative decline as more modern conifers arose. A very important extinct family was the Cheirolepidiaceae which were so diverse that the only reliable character which they all shared may have been the possession of a particular kind of pollen (*Classopollis*) (Watson, 1988). The range of morphological variation is very great encompassing three living families. They had a broad ecological range including even xerophytic or halophytic variants. They were abundant especially in tropical and subtropical regions in the Jurassic and early Cretaceous.

Of all the living conifers perhaps most similar to the Voltziales are the monkey puzzle trees, the Araucariaceae (Fig. 3.9). They have a very simple and primitive pattern of branching. The family originated in the Triassic with *Araucaria* itself present in the Jurassic. The Podocarpaceae were represented in the Triassic. Other families, the Cupressaceae, the Taxodiaceae, the Taxaceae, and the Cephalotaxaceae originated in the Jurassic.

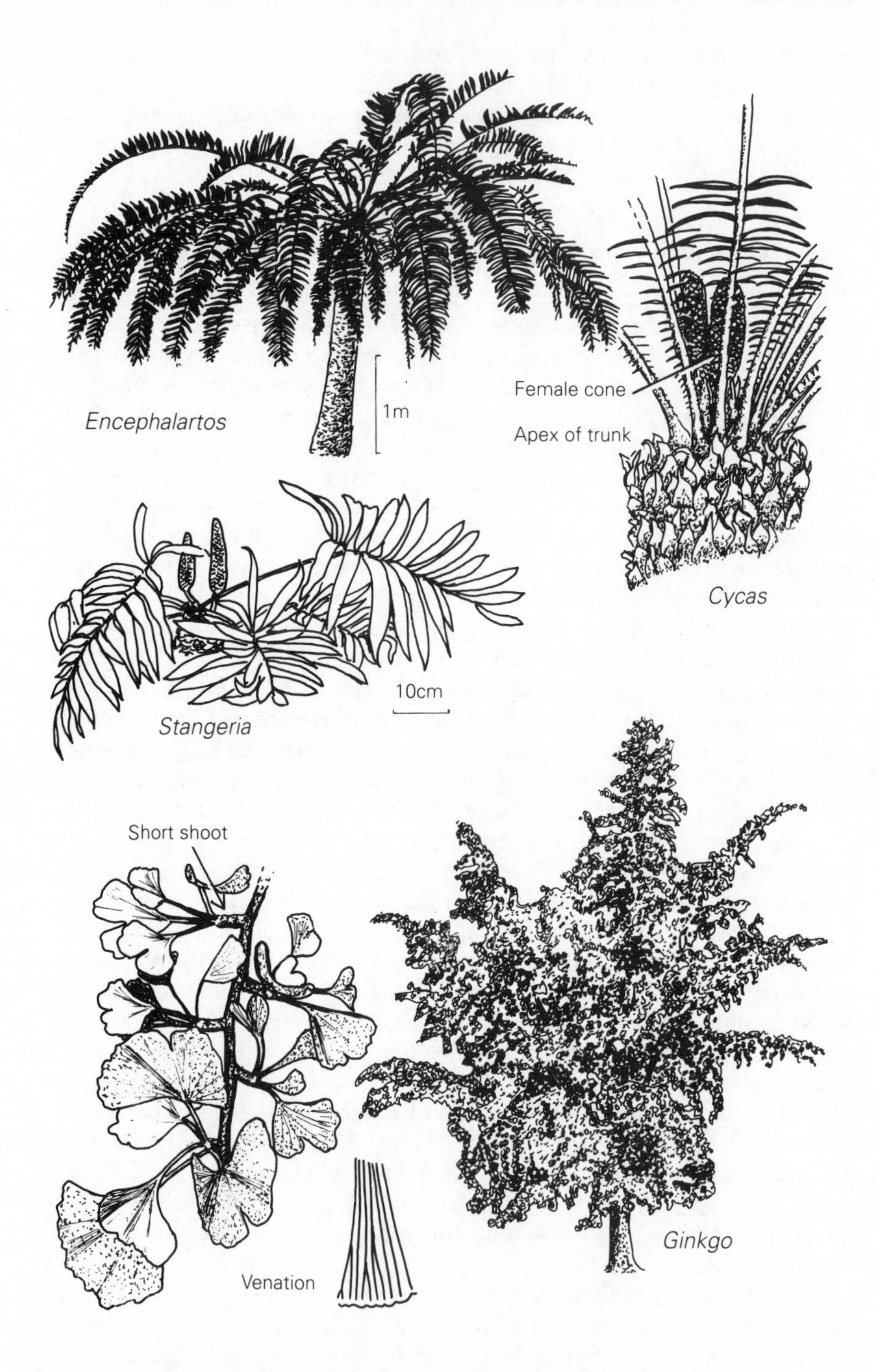

Fig. 3.8. Gymnosperms I: *Encephalartos, Cycas* and *Stangeria* are representatives of the three living families of cycad. *Ginkgo biloba* is a relict of a once diverse group.

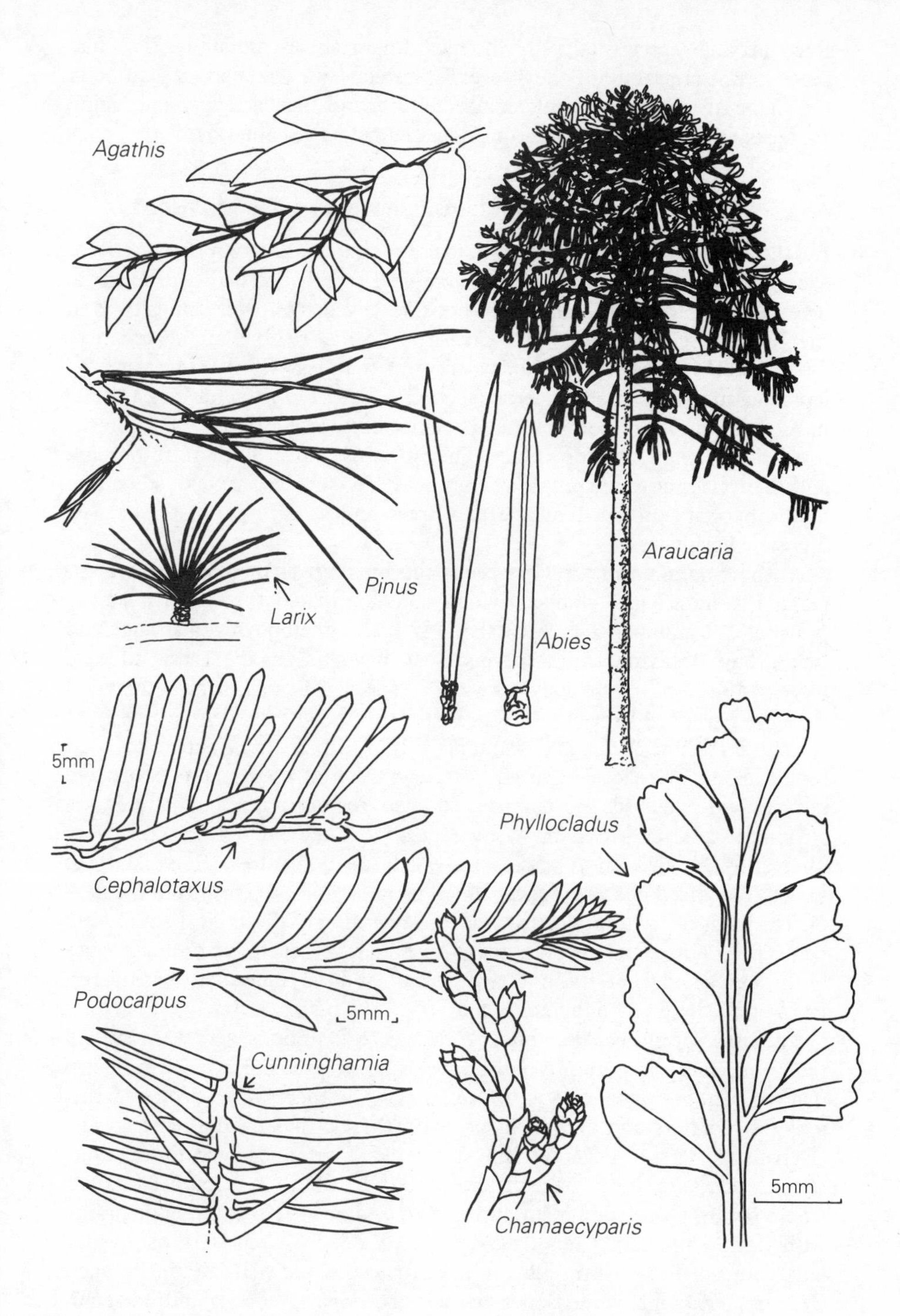

Fig. 3.9. Gymnosperms II: conifers. There is considerable diversity in the foliage of the conifers. *Araucaria* is a particularly primitive example.

Recognizable representatives of the most important living family, the Pinaceae do not appear until the Lower Cretaceous. Conifer families differ in the shape of leaf (broad, scale or needle-like) and their arrangement (spiral or decussate) and in many reproductive characters (Chapter 4).

3.6.2 Obscure origins: modern ferns, gnetophytes and angiosperms

By the end of the Cretaceous all major living plant groups had arisen. Most importantly the angiosperms were well established and had begun to diversify, but the seed plants were not the only plants diversifying. It is in the Jurassic (Harris, 1973) or even earlier (Sota, 1973) that the group of ferns which is most diverse today are first found as fossils. They are the leptosporangiate or modern ferns (section 4.2.2). They took a long time to become important or perhaps they have not been properly recognized in the fossil floras when they do occur. Their origins and evolution are obscure and their classification complex. They may have evolved to take advantage of the habitats provided by the new trees and as a response to the new dinosaur herbivores.

At the beginning of the Cretaceous the breakup of the continents dissipated the monsoonal climates which had dominated the world over the previous 250 million years (Parrish, 1987). The gnetophytes, a strange and tiny group of living gymnosperms, may have its origin at this time of increased dryness. It includes: *Gnetum*, 28 species of small tropical trees and lianes; *Ephedra*, 40 species of xerophytic shrubs with tiny leaves; and the strange xerophyte *Welwitschia mirabilis* (Fig. 3.10). *Ephedra* pollen has been found in the Triassic. The origins and evolution of these plants is obscure but **cladistic** analysis has emphasized their relationships though they are strikingly different in habit (Crane, 1985). They are nevertheless very interesting because they possess some advanced features. It is as if the various advanced characteristics, which were together to constitute the next major evolutionary leap forward with the evolution of the angiosperms, were being tried out in the gnetophytes. These include vegetative characteristics of the leaves and wood exhibited by *Gnetum*, and the reproductive characteristics of all three gnetophyte groups.

The angiosperms evolved into a world where conditions were warmer than today but more importantly latitudinal climatic gradients were much less marked. In one area of southwestern Canada there were diverse ferns, ginkgos, conifers and cycadophytes, with the last having the most species (Delvoryas, 1971). Together they formed a kind of gallery forest along streams or rivers. Ferns were generally abundant both in wooded situations where conifers were dominant and also in a kind of fern-savanna or prairie with scattered forests (Crane, 1987; Coe *et al.*, 1987). Some fern plains may have continued to exist throughout the Cretaceous and well into the Tertiary (Hickey, 1977) maintained open by periodic fires. Lycopods and horsetails were locally important, possibly in wetter areas. Cycadophytes were important plants of open areas.

The origin of the angiosperms, happened at some time in the Jurassic or

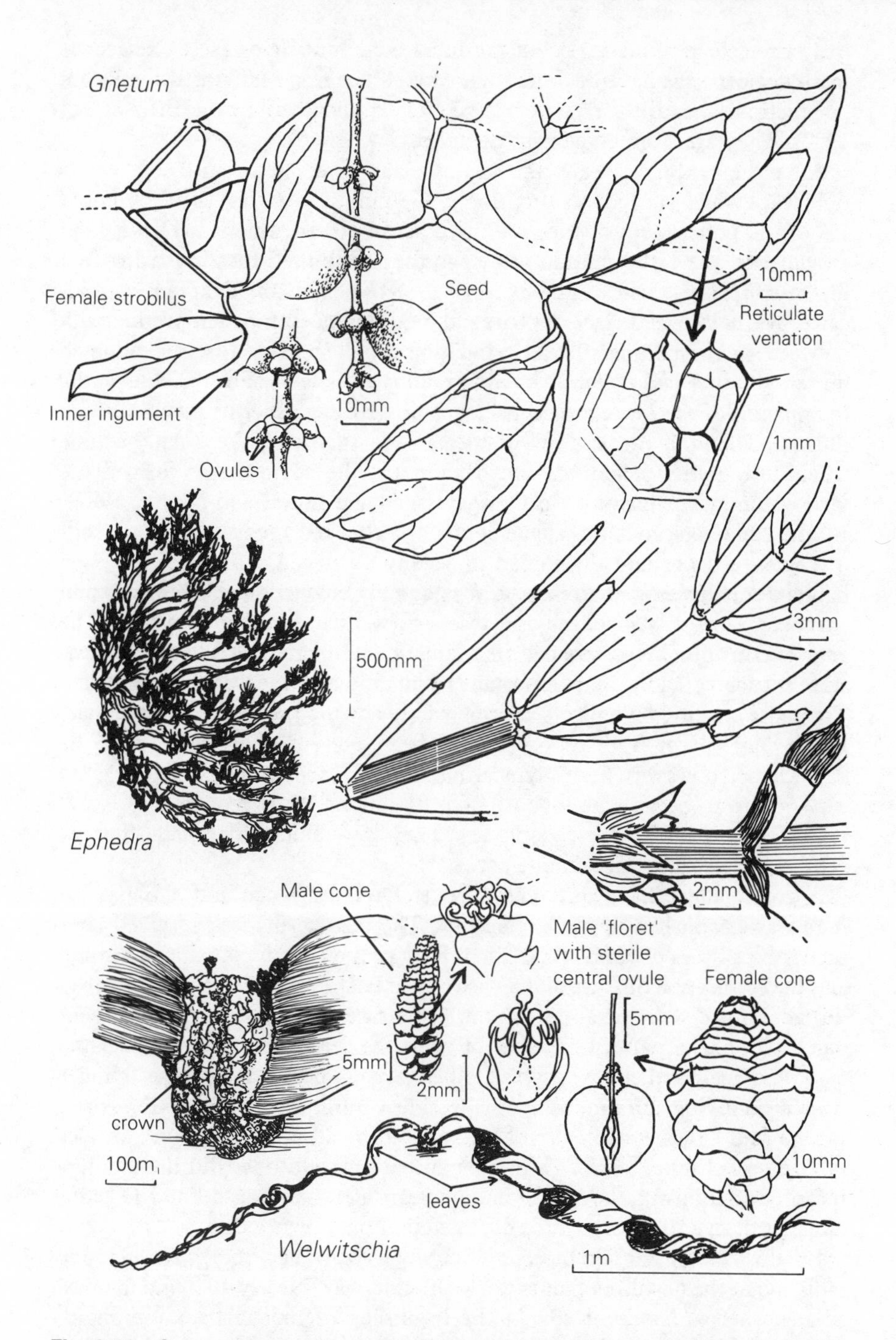

Fig. 3.10. Gymnosperms III: gnetophytes; *Gnetum* is a tropical tree or liane, *Ephedra* a xeromorphic shrub and *Welwitschia* a desert plant with a caudex and two continuously growing strap leaves. These three very different-looking genera are possibly related.

early Cretaceous (Fig. 3.11) but no undoubted Jurassic or early Cretaceous fossil angiosperms have been discovered yet. The first undoubted fossils are pollen from the late early Cretaceous, 125 million years ago (Friis *et al.*, 1987). They are from equatorial latitudes.

After the mid-Cretaceous the angiosperms diversified rapidly. It was a tidal wave which started as a tiny ripple. 113 million years ago they composed 1% of the pollen flora of the Atlantic coast of North America (Hickey and Doyle, 1977). By 100 million years ago they accounted for 20% rising in a few million years to 40%. The interpretation of pollen abundances is complicated by the different sizes of plant and by differences in pollen productivity between insect pollinated and wind pollinated species. Insect pollinated flowers produce less pollen than wind pollinated flowers. For this reason the domination of the late Cretaceous vegetation by angiosperms may be better illustrated by the proportions of megafossils such as leaves. At the time when 40% of the pollen was angiospermous, the angiosperms formed the 'overwhelming majority of leaf megafossils' (Upchurch and Wolfe, 1987). By the late Cretaceous the angiosperms had changed the world.

Angiosperms at first diversified to occupy disturbed and somewhat open habitats such as eroded stream-sides as quickly colonizing weedy herbs and shrubs (Crane, 1987). Ferns and lycopods went into relative decline. The angiosperms quickly diversified into aquatic and marshy habitats. This was an event which had a very important result; the evolution of the monocots. The most primitive monocots are the mainly aquatic group the Alismatanae. Large angiosperm trees may not have been present until the end of the Cretaceous; however, very atypical angiosperm trees, the palms, were an early group to become abundant, from 85 million years ago, in some ways taking the place of the cycadophytes. They were able to dominate the low scrub and perhaps even emerge above it.

A great geographical difference between a Gondwanan and a Laurasian flora was established in the Cretaceous. The floral provinces survived later successive phases of contact like the rafting of the Indian plate into Laurasia and the connection of Australasia with South East Asia. The very widespread distribution of some taxa indicates in some cases their great age. A present day geographical pattern of variation with the separation of closely related taxa, and the parallel evolution of other taxa, indicates how these families were diversifying at a time when the world's continents were breaking apart. By the Mid Cretaceous flowering plants had spread even into high latitudes in Laurasia (Crane, 1987). Only here and at high altitudes did the conifers remain as dominants though 80 million years ago there was a Polar Deciduous Broadleaved Forest, a type of vegetation unknown today.

By the end of the Cretaceous 65 million years ago the great adaptive radiation of the flowering plants was well underway. Nearly 40 living families of angiosperms had evolved. At the beginning of the Tertiary the angiosperms were adapting to temperate climates but there was increased wetness encouraging the development of modern tropical forests. Canopy closure provided more competition but the angiosperms diversified to occupy the new niches which became available, as tall buttressed trees, climbers,

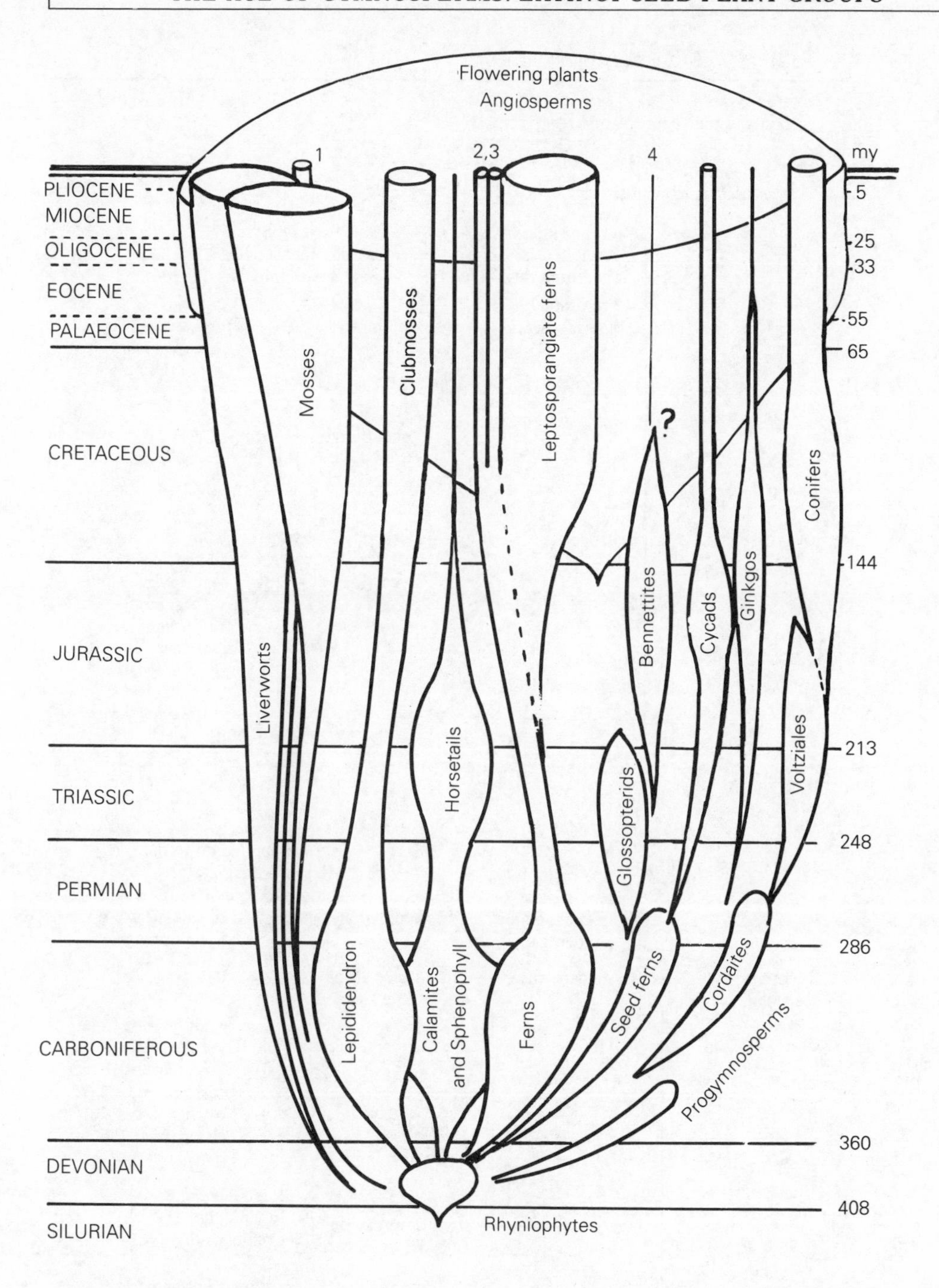

Fig. 3.11. Simplified evolutionary diagram of land plants; cross section of clades approximately proportional to the number of living species (information from Stewart (1983) and Niklas *et al.*, (1985)). 1. hornworts, 2,3. eusporangiate ferns (Ophioglossales, Marattiales), 4. gnetophytes.

Table 3.2 A hierarchy of names

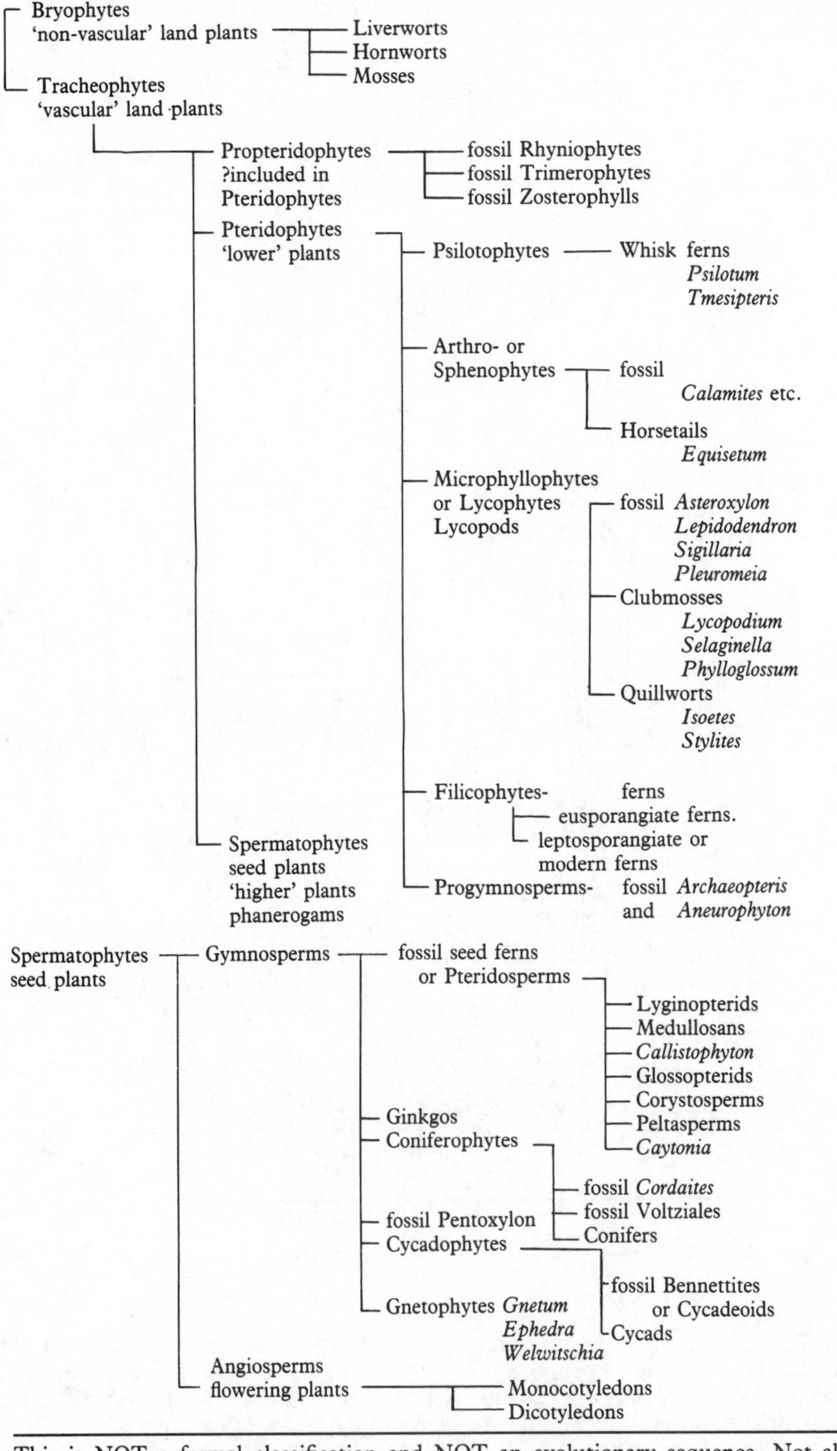

This is NOT a formal classification and NOT an evolutionary sequence. Not all alternative names are given.

epiphytes and understorey species (Upchurch and Wolfe, 1987). Most importantly the different conditions encouraged the development of new patterns of reproduction and dispersal.

4 Sex and dispersal: gametes, spores, seeds and fruits

4.1 THE ALTERNATION OF GENERATIONS
4.2 THE SPOROPHYTE
4.3 THE GAMETOPHYTE
4.4 HETEROTHALLISM
4.5 SEED PLANTS: THE OVULE
4.6 A CLASSIFICATION OF LAND PLANTS

SUMMARY

This chapter is about reproduction in land plants. The evolution of reproduction involves convergent evolution and the reduction of parts of the life cycle, as well as the origin of new structures to promote sexual outcrossing, multiplication and dispersal. The angiosperms with their diverse flowers are described in Chapter 5. Differences in reproductive characters are the signposts of major evolutionary events and are used in classification. A classification of living land plants is presented at the end of the chapter.

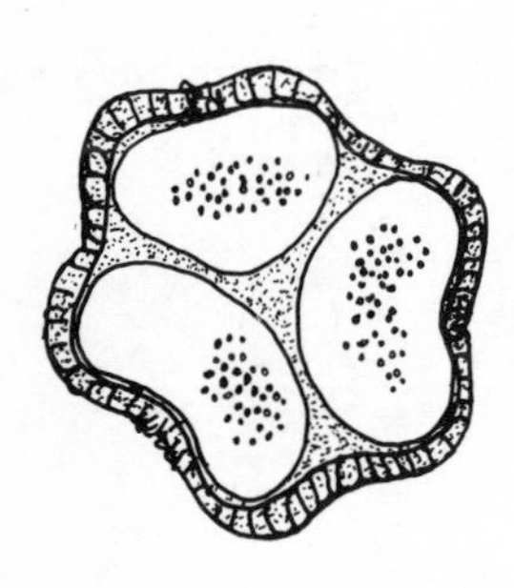

4.1 THE ALTERNATION OF GENERATIONS

There are two distinct purposes of reproduction. First, through sex and cross-fertilization, genetic diversity is maintained and new combinations of parental characteristics are produced in the progeny. Secondly, reproduction can lead to the multiplication and dispersal of the organism. This creates a potential for a newly adapted organism to arise and to spread to new areas. Without cross-fertilization there would be limited genetic diversity and thereby limited fuel for evolution. Differences in sexual reproduction also isolate species so that diversity can arise through the evolution of differently adapted and reproductively isolated species.

The two activities of cross-fertilization and dispersal overlap. Effective sexual reproduction requires the dispersal of the sperm. However the motility of sperm is extremely limited and some organisms produce a non-motile male gamete. In land plants the two activities of sex and dispersal are carried out by two distinct phases in the life cycle which alternate one after the other (Fig. 4.1). In living land plants there are two distinct kinds of plant, the **gametophyte** and **sporophyte**.

The gametophyte produces the male and female gametes in the process of **gametogenesis**. The gametophyte is **haploid**; cells have one set of chromosomes. As a result of mitotic divisions sperm are produced in organs called

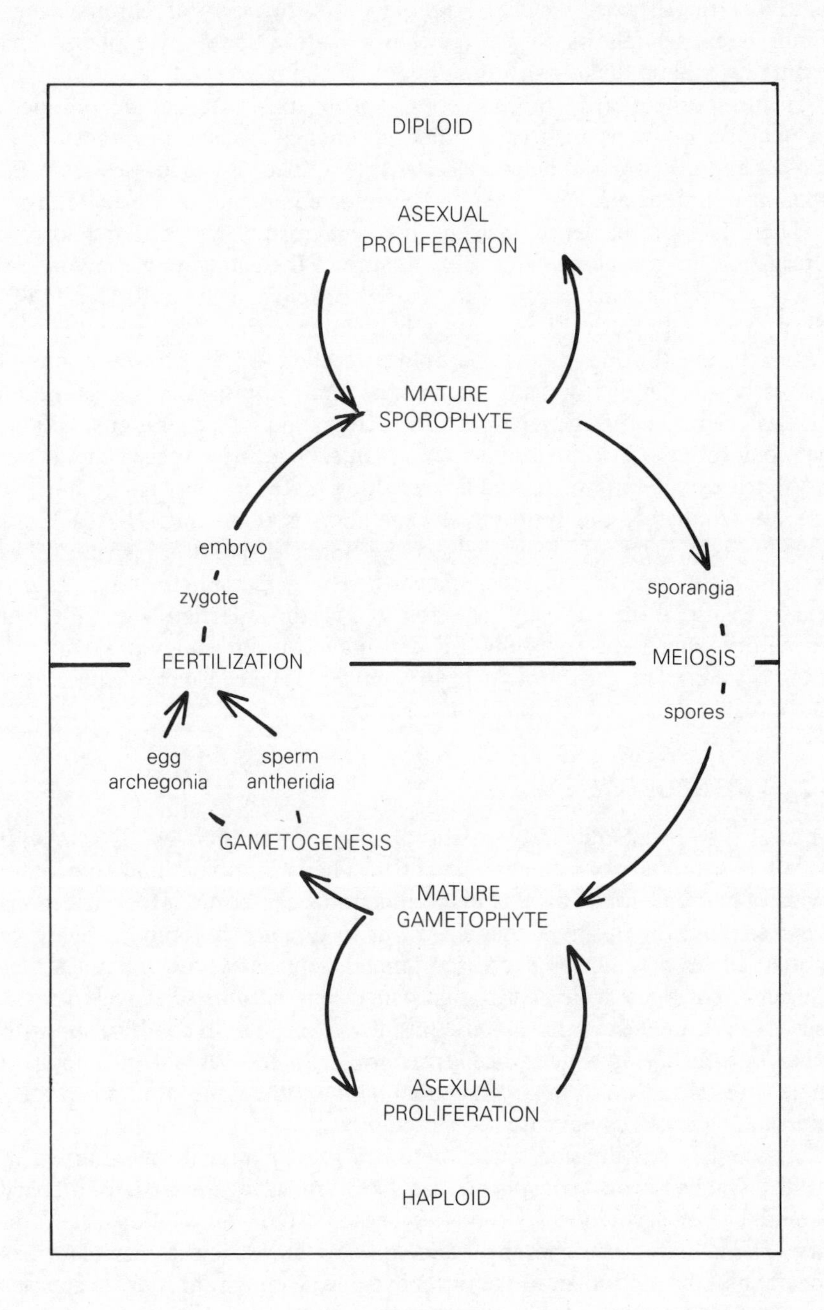

Fig. 4.1. The alternation of generations.

antheridia, and eggs in **archegonia**. Fertilization of an egg by a sperm produces the **zygote**, which is **diploid**, with two sets of chromosomes. **Embryogenesis** follows to give rise to a mature sporophyte plant. This occurs by mitotic divisions in the diploid sporophyte.

In the sporophyte, in the process of **sporogenesis**, spores are produced by **meiotic** divisions in organs called **sporangia**, often called capsules in mosses and liverworts. The spores are haploid. They are dispersed from the sporangia into the air. They land and germinate into a haploid gametophyte.

There is some evidence that the first land plants had a homomorphic alternation of generations with the structure of the gametophyte and sporophyte plants identical except for the reproductive organs (Remy, 1982). Even in some primitive living plants like *Psilotum* the gametophyte is very similar to the rhizome of the sporophyte (section 4.3.3). However, one of the early adaptations of land plants was surely the greater specialization of each reproductive generation. The alternation of generations became heteromorphic as it is in all living land plants. Vegetative specialization was restricted by the kind of reproduction taking place. In most land plants, in the tracheophytes, the sporophyte generation is dominant. It is a large, usually leafy plant. Tallness helped the aerial dispersal of spores. The gametophyte is small and inconspicuous. The gametophyte has remained small because of the necessity for water as a vector in fertilization. However in the bryophytes the sporophyte is dependent on the gametophyte and so it too is small. The gametophyte of bryophytes is the conspicuous plant; the moss, leafy or thalloid liverwort or hornwort.

4.2 THE SPOROPHYTE

Most of the obvious diversity of land plants is in sporophytes. Sporophytes have had a greater evolutionary potential. They are diploid and so contain twice as many genes. Twice as many mutations can occur. Most mutations are deleterious at the time they arise but in diploid sporophytes recessive mutant alleles may not be expressed immediately. However, mutant alleles can represent a reservoir of variation which may become selectively advantageous in future generations if circumstances change or in combination with other mutant alleles which arise. In contrast, in the haploid gametophyte there is restricted evolutionary potential because mutations are immediately expressed and thereby exposed to selection.

Meiosis is a cell division when there is a great potential for mutation to occur. Tracheophyte sporophytes are likely to have greater evolutionary potential than bryophyte sporophytes because spore production occurs in large plants with many sporangia and so many more meioses per plant per generation take place than in the bryophytes. In contrast, bryophyte sporangia (capsules) are produced in small numbers. The costs of producing spores carrying mutations which may be disadvantageous are less when very many spores are produced. There is greater leeway for potentially advantageous variation.

4.2.1 Spores and elaters

Spores are dispersal units, disseminules or diaspores. They are produced in large numbers and so propagate the plant, i.e. they are also propagules. They are protected by a spore wall which is resistant to desiccation. Spores arise directly from meiotic divisions within the sporangium, but they may undergo different degrees of development before spore release. Development of the gametophyte within the spore wall is said to be **endosporic**. Spores may be unicellular or multicellular at the time of release. Multicellular spores are released in some mosses, including the stone mosses Andreaeidae and some Bryidae. Most higher tracheophytes including *Selaginella* and it relatives, some ferns and all seed plants have endosporic development.

Accompanying the spores in the liverworts there are sterile cells called **elaters** (Fig. 4.5) which promote dispersal. They are long and have a band or bands of thickening which makes them twist as the humidity changes. *Boschia* has short irregularly thickened elaters. *Targonia* has branched elaters. In some species the elaters are attached to a small columnar **elaterophore**, either at the base of the capsule, as in *Pellia*, or at the apex, as in *Riccardia* (Schofield, 1985). In *Frullania* the elaters are fixed to both top and bottom of the capsule and act like springs when the capsule opens. Multicellular pseudo-elaters are present in the hornwort, *Anthoceros*. In *Equisetum* the wall of the spore itself partly peels back to form a strip-like elater which remains attached to the spore (Fig. 4.2).

Some plants produce two different kinds of spore in different sporangia, a condition called **heterospory** (see section 4.4.2).

4.2.2 The eusporangium and leptosporangium

The sporangium of tracheophytes, Tracheophyta, is usually a simple sac like organ. It may have a thick wall with several cell layers, **multistratose**. In this case it is called a eusporangium (Figs 4.3 and 4.4). The **eusporangium** has a multicellular origin. A group of sporangial initials on the surface of the plant divide **periclinally** to give rise to an inner sporogenous layer and an outer layer. The sporangium wall develops from the outer layer by subsequent periclinal and **anticlinal** divisions. It includes an inner layer (or layers) of **tapetum**, the nutritive layer for developing spores which breaks down at maturity. Eusporangia are present in all tracheophytes except the modern ferns. Hundreds even thousands of spores are produced in each eusporangium. They have no special adaptations aiding spore discharge but split open along a line of weakness called the stomium.

A different kind of sporangium called a **leptosporangium** is present in modern ferns, the Filicopsida, hence these are called leptosporangiate ferns. A leptosporangium arises from a single superficial sporangial initial cell. Periclinal and oblique divisions occur successively producing an apical cell and stalk cell, a jacket cell and internal cell among others. From these a **unistratose** (one-layered) sporangium wall, a two-layered tapetum, a sporogenous tissue and a stalk develop. Leptosporangiate ferns evolved well after

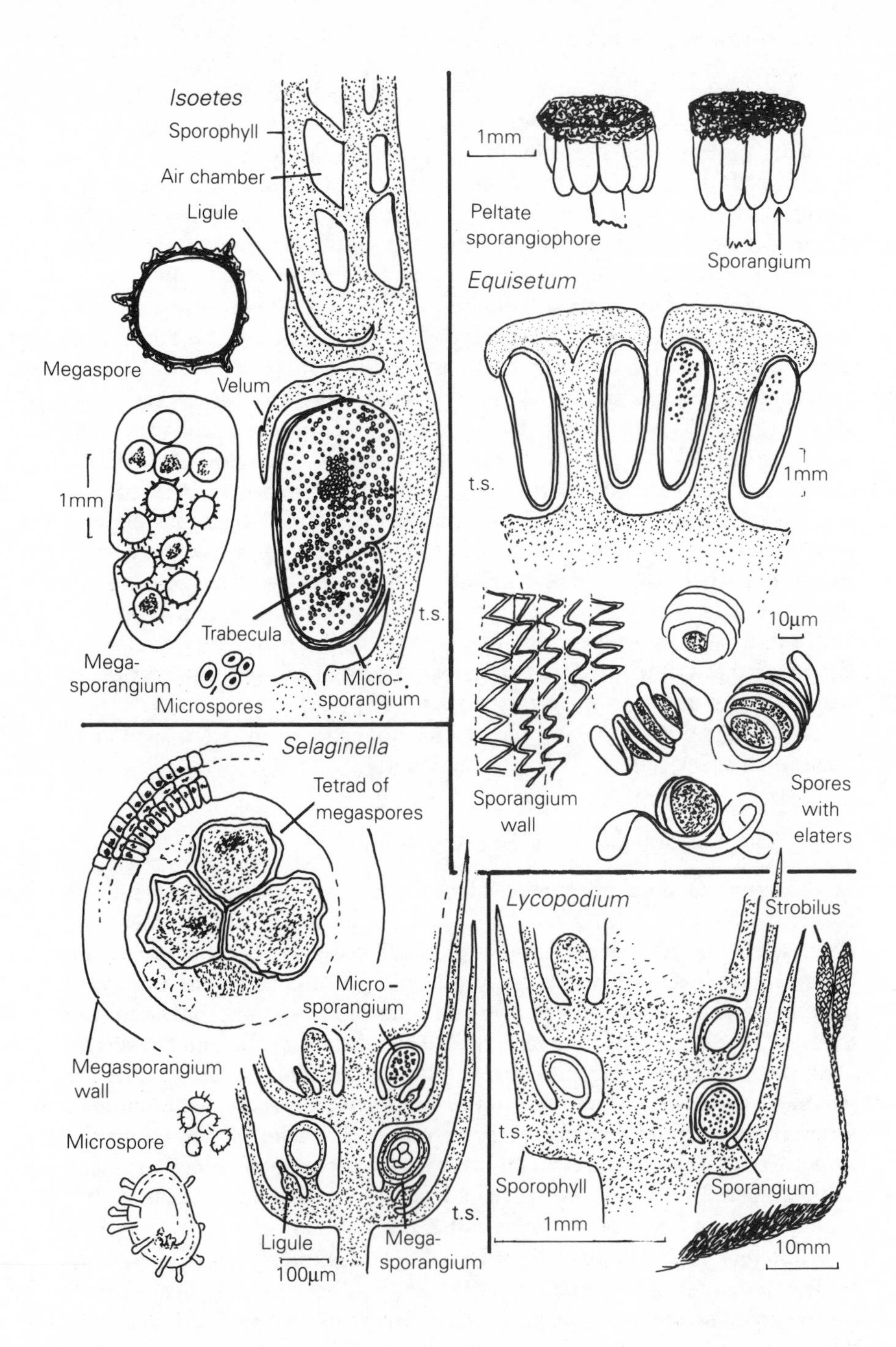

Fig. 4.2. Pteridophyte sporangia. *Lycopodium* and *Equisetum* are homosporous. *Isoetes* and *Selaginella* are heterosporous. The megasporangia of *Selaginella* only contain a single tetrad of spores.

most other pteridophytes. Leptosporangia differ at maturity from eusporangia in being small, having a single layered wall, in being long stalked, and in containing usually 64 spores or less (Fig. 4.4). Leptosporangia are produced quickly, in large numbers. They also have a special ring of cells, the annulus which helps to catapault the spores from the sporangium. *Osmunda* is a primitive leptosporangiate fern in which the annulus is not developed and a large number of spores, 256 or 512 per sporangium, are produced.

4.2.3 The arrangement of sporangia

Sporangia may be stalked or sessile. Frequently they are associated with leaves called **sporophylls** and then they may be **axillary** or **epiphyllous**, situated on the leaf itself. The sporophylls may look like vegetative leaves as in some clubmosses and the ferns or they may be modified in various ways. Sporangia may be single or fused together in a **synangium** as in *Psilotum* (Fig. 4.3), or compound on stalks called sporangiophores as in *Equisetum* (Fig. 4.2). The sporophylls may be concentrated in a particular part of a shoot or more commonly they are aggregated into a terminal compound structure, a cone or strobilus seen in clubmosses, conifers and cycads. *Equisetum* has a strobilus of closely packed peltate sporangiophores.

In ferns the sporangia are grouped in sori (singular = **sorus**) epiphyllous on the undersurface or at the margin of a frond. The sporangia arise from an area called the **receptacle** and, as in *Dryopteris*, they may be protected by a flap or pocket called the **indusium** (Fig. 4.4). Other ferns have the sporangia near the margin of the leaf, which is folded over as a 'false indusium' to protect them.

In some cones and strobili the sporangia mature at the same time. In others, sporangia at different stages of development can be observed. Similarly in ferns the fertile frond may bear sori at different stages of development though within each sorus the sporangia mature at approximately the same time. However, in the filmy ferns, Hymenophyllaceae, sporangia may be produced over an extended period at the base of a receptacle which elongates to carry sporangia out of the indusium (Fig. 4.4). Ferns like *Osmunda* have clearly differentiated fertile and sterile pinnae (Fig. 4.3). The peculiar ferns, Ophioglossopsida, do not have a sorus but have a bilobed frond, one lobe forming a fertile spike. In *Ophioglossum* there are two rows of sporangia forming an elongated synangium (Fig. 4.3).

4.2.4 The hornwort sporangium

The hornworts, are peculiar in having a sporophyte which is very different from other living plants and places them in a distinct subdivision of their own the Anthocerotophytina. The sporophyte has its base, the absorptive foot, embedded in the tissue of the gametophyte. The sporophyte first enlarges inside the thallus protected beneath a projection of the thallus, a kind of **calyptra**. It elongates into a sporangium which breaks out and the calyptra is left as a kind of collar or involucre (Fig. 4.5).

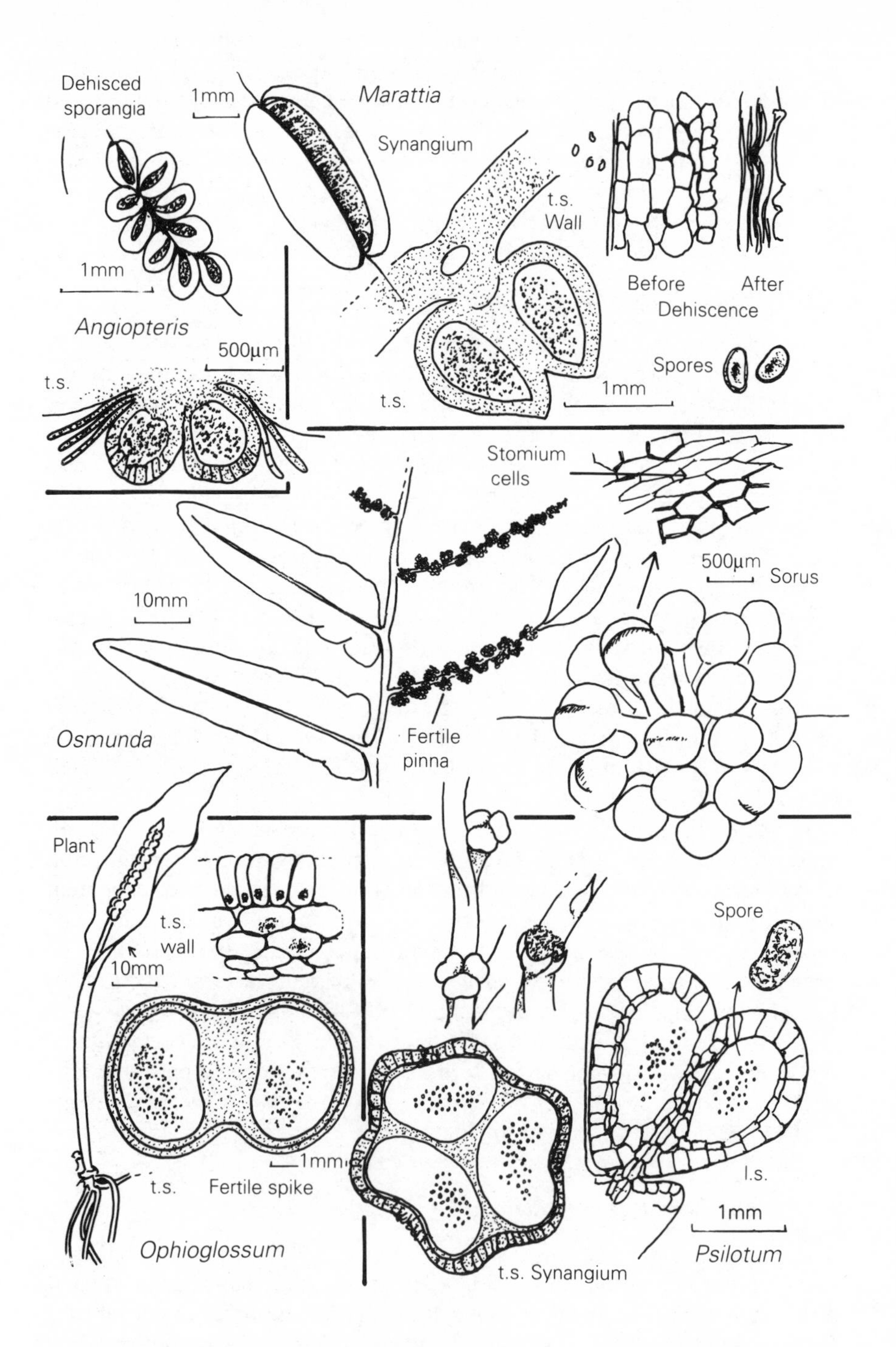

Fig. 4.3. Eusporangia: sporangia with a multiseriate wall which produce many spores. *Osmunda* has a primitive type of leptosporangium.

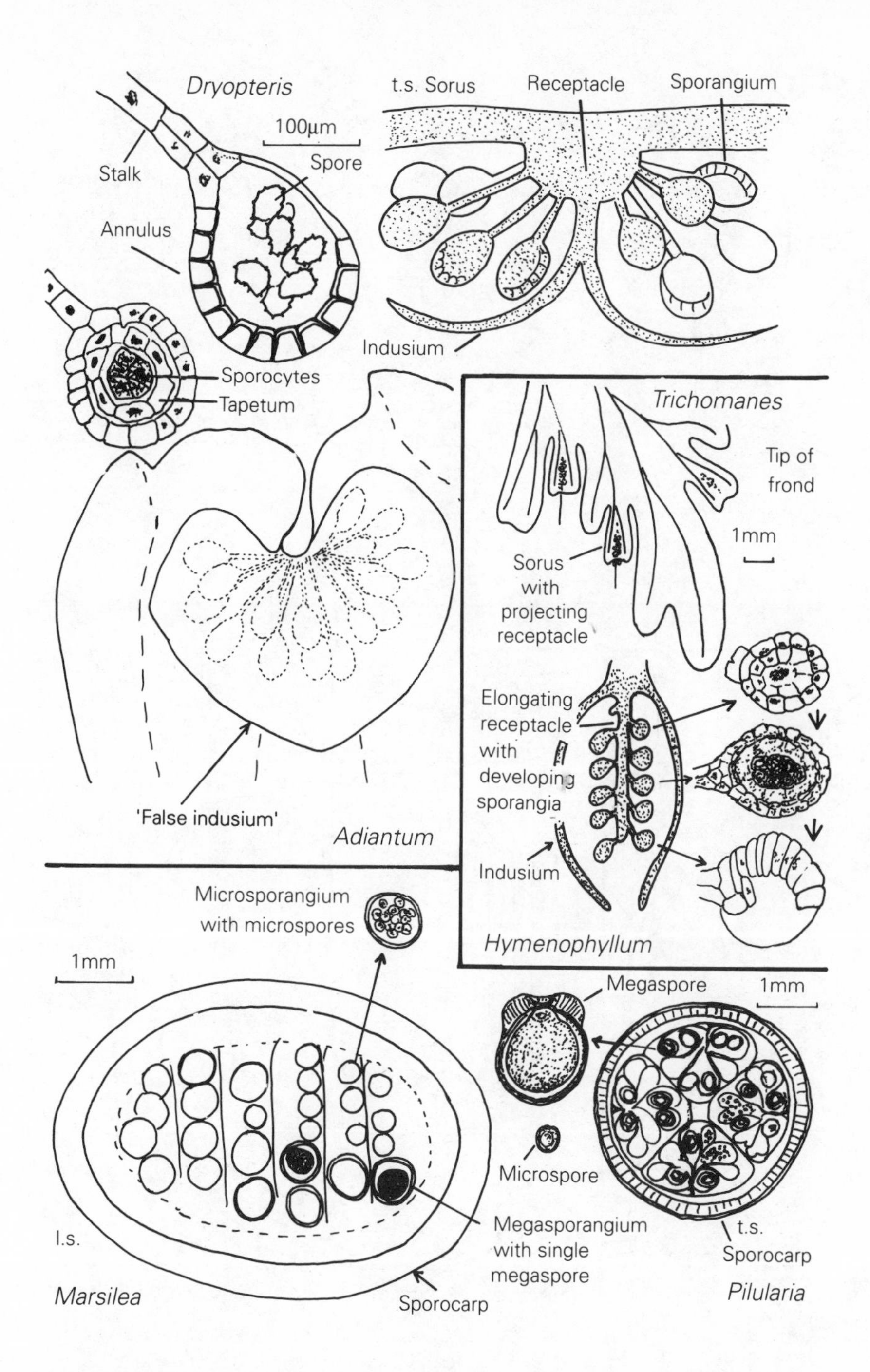

Fig. 4.4. Leptosporangia, the sporangia of 'modern' ferns. In *Hymenophyllum* and *Trichomanes* the receptacle elongates to carry the sporangia out of the indusium. *Marsilea* and *Pilularia* are heterosporous water ferns.

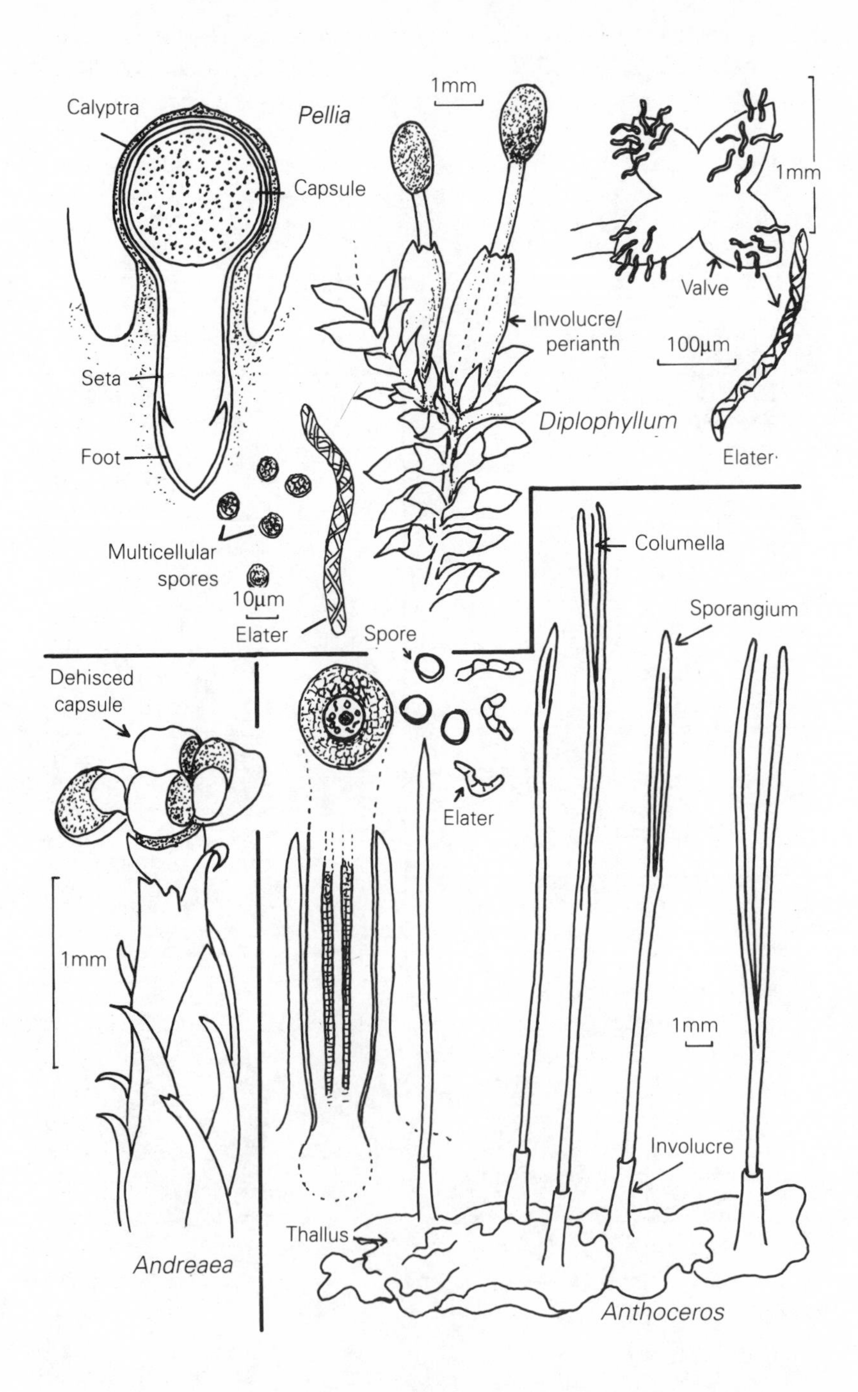

Fig. 4.5. Three different kinds of bryophyte sporangia (capsules). *Anthoceros* the hornwort is unique in having a zone at the base which produces spores over an extended period. *Andreaea* is an unusual kind of moss. *Diplophyllum* and *Pellia* are liverworts.

A basal meristem elongates the sporangium over an extended period. Inside the columnar sporangium there is a central sterile **columella** surrounded by sporogenous tissue. Spores and elaters mature from the apex downward as the sporangium elongates. The sporangium splits at its apex, in the region where the spores and elaters are mature, to release the spores. The sporangium of the hornworts is similar in some respects to the columnar sporangium of the fossil *Horneophyton* of the Rhynie chert, though this has a lobed structure.

4.2.5 The sporophyte of mosses and liverworts

There are some similarities between the sporophytes of mosses and liverworts but enough differences to indicate that they are not closely related and so they are placed in distinct subdivisions, the Bryophytina and Hepaticophytina respectively. In each case the sporophyte consists of only a foot, a stalk called the **seta** and a single sporangium called the capsule (Fig. 4.5). The foot is **haustorial** in the gametophyte. It develops from the zygote within the fertilized archegonium. The archegonium swells to accommodate it. Eventually the seta elongates breaking the archegonium and pushing the capsule out. This happens early in development in mosses but only when the capsule is mature in liverworts.

In mosses the archegonium splits open in such a way that the top part, called the **calyptra**, is carried up as a kind of a hat over the apex of the capsule. Here it continues to influence the development of the capsule until it shrivels and falls off. In liverworts the calyptra is broken and left as a tattered collar around the base of the seta. In some liverworts like *Marchantia* the capsules are elevated not just on the seta but because the whole sporophyte arises from a stalked platform, part of the gametophyte, called the **archegoniophore**. In some bryophytes the seta is absent or very short. In the bog mosses *Sphagnum* there is no seta but the capsule is raised up on a part of the gametophyte called the pseudopodium.

In most liverworts the capsule is spherical or ovoid (Fig. 4.5). The wall of the capsule may be one cell layer thick, unistratose, or have several cell layers, multistratose. In liverworts dispersal of spores is usually aided by the presence of sterile cells or elaters (sections 4.2.1 and 8.3). The capsule splits longitudinally either regularly or irregularly along lines of weakness, into valves, although in some an apical opening is produced. In *Asterella* a cap called the operculum falls off. In *Cyathodium*, tooth-like structures are present around an apical opening. In some liverworts like *Monoclea* the capsule is elongated and the capsule splits by a single longitudinal slit. In *Riccia* there is no seta and no elaters and the spores are released by the capsule disintegrating.

Moss capsules are very diverse (Fig. 4.6). Often they have a basal sterile photosynthetic region called the apophysis. Here there are stomata connecting with intercellular air-spaces between the photosynthetic cells. The apophysis can make a considerable contribution to the energy of the sporophyte. The sporogenous part of the capsule has a central core of non-sporogenous

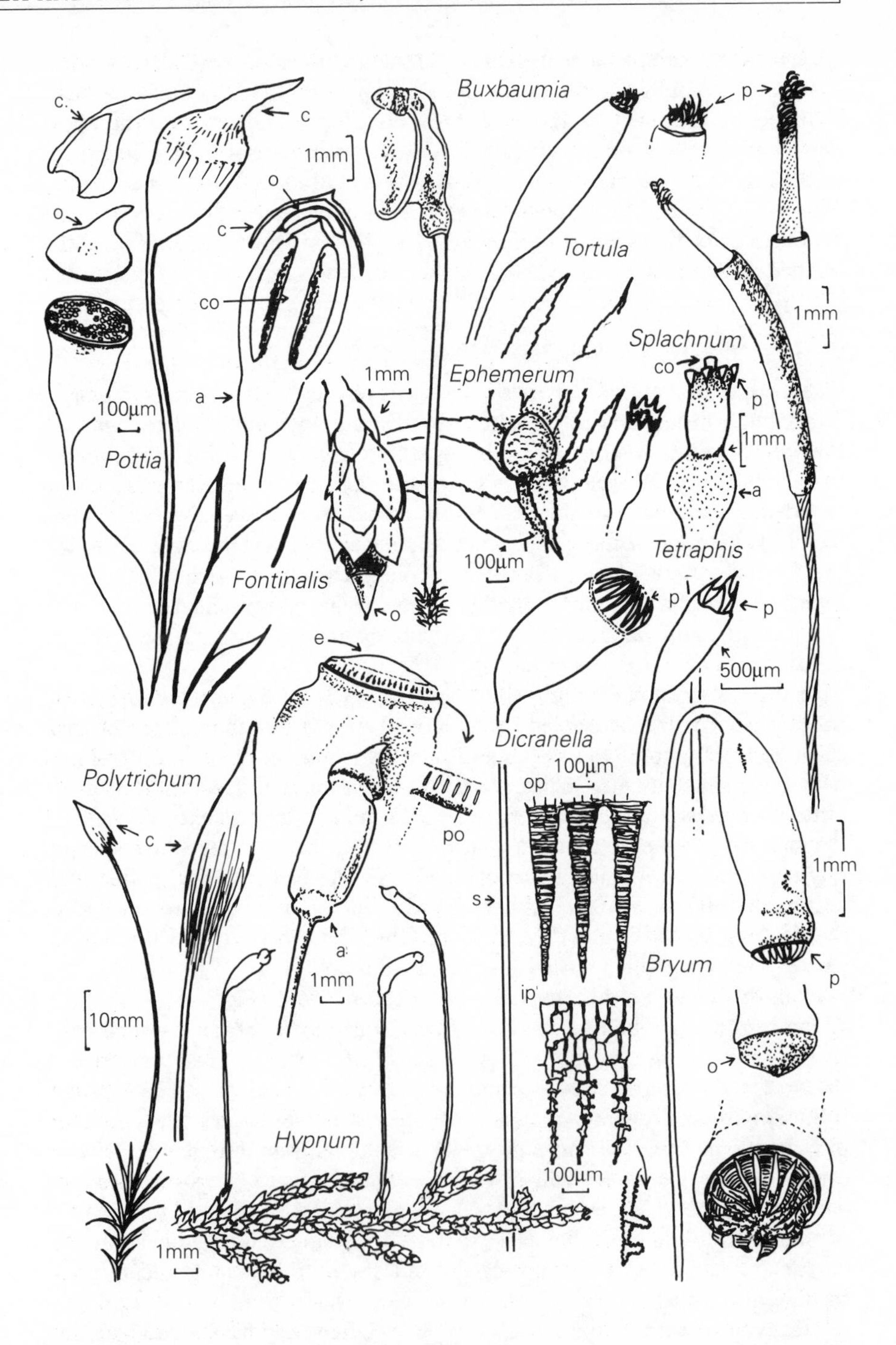

Fig. 4.6. Moss capsule variation. Variation occurs particularly in the form of the peristome. The capsule of *Ephemerum* only releases spores by disintegration. Key: a = apophysis, c = calyptra, co = columella, o = operculum, p = peristome, ip = inner peristome, op = outer peristome, po = pore, s = seta.

tissue called the columella, which runs through the centre of the spore mass (Fig. 4.6). At the apex of the capsule a ring called the **annulus** is present forming an obvious junction with the operculum. Below the operculum, in the class Bryopsida there is a simple or complex **peristome**. The peristome is a ring of 4 (e.g. *Tetraphis*), 8 (e.g. *Splachnum*), or 16 (e.g. *Dicranella*) teeth. The teeth may be multicellular and made of whole cells or constructed from the parts of adjacent cells which have broken down at maturity. Some mosses have a double peristome (e.g. *Bryum*). The peristome takes various forms. It may be twisted. The teeth may be forked or toothed or filamentous. In *Polytrichum* and its relatives a membranous epiphragm joins the teeth together and dehiscence is through gaps between the teeth.

In the bog mosses, class Sphagnopsida (one genus *Sphagnum*) the capsule is a spherical like that in the liverworts but unlike them it has a central columella. There is an operculum which is exploded off by the pressure which builds up inside as the rest of the capsule shrinks longitudinally, when drying. In the stone mosses, Andraeaopsida, (two genera including *Andreaea*) the capsule splits along four lines of weakness to gape open like a 'Chinese lantern', hence they are called lantern mosses.

4.3 THE GAMETOPHYTE

A gametophyte is a plant bearing the sex organs, antheridia and archegonia. A gametophyte arises from a germinating spore and is haploid. In most land plants the gametophyte is small and inconspicuous. In the bryophytes it is the dominant phase of the life cycle and the sporophyte is dependent on it. The archegonia and antheridia may be found together on the same axis. Then they either mature at the same time or at different times; **protandry** = male first, **protogyny** = female first. Alternatively the archegonia and antheridia may be found on different parts of the same plant (**monoecy**) in either male or female axes, or they are found on separate male and female plants (**dioecy**).

4.3.1 Archegonia

Archegonia are the female sex organs of land plants (Fig. 4.7). They are not present in algae which have instead a structure called an oogonium which differs from an archegonium in having no protective layer of cells around it. Each archegonium contains a single egg. Archegonia are strikingly similar throughout the lower plants. Each is made up of two parts; the lower chamber, called the **venter**, and the **neck**. Archegonia differ mainly in the number of cells they are composed of and in the relative size of the neck and venter. In some groups the archegonia develop from a superficial cell of the plant, and in others from a deeper cell. The wall of the archegonium is usually a single layer of cells, but in bryophytes the venter wall is two or more cells thick. An important stage in the development of the archegonium is the separation of the wall from the interior cells. The lowest of the interior cells becomes the single egg cell. Above it, there is a **ventral canal cell** and

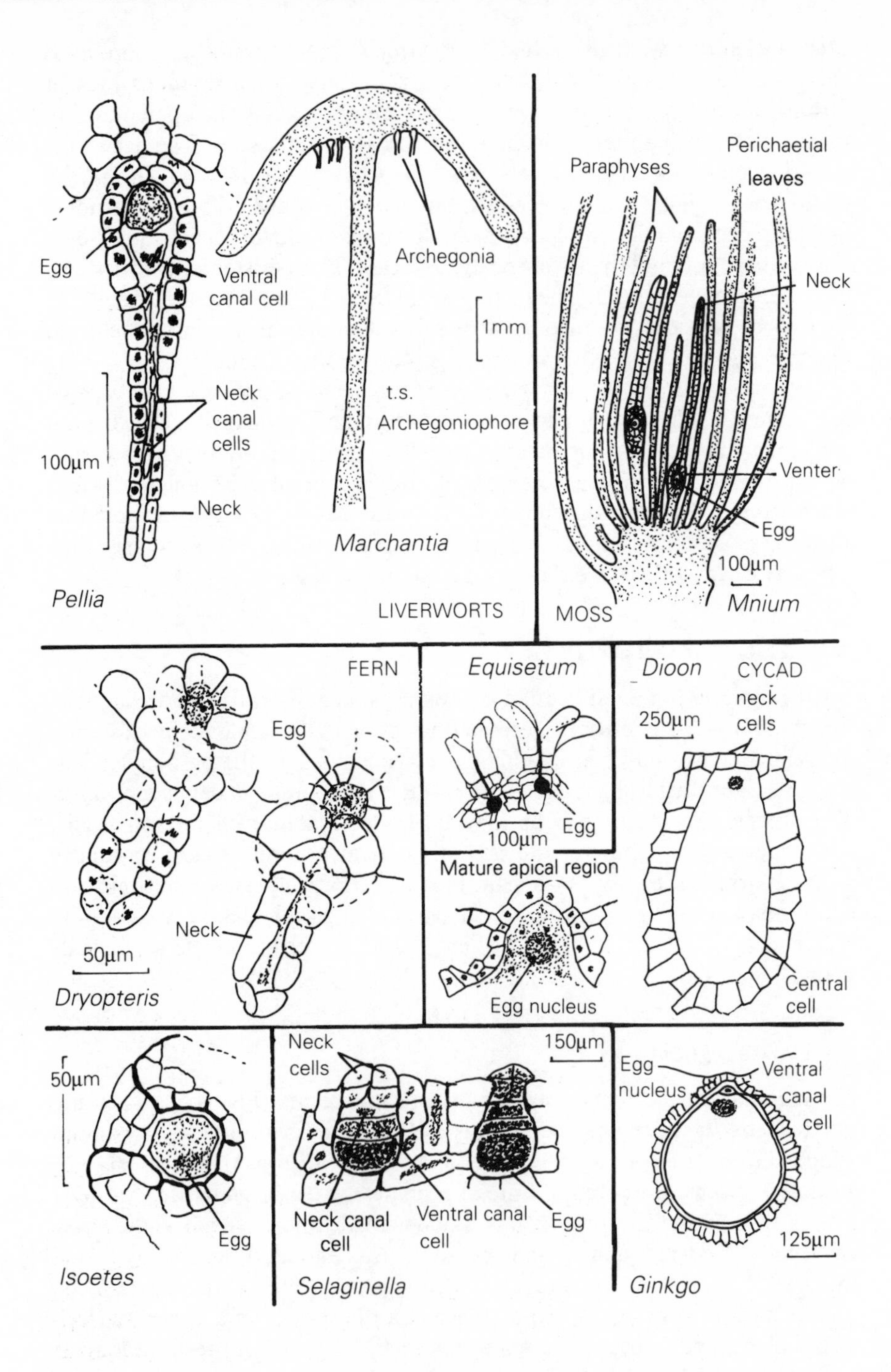

Fig. 4.7. Archegonia. (*Dioon* and *Ginkgo* from Coulter and Chamberlain, 1910.) The egg cells of cycads are very large and clearly visible to the naked eye.

several **neck canal cells** (e.g. *Pellia*, Fig. 4.7). The canal cells disintegrate at maturity to create the canal via which the sperm enters the archegonium.

In mosses and liverworts the archegonia are stalked and free, though they may be protected in an archegonial chamber. In all other land plants the archegonium is confluent with the body of the gametophyte though part of the neck may protrude above the surface (e.g. *Dryopteris*, Fig. 4.7). The degree of development of the neck and the neck canal cells varies in different groups. In ferns, for example, cell walls are not formed between the neck canal nuclei.

In seed plants there is an evolutionary trend of the reduction of the archegonium. A jacket layer of cells surrounds the central cells. The number of central cells varies. In *Ginkgo* (Fig. 4.7) a large egg cell, a ventral canal cell and four neck cells but no neck canal cells are present. In cycads (e.g. *Dioon*, Fig. 4.7) the pattern is similar but a cell wall is not produced between the egg nucleus and the ventral canal nucleus which is short lived. Conifers may follow either pattern but frequently have fewer neck cells than *Ginkgo*. In *Ephedra* there are 40 or more neck cells. In *Gnetum*, *Welwitschia* and the angiosperms no readily identifiable archegonia are recognizable. The egg nucleus or egg cell are found with other nuclei or cells which cannot be directly related to archegonium cells.

4.3.2 Antheridia, sperm and fertilization

The **antheridium** is the male sex organ within which the sperm are produced. It usually consists of a spherical or ovoid sac in which sporogenous cells give rise to sperm (Fig. 4.8). Antheridia vary greatly in size between different groups. In bryophytes the antheridium is stalked and either superficial or buried in antheridial chambers. The sterile antheridial jacket is a single cell layer. At maturity in mosses the cells at the tip of the antheridium break down and the jacket contracts squeezing out the sperm. As they mature the antheridia in chambers push against and rupture the roof of the chamber.

In ferns, *Equisetum* and whisk ferns, the antheridia are superficial. In clubmosses they are massive, buried and confluent with the gametophyte. In many cases, as in *Equisetum*, cap, opercular or cover cells at the tip of the antheridium can be recognized which break down to release the sperm. The jacket usually consists of only a single cell layer but in one family of ferns, the Ophioglossaceae, there are two layers. In many ferns the jacket consists of only three cells; two are ring-like and form the base, and there is a single cap cell. The cap breaks open as the contents of the antheridium swells. The number of sperm produced by each antheridium varies from hundreds, even thousands, in some bryophytes and ferns, to only four in *Isoetes*. In the seed plants there is no antheridial sac. The antheridium is reduced to only one or two cells and only two sperm or sperm nuclei are produced (section 4.5.5).

Several different kinds of male gametes have been described. In whisk ferns, horsetails, *Isoetes*, ferns, cycads and *Ginkgo*, a multiflagellate sperm is produced with the flagella arranged in a spiral around the pointed end of

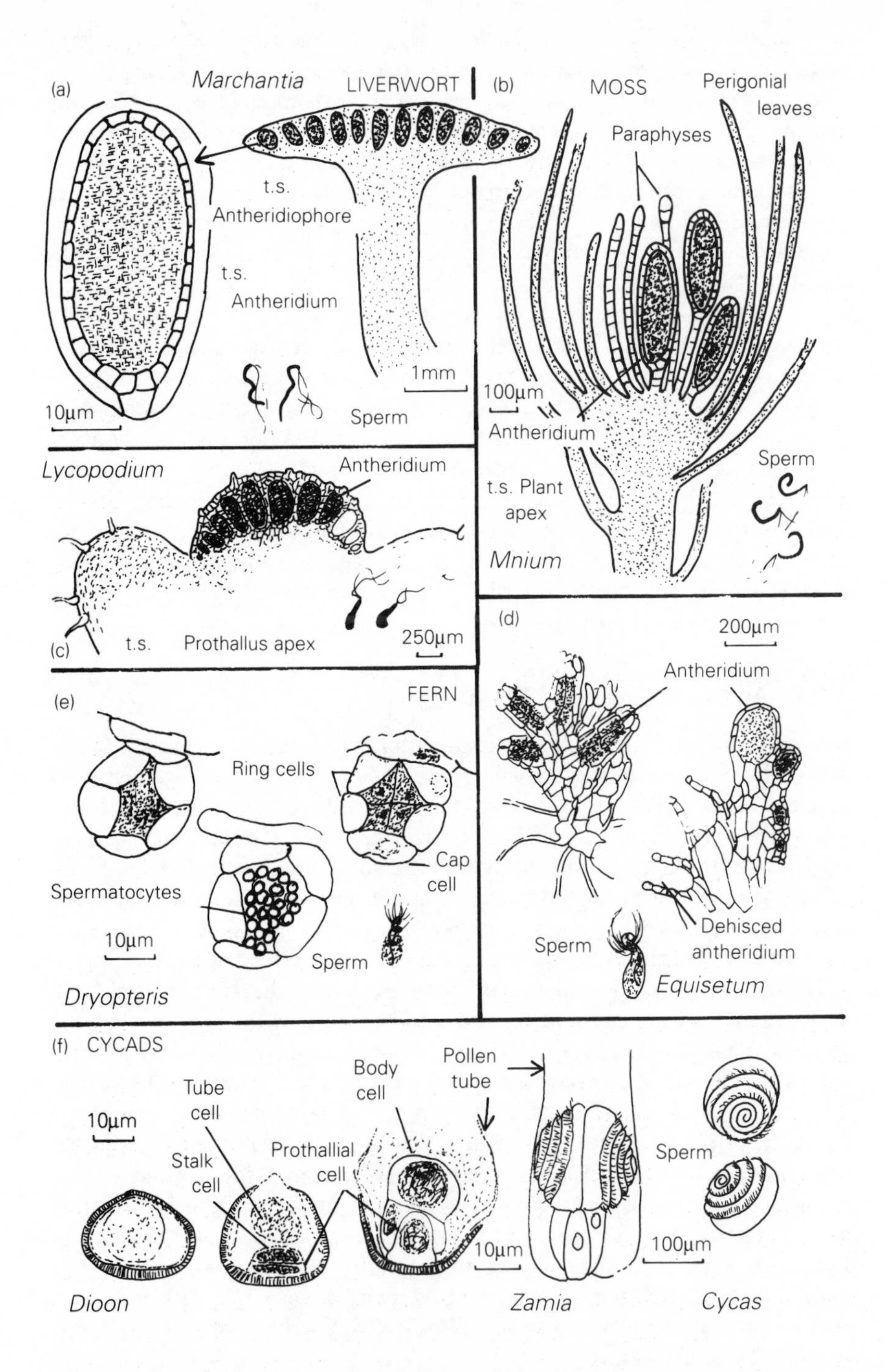

Fig. 4.8. Antheridia and sperm ((c) from Bruchman (1910); (d) from Duckett (1973); (f) from Coulter and Chamberlain (1910)). Sperm are not drawn to scale.

the rounded cone-like sperm. In the bryophytes, *Lycopodium* and *Selaginella* there is a biflagellate sperm. *Selaginella* and *Isoetes* are often placed in the same order, the Glossopsida, but the difference between their sperm is strong evidence for them not being very closely related. It is *Selaginella* and *Lycopodium* which are peculiar amongst tracheophytes, Tracheophyta, in having a biflagellate sperm. The possession of a biflagellate sperm is primitive and is shared with many algae as well as the bryophytes.

In the conifers, the gnetophytes, and the angiosperms the sperm is non-flagellate. The male cells or nuclei are conveyed towards the egg in a pollen tube in a process called siphonogamy. A pollen tube is also produced in cycads and *Ginkgo* but in these groups it has a haustorial function and is not directly involved in fertilization, and flagellate sperm are produced.

Flagellate sperm have very limited motilities. They may swim rapidly but only for short distances; however, sperm have been observed to be splashed distances up to 60 cm by rain. Archegonia produce a chemical sperm attractant which in ferns can be experimentally replaced by malic acid. In ferns the maturing prothallus produces a chemical signal, antheridiogen, as it produces archegonia. The antheridiogen encourages young neighbouring prothalli to produce antheridia. In this way cross fertilization is promoted.

4.3.3 Gametophyte variation in pteridophytes

There are three main growth forms of gametophyte in pteridophytes (Fig. 4.9). The gametophyte may be green and grow at the soil surface. Alternatively it may be a subterranean tuber like structure which is usually associated with a fungus and lives **saprophytically**. A third kind develops within the spore wall, that is **endosporically**, and is described in detail in section 4.4.3. All higher plants, the gymnosperms and flowering plants, have an endosporic gametophyte. In lycophytes (Lycophytina) all three kinds of gametophyte are found. *Selaginella* and *Isoetes* are endosporic while in the clubmosses, *Lycopodium sensu lato*, different species have either superficial and green or saprophytic tuberous gametophytes.

The gametophyte is often called a **prothallus**. *Lycopodium* has a large range of sizes and types of gametophyte. In some species it is a few millimetres long, lobed and bearing rhizoids below (Bruce, 1979a, b). The sex organs are produced between the lobes. It is monoecious bearing both male and female sex organs. In most ferns the prothallus is a little like a thalloid liverwort, a heart shaped or lobed flap of green cells, a few millimetres in length, with a thickened central region and rhizoids below. The prothallus may be infected by a mycorhizzal fungus. Near the notched apex archegonia are produced and further back amidst the rhizoids there are antheridia. Development is slightly protandrous, antheridia mature first.

The tuberous kind of gametophyte in *Lycopodium* is 1–3 cm long, and conical or disc shaped. There is a substantial mycorrhizal zone where the fungal hyphae penetrate between the cells. This apparently allows the plant to survive saprophytically in the absence of photosynthesis. Sex organs are produced buried in a kind of apical crown. A similar saprophytic growth

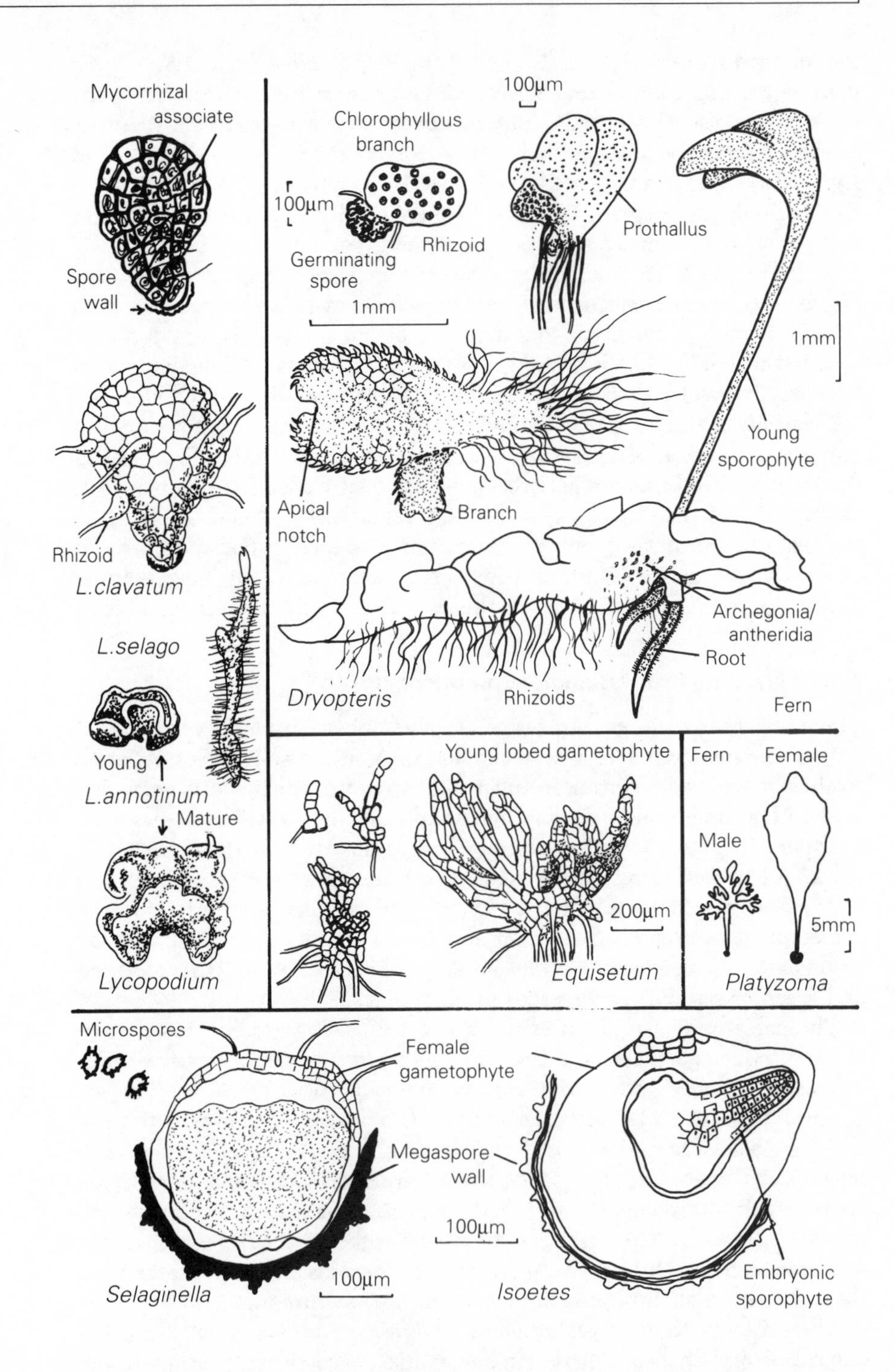

Fig. 4.9. Gametophyte variation in pteridophytes: prothalli, mycotrophic tubers and endosporic development. (*Lycopodium* from Bruchman (1910), *Equisetum* from Duckett (1973), *Platyzoma* from Duckett and Pang (1984).)

form is found in the whisk fern *Psilotum*. In this case the gametophyte is cylindrical, up to 2 mm in diameter and branched dichotomously. It looks rather like a part of the rhizome of the mature sporophyte, which is evidence for a **homomorphic alternation of generations** being the primitive condition of land plants. Antheridia are superficial and archegonia are partially sunken in it. Rhizoids are present as well as a mycorrhizal fungus. Similar branched cylindrical gametophytes or globose or hemispherical ones are found in the fern class Ophioglossopsida. Small amounts of chlorophyll may be found at the apex of these mainly saprophytic fern gametophytes if they are exposed to the light.

Dioecism is uncommon in **homosporic** pteridophytes. *Equisetum* is the exception in being **heterothallic**, producing male and hermaphrodite prothalli. The green prothalli grow on the soil surface, have rhizoids and are very small (1–8 mm) (Duckett, 1973). Male gametophytes are lobed with a few sterile lamellae. The other kind of gametophyte is protogynous: at first they have many sterile lobes between which the archegonia are situated; later, after several weeks, female development ends and antheridial lobes are produced.

4.3.4 Gametophyte variation in the bryophytes

Only in the bryophytes has the gametophyte become a diverse and well developed phase of the life cycle. However, even in the bryophytes a sexually effective gametophyte has climatic limits. The bryophytes of arctic and antarctic regions reproduce mainly asexually (Schofield, 1985). The largest bryophyte forms are aquatic, procumbent or pendulous forms up to a metre in length. The structure of the mature vegetative gametophyte is described in detail in Chapter 7. Unlike the tracheophytes the gametophyte in bryophytes is the dominant generation of the life cycle.

In most bryophytes the spore germinates to produce a juvenile filamentous stage called the **protonema**. The filaments are usually uniseriate. In mosses, the protonema can form an extensive branched axis with several kinds of filaments. Prostrate creeping filaments are green and have either transverse cell walls (**chloronema**) or oblique cell walls (**caulonema**). The latter are identical in structure to the non-green rhizoids which are also produced. Chlorophyllous upright filaments have transverse cell walls. In the stone moss, *Andreaea*, the protonema is multiseriate. Cells in the caulonema round up and by oblique cell divisions produce the apical cell of a bud which gives rise to the mature gametophyte. In the bog moss, *Sphagnum*, the filamentous protonema soon enters a thalloid phase from which the mature gametophyte then arises.

The reduction of the juvenile protonemal phase appears to be an evolutionary advanced feature in bryophytes. Development of the mature gametophyte directly from a spore in liverworts and hornworts can therefore be regarded as advanced; the juvenile filamentous phase consists of only two or three cells. In thalloid liverworts the apical cell of the protonema divides to produce a quadrant of cells, giving rise to a plate of cells in which the apical cell of a growing thallus appears. In leafy liverworts the protonema produces a clump of cells sometimes called the sporeling from which the mature liverwort arises. Some hornworts enter a thalloid growth phase directly. In the moss *Dicnemon* and the thalloid liverwort *Conocephalum* development of the gametophyte is direct from the multicellular spore.

In mosses the archegonia and antheridia are often found at the apex of the leafy shoot, sometimes called the **gametophore**, intermixed with sterile hairs called **paraphyses**. In **acrocarpous** mosses the inflorescence terminates the unbranched or weakly branched shoot. In **pleurocarpous** mosses the gametophores are short lateral branches of the main vegetative, usually profusely branched, shoot. The majority of mosses are **dioecious** or where **monoecious** they are often either **protandrous** or **protogynous**, or **diclinous**. The antheridia and archegonia are usually found in a cluster surrounded by an involucre of leaves. A male involucre is called a **perigonium**. A female involucre is a **perichaetium**. These involucres function as rain splash cups. Rain drops splash sperm from the perigonium into the perichaetium. The paraphyses help to hold a reservoir of water within the splash cup. Paraphyses are usually uniseriate filaments with the apical cell swollen. In some genera they are multiseriate and club shaped. Similar kinds of splash cups are found in the fossil *Lyonophyton* from the Rhynie chert.

In the bog moss *Sphagnum* paraphyses are absent. Perichaetia are present but antheridia are found individually at the base of each leaf in the upper third of a male branch. The stone moss *Andreaea* has perigonia on short lateral branches, and perichaetia at the shoot apex.

In leafy liverworts the sex organs are either lateral in axils of leaves as in the Calobryales or the archegonia are apical, with a kind of perichaetium, called the perianth surrounding them. Various kinds of perianths are found in different species. A few genera such as *Calyopogeia* have a subterranean pouch, called the marsupium which protects the archegonia.

In the thallose liverworts the antheridia and archegonia may be exposed or buried in the surface of the thallus in a chamber or a pouch with a pore to the exterior. Some have the sex organs protected by leaf-like flaps, or have an involucre called a **perigynium** protecting the archegonia. In *Marchantia* the archegonia and antheridia are raised on upright umbrella- or mushroom-shaped projections of the thallus, called the archegoniophore and the antheridiophore respectively. The archegoniophore, undergoes considerable growth so that the archegonia which are at first exposed on its upper surface are carried onto the under surface by its growth. The antheridia are buried in the top surface of the antheridiophore.

In hornworts the antheridia are in chambers, in some cases several to a chamber. They arise in groups behind the growing point. Later the archegonia arise in a similar position.

Smallness in bryophytes is a specialization which has advantages in those habitats and niches which they occupy. One subclass of mosses, the Buxbaumiidae, is interesting in that it shows a great reduction of the gametophyte. *Buxbaumia* (Fig. 4.6) is only obvious when the sporophyte is present. The gametophyte is tiny. The protonema produces side branches which bear perigonia or perichaetia.

The whole bryophyte life cycle may be very rapid. They are adapted to take advantage of transient but abundant water. Many can then dry out to rehydrate rapidly when water becomes available again. Many bryophyte habitats have been created by the evolution of large plants; in their shade or with the bryophyte growing as their epiphytes and epiphylls. These habitats all require smallness. It is intriguing that evolutionary reduction in the three groups, liverworts, hornworts and mosses, which have long separate evolutionary lineages, has lead to evolutionary convergence, so that the leafy gametophytes of mosses and liverworts resemble each other and the thalli of the hornworts and the thalloid liverworts resemble each other.

4.4 HETEROTHALLISM

With the evolution of dioecy (separate male and female plants) or gynodioecy (separate male and hermaphrodite plants) one evolutionary innovation of great importance was **heterothallism**, the specialization of the gametophyte for its particuliar sex, either male, female or hermaphrodite. In *Equisetum* spores of only one size are produced but they may germinate and grow into two different kinds of gametophyte. Small gametophytes are male while large ones are hermaphrodite but at first female and then male (protogynous). The larger size of hermaphrodite or female gametophytes is readily understood because eggs are larger than sperm, and produced individually in archegonia: a larger female gametophyte is required to produce eggs in large numbers. In all land plants the female or hermaphrodite gametophyte also supports the development of the sporophyte at least in the early stages of embryogenesis.

4.4.1 Anisospory

In all heterothallic plants, apart from *Equisetum*, heterothallism is associated with the production of spores of different sizes, either within the same sporangium, a condition called **anisospory**, or in different sporangia, a condition called **heterospory**. The observation of different sizes of spore in some fossil taxa has been interpreted by some workers as being evidence for either the production of functional and abortive spores (eg. *Chaleuria*, Andrews *et al.*, 1974) or as anisospory. Taylor and Brauer (1983) claim from ultrastructural comparisons that the spores of the fossil *Barinophyton*, which fall in two size classes, 30–50 μm and 650–900 μm, were all functional. Anisospory gradually increased in the Devonian. In contrast heterospory became suddenly important at the beginning of the Upper Devonian.

Anisospory is uncommon in living land plants and unknown in tracheophytes. Variation in spore size may be analogous to variation in seed size within some species of flowering plants; part may result from competition between spores for the same resources in development within the same sporangium, and perhaps has adaptive value where the environment is very heterogeneous, favouring two different strategies for dispersal and germination. Anisospory is confined in the living land plants to a few mosses and here it is associated with heterothallism. Although a large proportion of bryophytes are dioecious (over 60% of the Bryidae), there is usually no heterothallism. In a few however dioecy is accompanied by striking anisospory and heterothallism (Ramsay, 1979). In *Macromitrium* spores of two sizes are produced. Two of the spores of each tetrad remain small giving rise to male plants, while the larger spores produce female plants. The male plants are dwarf and grow epiphytically on the female plants. Under the chemical influence of the female host the males consist only of a perigonium and a few rhizoids. In this peculiar genus, males growing separately may be either dwarf or identical vegetatively to the female.

4.4.2 Heterospory

Heterospory is a much commoner phenomenon than anisospory. Heterospory has evolved separately in several different groups. The first land plants were homosporous. In living plants, with a few exceptions, all bryophytes, whisk ferns and horsetails are homosporous. Lycophytes and ferns contain some heterosporous taxa: *Selaginella*, *Isoetes*, *Stylites* in the former and aquatic ferns in orders Marsileales and Salvineales in the latter. All gymnosperms and flowering plants are heterosporous. Two different size of spores, **megaspores** and **microspores**, are produced in different sporangia, called megasporangia and microsporangia respectively. The megaspores germinate into female gametophytes, with the exception of the fern *Platyzoma* where they are hermaphrodite (Fig. 4.9). The microspores germinate into male gametophytes. Megaspores are produced in smaller numbers than microspores.

The fossil relative of the horsetails, the calamites may show an evolutionary trend from homosporous forms to heterosporous forms. *Calamostachys binneyana* is apparently homosporous, though this fossil may be the microsporangiate cone of a heterosporous species where no fossils showing connection to a megasporangiate cone have yet been discovered. *Calamostachys americana* is certainly heterosporous with megasporangia containing spores three times as large as those in the microsporangia in the same cone. In *Calamocarpon* each megasporangium contains only one very large functional megaspore. It is likely that dioecy was the first step in the evolution of heterospory. There are many dioecious homosporous taxa but no monoecious heterosporous non-seed plants. There is a strong relationship between heterospory, dioecy and heterothallism. This is illustrated in the anomalous **gynodioecious** leptosporangiate fern *Platyzoma*, which grows in north-eastern Australia. It is heterosporous but not very markedly so. It is quite strongly heterothallic. It produces 32 microspores (diameter 71–101 μm) in each of its microsporangia and 16 megaspores (diameter 163–83 μm) in each of its megasporangia (Duckett and Pang, 1984). The microspores produce a filamentous male gametophyte which produces only antheridia. The megaspores produce a spathulate hermaphrodite gametophyte which initially produces archegonia and later antheridia.

Heterospory may be considered as an evolutionary specialization whereby one class of spores, the microspores, enhance the potential for cross fertilization since they are small and produced in large numbers and are easily dispersed. The other class of spore, the megaspores, enhance the survivability of the female gametophyte, maximize the production of eggs and enhance the early life of the sporophyte after fertilization. Separation of the sexes precludes any advantages small spores confer for the dispersal and colonization of new geographical areas. The larger megaspores are less easily dispersed in the air. Microspores and megaspores have to settle in close proximity for fertilization to be effected. If air dispersed this is very unlikely to occur away from the original range of the species. Heterospory is particularly advantageous, or least disadvantageous, in an aquatic environment where the megaspore is possibly as easily dispersed in water as the microspores. This helps to explain the presence of heterospory in aquatic ferns.

4.4.3 Endospory

Endosporic development is development of the gametophyte which occurs within the spore wall, so that, when dispersed, spores may be multicellular. In lower plants the spore germinates and the gametophyte grows out of the spore wall. However, higher plants have a tiny gametophyte which develops almost entirely within the spore wall. These plants are all heterosporic and heterothallic.

Selaginella and *Isoetes* are pteridophytes with endosporic development. Within the microspore the first cell division produces a sterile prothallial cell and an antheridial cell. The prothallial cell does not develop any further. The antheridial cell divides to produce a single antheridium. The antheridium consists of a jacket layer of cells surrounding either 128 or 256 **spermatocytes**. Each spermatocyte gives rise to a single sperm. When the antheridium is mature the prothallial cell and jacket cells disintegrate and the spore wall ruptures to release the sperm. The female gametophyte develops within the megaspore wall (Fig. 4.9). There is a period of free nuclear divisions and then apical nuclei become separated from each other by cell walls. At the tri-radiate apical scar the spore wall splits open and rhizoids grow out. The female gametophyte protrudes from the megaspore but usually it remains achlorophyllous, relying on the nutrient reserves in the undifferentiated mass below. Archegonia develop in the surface layer of the exposed region.

The reproduction of *Selaginella* differs from *Isoetes* in a number of ways. In *Isoetes* the megasporangium is at the base of the sporophyll and protected by a flap called the velum (Fig. 4.2). It produces between 100 and 300 megaspores. In *Selaginella* a single tetrad of megaspores is produced. The next evolutionary step was a further reduction in the number of megaspores which develop in each megasporangium so that only one megaspore from a single tetrad develops. This trend has already been noted for fossil calamites ending with the single-megaspored *Calamocarpon*. In living aquatic ferns like *Marsilea* and *Pilularia* (Fig. 4.4) only a single megaspore develops in, and is released from, each megasporangium.

Selaginella differs from *Isoetes* in another way. Development of the female gametophyte occurs within the megaspore while it is still in the megasporangium. It has been observed that, in some cases, archegonia are present at the time of megaspore release. It is imaginable that fertilization could occur before release if a microspore was to land close to the megasporangium.

The combination of a reduction to one megaspore per megasporangium, and its retention within the megasporangium, along with the development of the female gametophyte endosporically was the next step in the evolution of reproduction in land plants. It probably happened at least twice in different evolutionary lines giving rise to different groups of seed plants.

4.5 SEED PLANTS: THE OVULE

An **ovule** is the megasporangium of seed plants containing a single megaspore surrounded by an extra structure, the **integument** (Fig. 4.10). The megasporangium is usually called the **nucellus**. One or more integuments may be present. The female gametophyte develops endosporically in the surviving megaspore of a tetrad. The tetrad is linear and not tetrahedral as it is in non-seed plants. There is a close connection between the megaspore and the nucellus. The haploid female gametophyte which develops is almost entirely submerged in the diploid sporophyte which gave rise to it. The ovule has a **micropyle** at one end, a passageway between the lobes of the integument. The other end is called the **chalaza**.

The probable course of evolution of the ovule has been traced in several groups of progymnosperms and seed ferns. A series of fossil pre-ovules shows the envelopment of the megasporangium by sterile branches and their gradual fusion to form an integument. This has been accompanied by a reduction in the wall or exine of the megaspore, its function now superseded by the integument. A further stage of envelopment is provided, in some fossils, by the presence of a cupule, like an outer integument, and derived from a ring of sterile branches in the same way as the integument. In some evolutionary lineages the inner integument becomes fused to the nucellus.

Endospory and the retention of a single megaspore was associated with an increase in the size of the megaspore. The integument may have evolved for protection, to promote efficient pollination or for dispersal. The ovule represents a concentrated, highly nutritious energy source for predators. After fertilization the ovule develops into the seed. In living plants very high loss of seed has been demonstrated prior to dispersal (Louda, 1982). The presence of glands on the cupule of the fossil *Lagenstoma* suggest that predators have always been a problem for plants. It is possible that the integuments and cupules evolved to conceal and protect the developing female gametophyte and later the developing seed from predation. Another possibility is that they provided protection from the environment, perhaps drought. This is especially important in large plants where the ovules are produced high up and in an exposed position.

The integument also may have evolved to help promote the efficient reception of the microspores, which now had to land not on the soil but on or near the nucellus. Some of the earliest ovules, in a fossil called *Genomosperma* from the Lower Carboniferous, had a nucellus with an elongated funnel shaped tip called a salpinx which probably functioned to trap pollen and direct it to a pollen chamber above the archegonia (Long, 1975). Projecting integuments are found in the living gnetophytes (Fig. 3.10). The lobed structure of early integuments would have modified patterns of airflow to allow pollen to settle on the nucellus (Niklas, 1985). Alternatively the integuments may have prevented rain interfering with pollination in the pendulous ovules.

A third, and eventually all important adaptation, was the protection the integuments gave the seed in dispersal. Some integuments are winged as an aid for dispersal.

In gymnosperms the food reserves of the seed are derived from gametophytic tissue existing before fertilization and are potentially wasted if fertilization does not occur. In angiosperms food reserves are transferred to the developing seed only after a successful fertilization.

4.5.1 The female gametophyte in seed plants

An evolutionary reduction of the gametophyte can be traced in living plants. In seed plants a linear tetrad of megaspores is produced of which three disintegrate. Development of the megaspore and female gametophyte involves the production of a megaspore wall in most gymnosperms but not in angiosperms. Megaspore wall formation is particularly conspicuous in cycads but is not or only scarcely visible in some conifers.

Female gametophyte development usually begins with a free nuclear phase and then cell walls are formed. The size and number of nuclei which are produced varies between groups. In some cycads the ovules are up to 6 cm in length and between 1000 and 3000 nuclei have been counted. In *Ginkgo* there are up to 8000 nuclei, 7200 in *Gnetum* and 2000 in the conifer *Pinus*. This free-nuclear phase may last for a considerable time. In *Pinus* there is a halt at the end of one season at the 32 nuclei stage and nuclear division resumes in Spring. Cell wall formation begins at the periphery of the coenocytic gametophyte and proceeds centripetally. In the cycads and *Ginkgo* regular hexagonal shaped cells called alveoli are produced. After cellularization some of the cells at the micropylar end become archegonial initials. The number of archegonia produced varies considerably. Only two are produced in *Ginkgo*, usually between 2 and 6 in cycads, and sometimes many in conifers.

The process of female gametophyte development in the gnetophytes is rather like other gymnosperms in *Ephedra*. The number of nuclei varies between 256 and 1000 in different species. However in *Gnetum* the female gametophyte remains free-nuclear and no archegonia are formed. A free nucleus at the micropylar end functions as an egg. In *Welwitschia* multinuclear cells are formed. In fertilization these produce tubes which grow up through the nucellus to meet the pollen tube and one of the nuclei functions as an egg.

Most angiosperms, with **monosporic** development, produce a tetrad of which three nuclei disintegrate, like gymnosperms, but no megaspore wall is produced. In other angiosperms, with tetrasporic development, no cell wall of any sort is formed between the four nuclei of the megaspore tetrad and there is no subsequent disintegration of the nuclei, which are all involved in the development of the female gametophyte. This evolutionary specialization avoids the waste of three quarters of the tetrad. In other angiosperms development is bisporic with two megaspore nuclei going on to produce the female gametophyte. In angiosperms the female gametophyte, called the **embryosac,** is very small. It consists of between four and sixteen nuclei in different taxa and is only partly cellularized. One pattern with eight nuclei is illustrated diagrammatically in Fig. 4.15. Three nuclei at the chalazal end develop cell walls and are called antipodal cells. Two polar nuclei which migrate to the centre from opposite poles of the embryosac may fuse before fertilization to form a secondary nucleus or central cell. The remaining cells are two synergids and an egg.

4.5.2 Cones, cupules and ovaries

A range of structures bearing and protecting the ovules are found (Fig. 4.10). The gymnosperms are so named because they are supposed to bear their seeds unprotected (*gymnos* = naked in ancient Greek). This is true for some of them, so for example in *Ginkgo* and *Taxus* the ovules are exposed on peduncles, in pairs in *Ginkgo* and singly in *Taxus* (Fig. 4.10). In *Phyllocladus* they are borne on lateral shoots called phylloclades surrounded by scales. In *Cephalotaxus* they are borne on short lateral fertile shoots again surrounded by bracts but still rather exposed. *Cycas* has loosely arranged pinnate megasporophylls with a number of ovules at the base, all with a dense covering of hairs. Another cycad *Ceratozamia* has just two ovules on each megasporophyll.

However, a common pattern in conifers is for megasporophylls to be arranged in a tightly packed cone. The cones protect the developing ovules but gape open in the right season to allow the pollen to reach the ovules. In the conifer family Pinaceae the ovules are situated on the adaxial side and near the base of a tough woody ovuliferous scale (Fig. 4.11). Each megasporophyll has an accompanying bract scale in these conifers. Some fossil gymnosperms like the Caytoniales have a cupule, a spherical structure containing ovules. Cupules were arranged opposite each other, like pinnae on a sporophyll.

In the gnetophytes *Ephedra* has a rather conifer-like female cone though with usually only two ovules. In *Gnetum* the ovules are exposed on a kind of female catkin and in *Welwitschia* they are surrounded by a '**perianth**' of fused bracts each '**floret**' hidden within the cone (Fig. 3.10).

In angiosperms the ovules are enclosed in a **pistil**, a feature which gives them their name (*angion* = vessel in ancient Greek). Unique characteristics

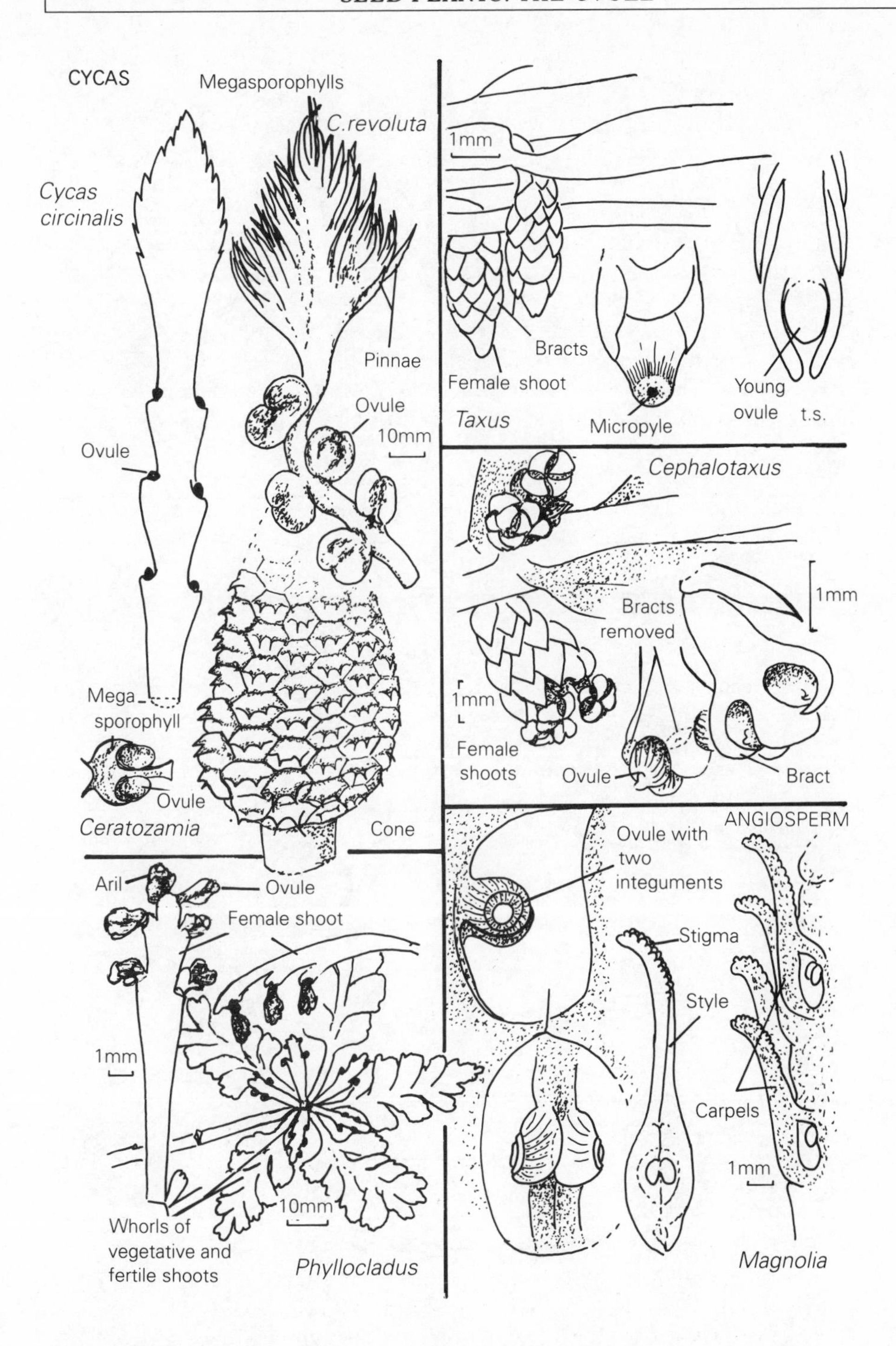

Fig. 4.10. The gynoecium in seed plants. There is considerable variation in the conifers (*Taxus, Cephalotaxus, Phyllocladus,* and see Fig. 4.11), in the cycads (*Cycas* and *Ceratozamia*) and the angiosperm *Magnolia.*

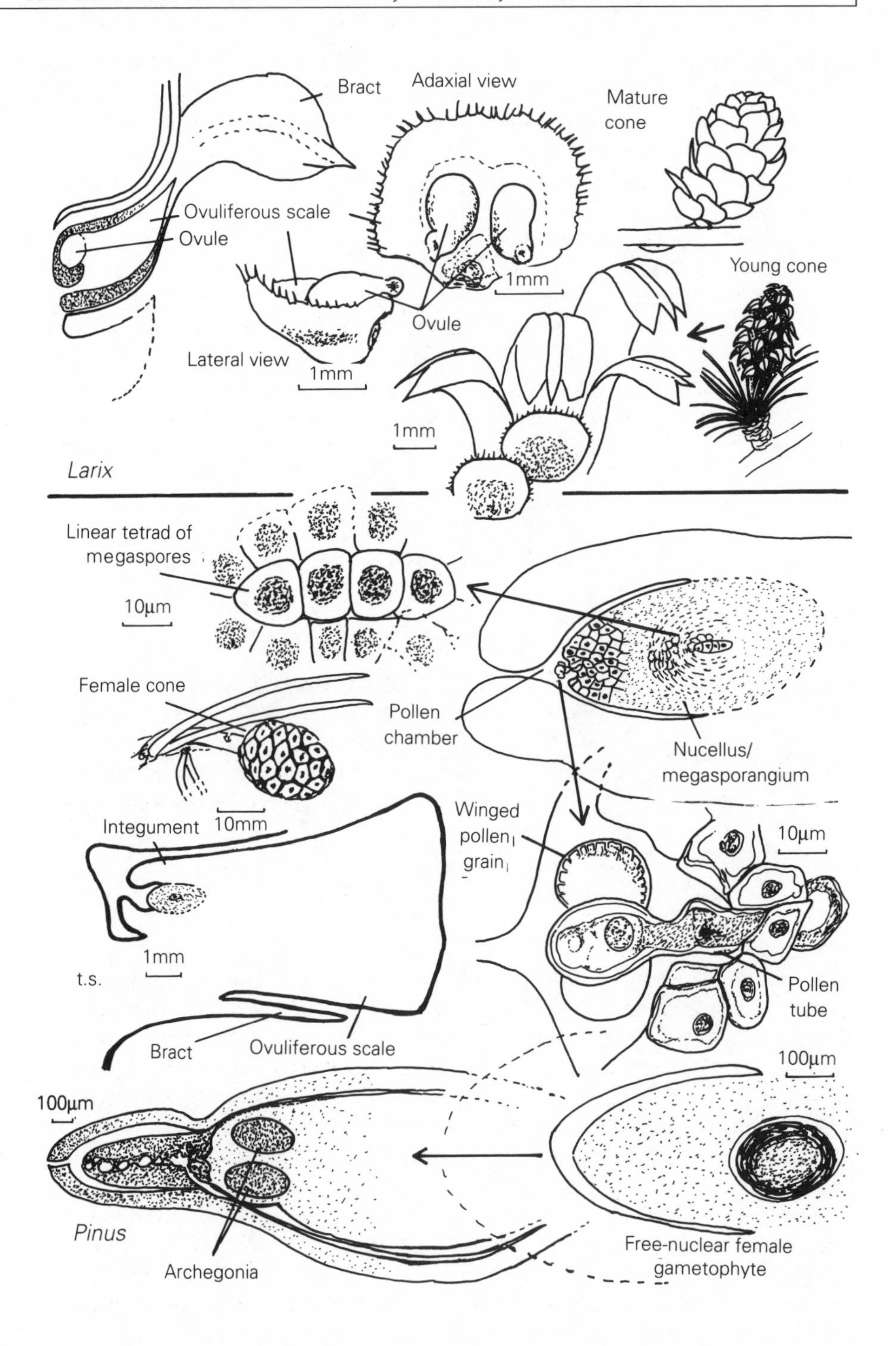

Fig. 4.11. The ovuliferous cone and ovule development in two species of the Pinaceae: *Pinus* and *Larix*.

of the angiosperms are the accessory structures of the pistil; the **stigma** and **style**. The stigma receives pollen and the style conducts the pollen tube to the vicinity of the ovules. The basal chambered part of the pistil, containing the ovules, is called the **ovary**. Each ovule has a stalk called a funiculus which connects it to the ovary wall. The ovules are arranged either singly or in groups in several different ways (section 5.4.2). In many angiosperms the ovule is protected by two integuments. It is possible that the outer integument is derived from a cupule-like structure. Collectively a set of pistils in a flower is called the gynoecium. They are often fused together creating a compound ovary or pistil. The pistil may have evolved to protect the ovules from insect pollinators with biting mouth parts.

One kind of primitive pistil, called a carpel, is like a folded fertile leaf, with ovules arising on the adaxial surface but appearing marginal because the distal surface are adpressed along the line of the suture. The stigma is more or less sessile. In fact the fundamental form of the carpel is as a peltate more or less tubular leaf (Weberling, 1989) as if the megasporophyll has first become concave and then grown up around the ovules. This helps to explain the orientation of the ovule and the origin of the stylar canal.

4.5.3 Microsporophylls, microsporophores and stamens

In most gymnosperms the **pollen sacs** (microsporangia) are grouped together on the abaxial surface of special scale leaves called **microsporophylls** (Fig. 4.12). In some cycads like *Cycas* there are over a thousand pollen sacs on each microsporophyll. In the cycad *Zamia* there are only five or six. Similarly in the conifers each sporophyll may bear many pollen sacs as in the Araucariaceae, but it is normal to have between two and nine pollen sacs to each scale: two in *Pinus*, four in *Cephalotaxus*. In both cycads and most conifers the microsporophylls are arranged spirally in a male cone or strobilus (*Pinus*) or on a pendulous catkin as in *Podocarpus*. In *Taxodium* the male cones are arranged in a pendulous compound microstrobilus.

A number of other arrangements are found. The Taxaceae (*Taxus*) are peculiar in the conifers in having a stalked structure, a peltate microsporangiophore, reminiscent of the **sporangiophore** of *Equisetum*, bearing 6–8 pollen sacs (Fig. 4.13). The microsporophylls of the Taxodiaceae (including *Taxodium* and *Sequoia*) are somewhat similar to those of *Taxus* though their derivation from the more usual microsporophyll is more obvious. In *Ginkgo* the microsporangia are also borne on microsporangiophores arranged in a catkin.

The microsporophyll of angiosperms is called a **stamen**. The pollen sacs are fused together in pairs (**thecae**) to form the anther, a kind of **synangium** (Fig. 4.12). Each chamber of the anther is called a **locule**. At maturity the wall between the pollen sacs may break down. Pollen sacs are joined in pairs by the connective to the filament. In the gnetophytes, *Ephedra*, *Welwitschia* and *Gnetum* various different microsporangiate structures are present which at least in *Ephedra* have a superficial resemblance to stamens.

In the development of the pollen sac, cell divisions give rise to several

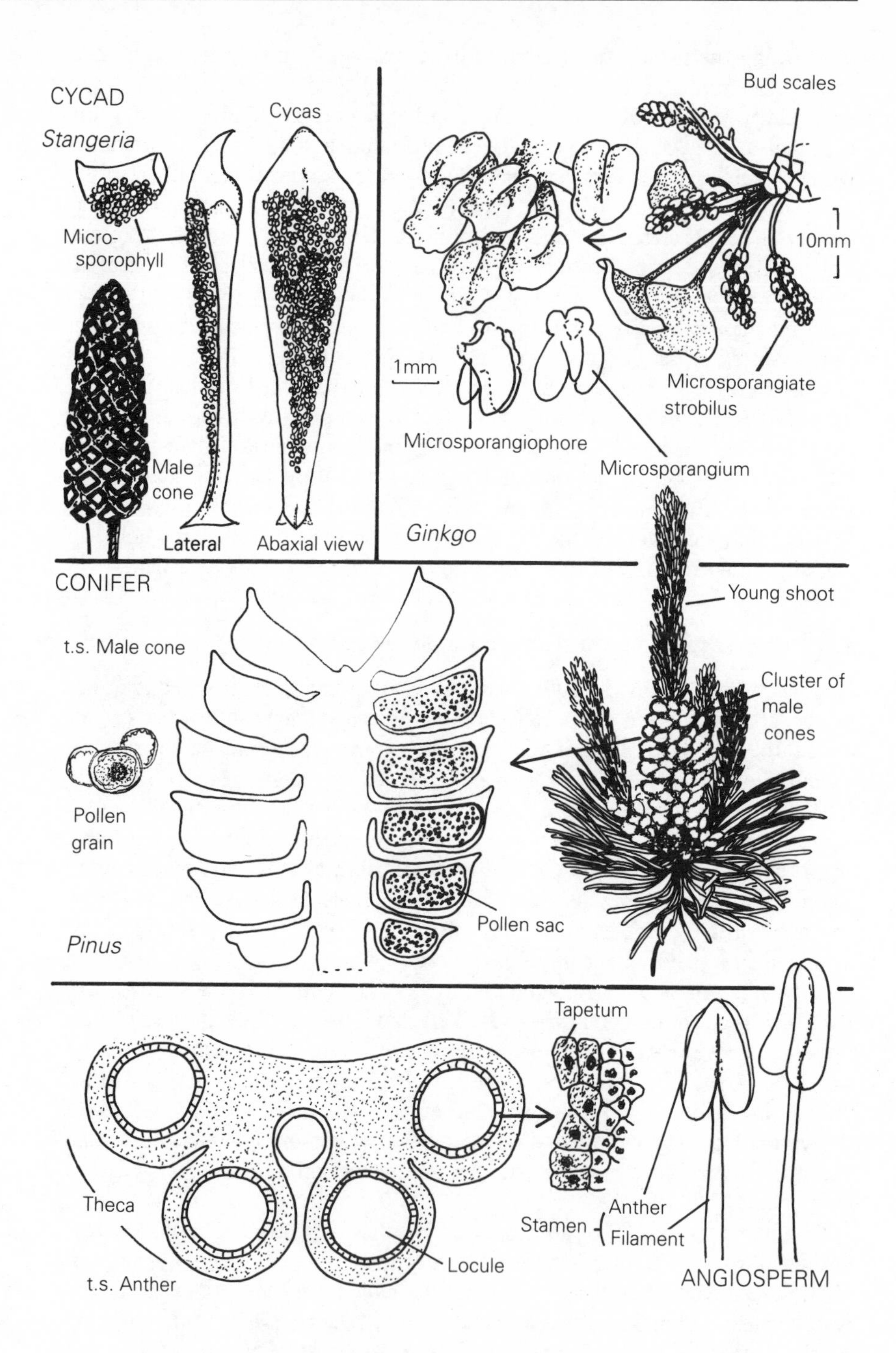

Fig. 4.12. The androecium in the four of the five main groups of living seed plants (the gnetophytes are illustrated in Fig. 3.10).

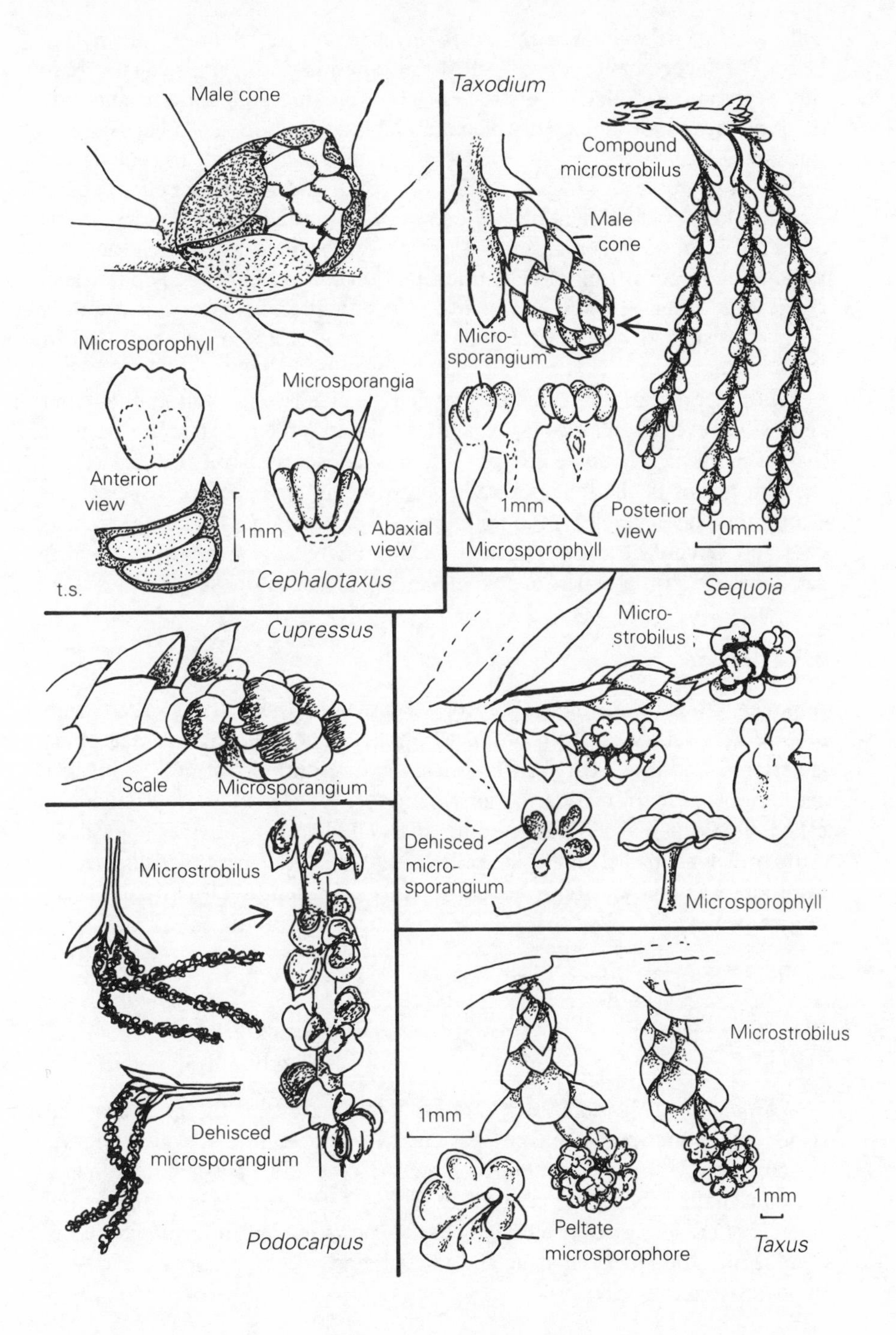

Fig. 4.13. Variation in the male axis of conifers: some have a compact cone but in others there are structures more like catkins.

cell layers in its wall. The innermost layer gives rise to the pollen mother cells. The layer outside this is called the **tapetum** and has a special role in the formation of pollen. It nourishes the developing pollen grains and may determine the morphology of the outer wall of the grain. Cells here may be multinucleate and polyploid. One kind of tapetum, called amoeboid, has cells with protoplasts which penetrate between the developing pollen grains. Another kind, called glandular or secretory, eventually breaks down and is absorbed into the pollen grains. Below the epidermis of the pollen sac is an **endothecium** consisting of cells thickened on the anticlinal and inner periclinal wall. When the anther is mature the endothecium loses water and its outer wall shrinks pulling the pollen sac open at a line of weakness called the **stomium**.

Meiosis occurs in the pollen mother cells to give rise to a tetrad of haploid cells which eventually break apart and develop into the pollen grains without increasing in size. In some groups, like the angiosperm family the Ericaceae, the four grains in the pollen tetrads do not break apart before dispersal. In *Acacia* the pollen is released in **pollinia** of 16 or 32 grains. In the Asclepiadaceae and Orchidaceae the pollen is further aggregated into pollinia which can combine all the pollen of a single anther locule into a single mass.

4.5.4 Pollen

A pollen grain is the microspore of seed plants. It is distinguished from other microspores because it produces a pollen tube. The male gametophyte is very reduced. No clearly differentiated antheridium is formed. No jacket cells are present. In cycads the male gametophyte is a single vegetative cell called the prothallial cell and an antheridial cell. The latter divides to produce a tube nucleus and a generative cell. In its turn the generative cell divides to produce a stalk and body cell. The body cell then divides at pollination to give two sperm cells i.e. a total of five nuclei are produced.

microspore→prothallial cell
→antheridial cell→tube cell
→generative cell→stalk cell
→body cell→2 sperm

In *Ginkgo* and the conifers, there are two divisions at the prothalial cell stage but the first prothallial cell has disintegrated at time of pollen release. In some conifers from the families Araucariaceae and Podocarpaceae many more prothallial cells are produced. In *Ephedra*, *Welwitschia* and *Gnetum* the generative nucleus gives rise directly to the sperm nuclei. In the angiosperms there is no prothallial cell stage and the generative cell gives rise directly to the two sperm nuclei.

microspore→tube cell
→generative cell→2 male nuclei

In angiosperms division of the generative cell may occur either before or after the pollen is shed and so the dispersed pollen grain may contain either two or three nuclei.

The wall of the pollen grain is a complex structure with two main layers, the **exine** and **intine**. The outer exine may be sculptured in many different ways, with spines and ridges. The exine is not continuous but broken by various slits, furrows (colpi) and pores. The pollen germinates through these apertures. The sculpturing and shape of the pollen grain is so varied that many different species can be identified from their pollen grain alone.

The structure of the pollen exine and intine probably has a role in the functioning of the sexual incompatibility systems of angiosperms which promote out-crossing. In angiosperms it often it has a chambered structure with a roof, the **tectum**, above a layer of supporting **columellae**. The exine has a particularly complex sculptured structure in many insect-pollinated groups. It is complex in the fossil Bennettitales (Crepet and Friis, 1987) and in a very diverse family of Mesozoic conifers the Cheirolepidiaceae. The latter are identified from their pollen, called *Classopollis*. They are astonishingly diverse which has led some workers to suggest that this diversity is a result of having a sexual-incompatibility mechanism, which the posession of a complex pollen exine also suggests.

4.5.5 Pollination

The spores of non-seed plants need only any moist surface where they can rehydrate and germinate. Seed plants have specialized receptive surfaces for trapping pollen so that it can germinate.

Most gymnosperms and many angiosperms are wind pollinated. A variety of structures have evolved to trap pollen from the wind. In wind pollinated angiosperms the stigma is often feathery like that in the grasses. It acts like a net, sieving a large volume of air as the inflorescence vibrates backwards and forwards in the wind. Some wind pollinated flowers have a relatively simple stigma. However vortices, and doldrums are created in the airstream in the area of the stigma, because of the shape of surrounding bracts, so that pollen grains are caught in a vortex and settle onto the stigma. The shape of the surrounding bracts and scales is equally important in the gymnosperms. There is complicated recirculation of pollen bearing air around the receptive areas.

In most of the wind pollinated gymnosperms, including cycads and conifers, pollination is aided by the production of a drop of fluid, the pollination drop, which captures the pollen grain. It is particularly obvious in *Taxus* and other species with more or less exposed ovules. The viscous drop first catches the pollen grains and then as it is reabsorbed they are drawn through the micropyle into the **pollen chamber** above the nucellus. In *Pinus* the pollen is at first trapped on the lobes of the integument and then drawn into the pollen chamber by fluid secreted by the ovule. In the conifer *Pseudotsuga* pollination is dry. This is also the case for the vast majority of angiosperms (but see 'wet' stigmas described below). In a very few flowering plants pollen is caught by a liquid trap though this arises from the stigma. In the water lily, *Nymphaea*, pollinating insects are caught and drowned before their bodies are brought into contact with the stigma as the liquid trap dries out.

A critical stage in the evolution of the ovule was the evolution of a pollen tube. In gymnosperms, once the pollen has been drawn into the pollen chamber the pollen grain germinates and a pollen tube is produced which grows into the nucellus. In cycads and *Ginkgo* the pollen tube is branched and purely **haustorial**. The haustorial pollen tube provides the male gametophyte with nutrients.

Competition between microspores for fertilization of the egg may have encouraged early pollination because the first microspore to arrive had a better chance of being the one to effect fertilization. The early release of pollen before the maturity of the female gametophyte may then have resulted in a delay between pollination and fertilization thereby encouraging the evolution of a haustorium to allow the male gametophyte to survive the intervening period. In more advanced groups the pollen tube gained the function of transferring the sperm nuclei to the archegonium (siphonogamy).

In *Ginkgo* and the cycads the pollen tube gradually digests the tissue of the nucellus. Finally it breaks into the **archegonial chamber**, a space lying above the archegonia, where it releases the motile sperm. The presence of motile sperm in seed plants is primitive. The advanced gymnosperms and the angiosperms are siphonogamous. There is no archegonial chamber and the pollen tube conveys the male gametes directly to the egg. In conifers the pollen tube is initially branched and haustorial but develops eventually to transfer the sperm nuclei to the archegonium. In some species of *Pinus* there is a period of over a year between pollination and fertilization. In *Pseudotsuga* the pollen tube conveys the male gametes from the integument lobes all the way to the archegonium.

Adaptations for the reception of wind-borne pollen also exposed plants to an increased risk of collecting pathogenic wind-borne fungal spores. The development of a pollen tube may have allowed the nucellus to act as an antipathogen screen, recognizing the pollen tube with its sperm nuclei and allowing it to grow (Haig and Westoby, 1989). Perhaps it is in this self-recognition process that sexual-incompatibility between different species and self-incompatibility within a species has its origins (see below).

Angiosperms are siphonogamous. In most of them the pollen tube is not haustorial. The pollen tube has to grow from the stigma, down the style, before reaching the ovule (Fig. 4.14). The pollen grain germinates directly on the stigma, usually producing only a single pollen tube. In the angiosperm family Malvaceae both haustorial and absorptive pollen tubes are produced. Selection has been for rapidity in sexual reproduction and usually only 12–48 h may intervene between pollination and fertilization. However in some angiosperm trees from seasonal habitats like *Quercus* and *Corylus* pollen tube development may be arrested for weeks or longer after a period of initial growth.

Two types of stigma have been described in angiosperms, called 'wet' and 'dry'. A'wet' stigma produces an exudate (cf. pollen drop of gymnosperms). The exudate does help to attach the pollen grain to the stigma but seems

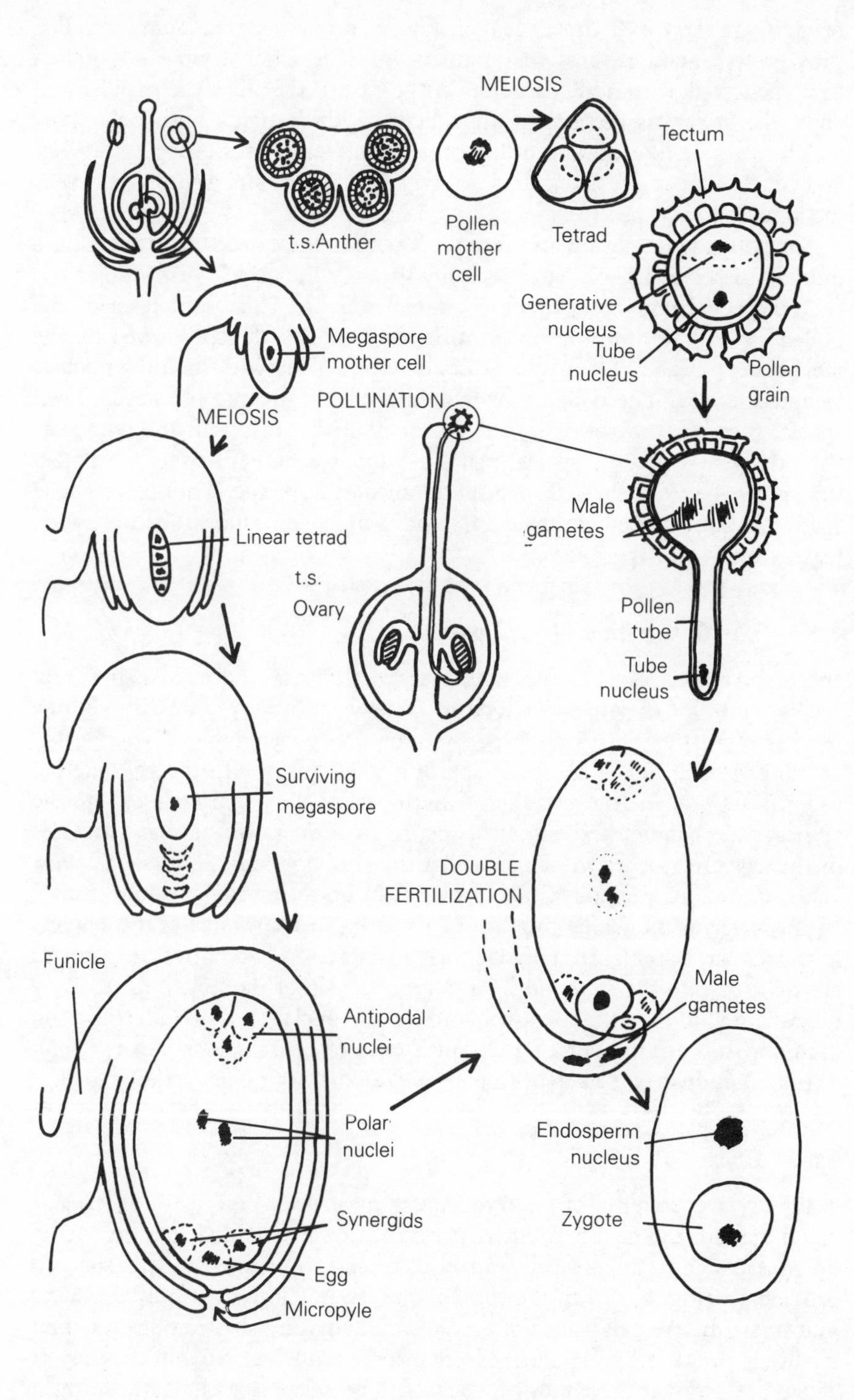

Fig. 4.14. Reproduction in a generalized angiosperm.

primarily to serve to hydrate the pollen grain, providing a medium for pollen tube growth and prevents the stigmatic papillae from drying out. Pollen tubes penetrate the **middle lamellae** at the base of the stigmatic papillae and enter the stylar tissue. Pollen attaches to a 'dry' stigma with pollenkitt, which is a glue like coating of lipoprotein derived from the tapetum.'Wet' and 'dry' stigmas are associated with gametophytic and sporophytic self-incompatibility respectively (section 5.3.1).

The pollen grain germinates through a pore in the exine; the intine bulges out and dissolves its way enzymatically through the cuticle of the stigmatic cell, thereby being able to absorb water from it. The protoplasm of the pollen grain flows into the pollen tube, tube nucleus first, followed by the male gametes. The pollen tube grows down the style with the tube nucleus leading the way. The pollen tube is guided chemically and physically down special tracts in the style into the chamber of the ovary. In the monocots the tube grows down a special glandular stylar canal. In dicots the pollen tube grows between the cells of a solid transmitting tissue. The sperm nuclei follow the tube nucleus to the embryosac. The pollen tube enters the ovule, usually by way of the micropyle.

4.5.6 The fertilization of seed plants

In gymnosperms each of the several archegonia may be fertilized so that several embryos may arise in each ovule (polyembryony) though only one reaches maturity. Fertilization occurs after the pollen tubes discharge the sperm or sperm nuclei. A single sperm nucleus fuses with the egg nucleus to form the zygote. In cycads and *Ginkgo* part of the egg protrudes into the archegonial chamber and engulfs a sperm. In *Welwitschia* tubular processes of the egg grow up to meet the pollen tube. In conifers, *Ephedra*, *Gnetum* and angiosperms the pollen tube grows right up to the egg (**siphonogamy**). In angiosperms the pollen tube enters the embryo sac via one of the **synergids** which has a special receptive filiform apparatus. Then both male gametes are released into the synergid by a pore in the tip of the pollen tube. They migrate, one to the egg, the other to the polar nuclei. A double fertilization occurs with the formation of a diploid zygote (sperm nucleus plus egg) and a triploid **endosperm** nucleus (male gamete plus two polar nuclei).

4.5.7 Embryos

Many diverse patterns of **embryogenesis** have been described. In *Ginkgo* and the cycads there is an extensive period of free nuclear division, producing up to 256 nuclei, before cell walls are formed to create a **suspensor** and **proembryo** (Fig. 4.15). In the conifers the free nuclear stage only produces four nuclei before cell walls are produced. Transverse division of these first four cells produce two or more layers of cells with four cells in each layer. These are a proembryo and a suspensor. The suspensor elongates to push the tiered proembryo into the tissue of the female gametophyte. The gametophyte is digested by the growing embryo.

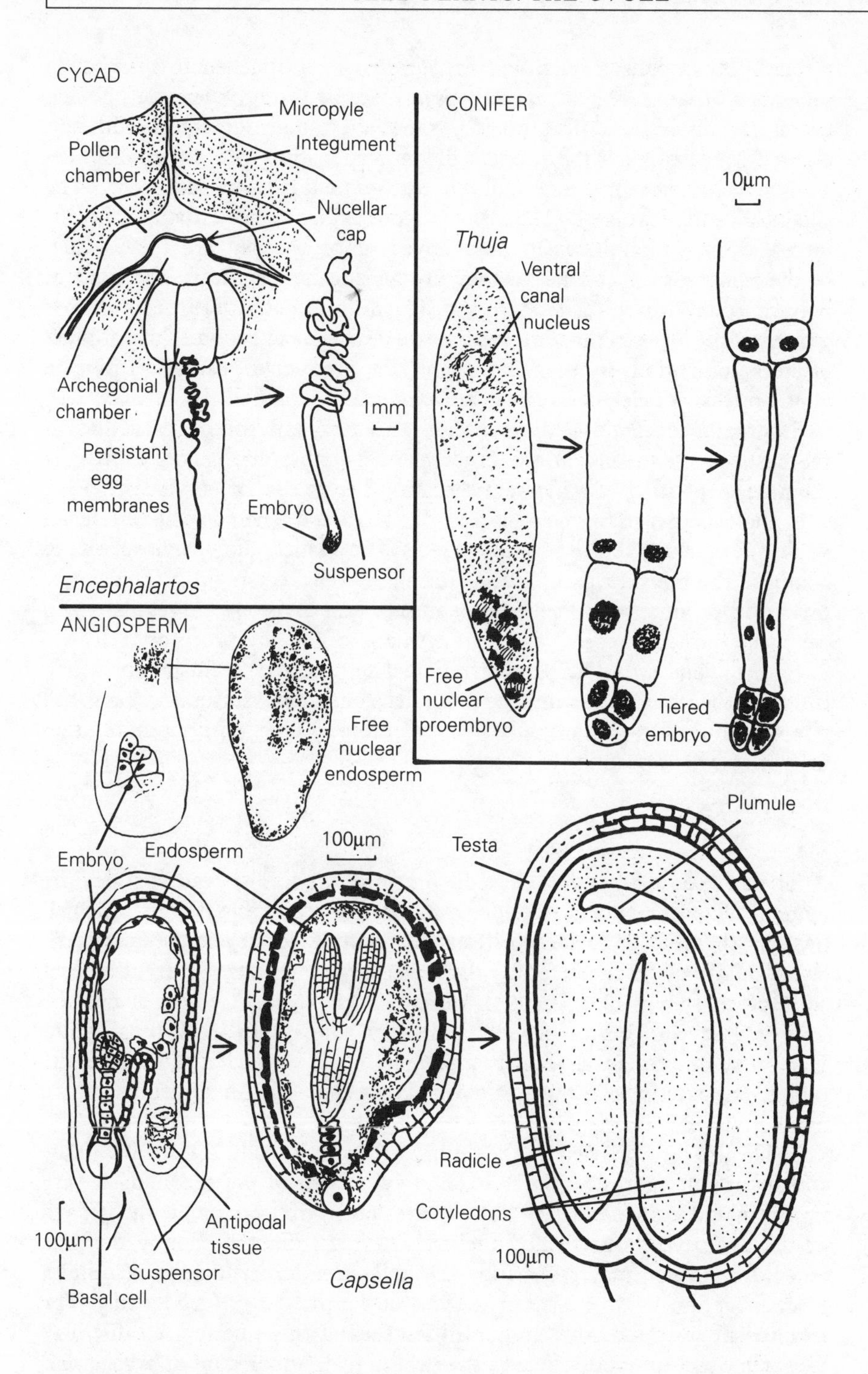

Fig. 4.15. Embryogenesis in seed plants: *Encephalartos* is a cycad; *Thuja* (from Coulter and Chamberlain (1910)) is a conifer; and *Capsella* is an angiosperm.

The four suspensor cells, and the embryo cells attached to them, may separate so that several separate embryos may be started (cleavage polyembryony). This is in addition to the potential existence of several different embryo systems, each the product of the fertilization of a different archegonium. However only one embryo reaches maturity in each seed. The significance of cleavage polyembryony is not evident since although there is intense competition between the derived embryos they are the product of the same zygote and hence they are genetically identical. The kind of polyembryony where the embryos arise from different zygotes may provide a mechanism for selection of a vigorous heterozygous genotype, the product of cross-pollination; by selection against the products of self-pollination in which recessive deleterious genes are expressed.

Eventually the embryo differentiates an embryonic root, the **radicle**, at the suspensor end and at the other end an embryonic shoot, called the **plumule**, consisting of a **hypocotyl** bearing two to several cotyledons.

In angiosperms embryogenesis is said to be direct. There is no free-nuclear phase. This is one of the several characteristics which allow angiosperms to complete the reproductive phase of their life cycle very rapidly. The zygote divides to produce a two celled proembryo. Cell divisions are regular, with cell walls formed at each stage. The terminal cell of the proembryo divides to form the embryo. The basal cell divides to produce the suspensor which connects the embryo to the seed wall. In some angiosperms the basal cell also contributes to the formation of the embryo. This happens more often in species which produce large seeds.

4.5.8 Seeds

After fertilization the ovule develops into the **seed**. The seed contains the embryo. The integument develops into a tough seed coat, the **testa**, which protects the seed. The mature testa is a complex of several or many cell layers. Hardness is conferred by the presence of sclereids. Several different sorts of cells may be present in different cell layers. Cell walls may be impregnated with cutin, phenols and other compounds as well as lignin. Hairs may be present, as in cotton seeds. On the surface of the seed is the hilum, the scar where the ovule was connected by the funiculus to the ovuliferous scale or pistil.

Inside the testa as well as the embryo there is usually a nutritive tissue which supports germination and the early growth of the seedling. A very important difference between angiosperms and gymnosperms is the growth of the nutritive tissue, called the endosperm, from the rapidly dividing triploid endosperm nucleus (Fig. 4.16). The endosperm may be cellular from inception or free nuclear at first or for the length of its life. The endosperm surrounds the embryo in the seed at the time of seed dispersal except in exalbuminous angiosperms where it disappears in embryogenesis as its nutritive reserves are absorbed into the embryo. In this case food reserves are present in the **cotyledons**, and directly available to the growing apices of the seedling. Seeds with an endosperm require a longer period of

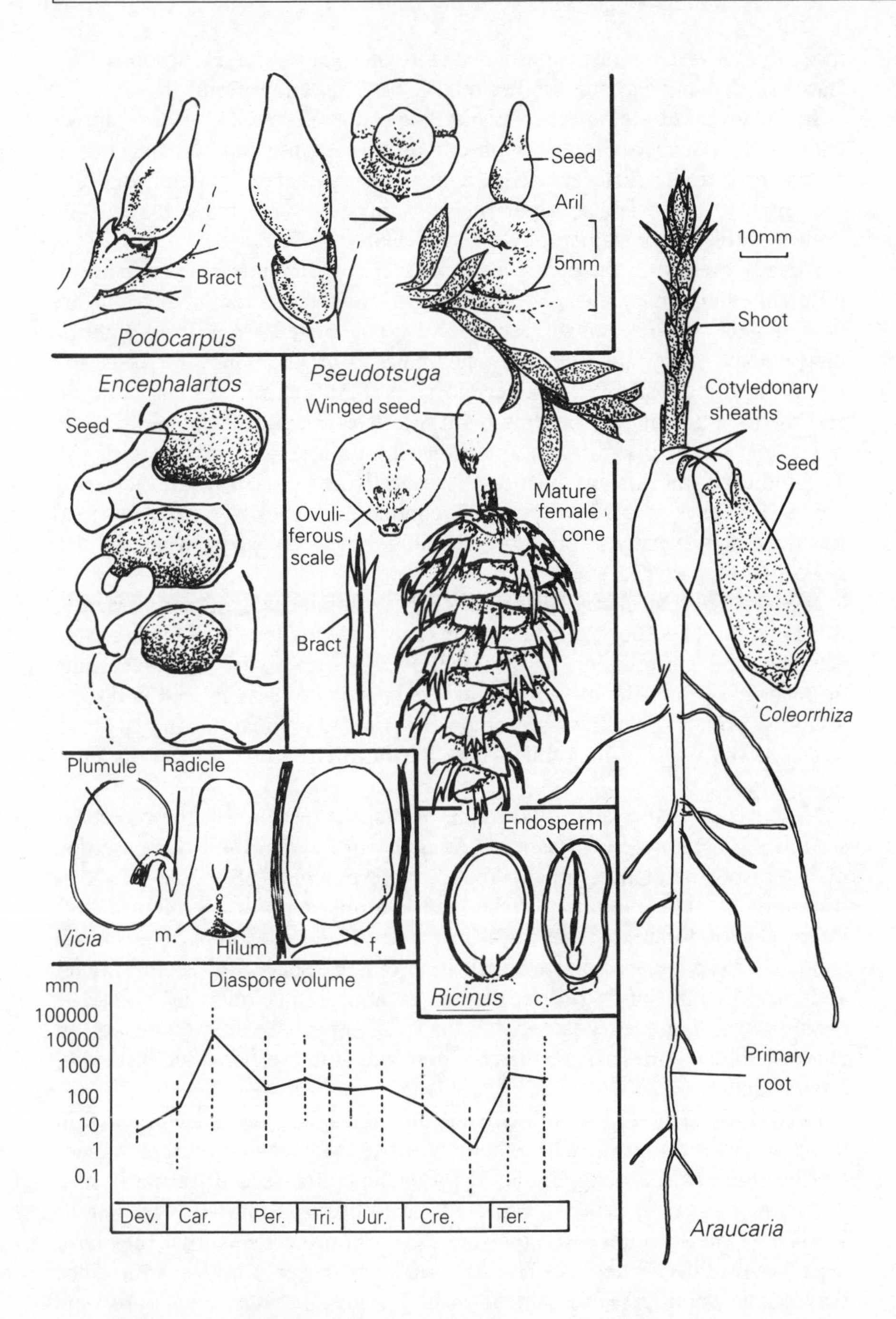

Fig. 4.16. Seeds and seed bearing structures: m.=micropyle, f = funicle, c = caruncle. Diaspores, of which seeds are one kind, have changed markedly in volume over geological ages (from Tiffney (1984)) possibly giving some indication of the levels of competition that exist – large diaspores mean high levels of competition.

incubation in warm moist conditions before they germinate (Hodgson, 1989). Large seeded angiosperm families tend to have an endosperm.

In one group of the flowering plants, the centrosperms, (order Caryophyllales), the endosperm is either absent or very scanty but the nucellus is persistent forming a storage tissue called the perisperm. In gymnosperms, the nutritive tissue is also a **perisperm**, supplied by the massive surviving tissues of the female gametophyte and nucellus.

Already present in the oldest fossil seeds is a distinction which marks out different evolutionary lineages (Meyen, 1987) and shows that seed plants are polyphyletic. Two kinds of seed can be recognized from their external morphology. They are bilaterally symmetrical or **platyspermic** seeds and radially symmetrical or **radiospermic** seed. Platyspermic seed is flattened and the integument may be drawn out into two wings, perhaps as an adaptation for wind dispersal. The platyspermic lineage includes the ginkgos. The radiospermic lineage includes the cycads and probably the flowering plants. There is a third lineage, the conifers, which has a flattened seed but their platyspermy is probably secondary and derived from a primitive radiospermic pattern.

It is strange that overall there are very few fossil seeds in comparison to fossil ovules. This shortage could mean that following fertilization abscission occurred and germination of the seedling was direct with no intervening dormancy. Perhaps the integument evolved not to protect the seed in dispersal but to protect the ovule in development or to aid pollination. In platyspermic seed the winged integument may have evolved to aid in the dispersal of the seeds.

Nevertheless, especially in flowering plants, the testa serves to protect the seed as a disseminule or **diaspore**. As noted in section 4.4.2 the evolution of heterospory and heterothally reduced the potential of spores to act as disseminules. The evolution of the seed is a milestone in a profound shift in function of different stages of the life cycle. The evolution of heterospory, and the heterothally associated with it, promoted successful establishment and cross-fertilization at the expense of the ability to colonize new areas. A megaspore is not well adapted for dispersal especially since it is not an independent colonizer. It requires the presence of microspores for successful establishment.

In contrast each seed is an independent colonist and seeds are adapted in many ways for dispersal. The evolution of the seed provided a replacement for the spore as the disseminule. Early evidence for seed dispersal is seen in the fossil *Caytonia*, where abundant isolated platyspermic seeds are found. Seeds have the advantage that they are very well protected so that they have high survivability. They can be dispersed in both space and in time since they can in some cases remain dormant for long periods. Seeds are time capsules because a tough resistant testa allows some seeds to survive dormant in the soil for many years before germinating.

Accessory structures may be modified to promote dispersal. In *Pinus*, part of the ovuliferous scale becomes a wing for the seed. In *Taxus* and *Podocarpus*, an **aril**, part of the stem below each ovule, expands and becomes a brightly coloured fleshy attractant for animal dispersal agents. In flowering plants after fertilization the ovary wall develops into a fruit wall, called the **pericarp**. The pericarp is adapted in many different ways to promote the survival and dispersal of the seed and the successful establishment of the seedling (section 8.3.5). There are many remarkable analogies between spore, seed and fruit release and dispersal. Diaspores of all sorts, spores, seeds and fruits, cover several orders of magnitude in size.

Seeds tend to be much larger than the dispersed spores of lower plants. There is a potential conflict between being large for successful establishment and being small for successful dispersal. However in seed plants the presence of accessory dispersal adaptations counteracts the disadvantages of largeness. Seeds and fruits themselves show a remarkable range of size from the double coconut of *Lodoicea* which weighs up to 27 kg to the seed of some orchids which weighs about 1 μg. Large diaspores allow a quicker and less risky germination and establishment of a young plant, although they cannot be produced in as large numbers as small diaspores. The nutritive reserves a seed carries are particularly important where the seedling is stressed, by for example shade or drought.

It is potentially wasteful to put large amounts of food reserves into spores. The reserves are only weakly protected from predation by the spore coat and they are only indirectly used for the establishment of a young sporophyte, and then only if the gametophyte is successfully fertilized. In seeds the reserves are well protected within the integument and they are certain to be used by the young sporophyte, the germinating seedling.

However, the ovules of some gymnosperms, especially the cycads and *Ginkgo*, reach mature size after pollination and before fertilization. This prior provisioning of the seed may be wasted if fertilization is not achieved. Unpollinated ovules are aborted. Conifers are more efficient than cycads or *Ginkgo* by either reallocating the food reserves of unpollinated ovules or by increasing the size or quality of the food reserves, as measured by dry weight, only after fertilization (Haig and Westoby, 1989). In angiosperms the food reserve is created very rapidly post-fertilization by the growth of the endosperm. The triploid condition of the endosperm may allow it to compete very effectively with the diploid maternal parent for food reserves.

Seed plants, where cross fertilization is promoted by the production of different sexed spores and dispersal by the production of seeds, contrast with homosporous lower plants, where the spore is the dispersal unit and cross fertilization is promoted by having a motile sperm. Although the alternation of generations is fundamentally unchanged, the function of the two critical transitions between the haploid and diploid phases has been swapped.

The four main different life cycle patterns exhibited in land plants are summarized in Fig. 4.17.

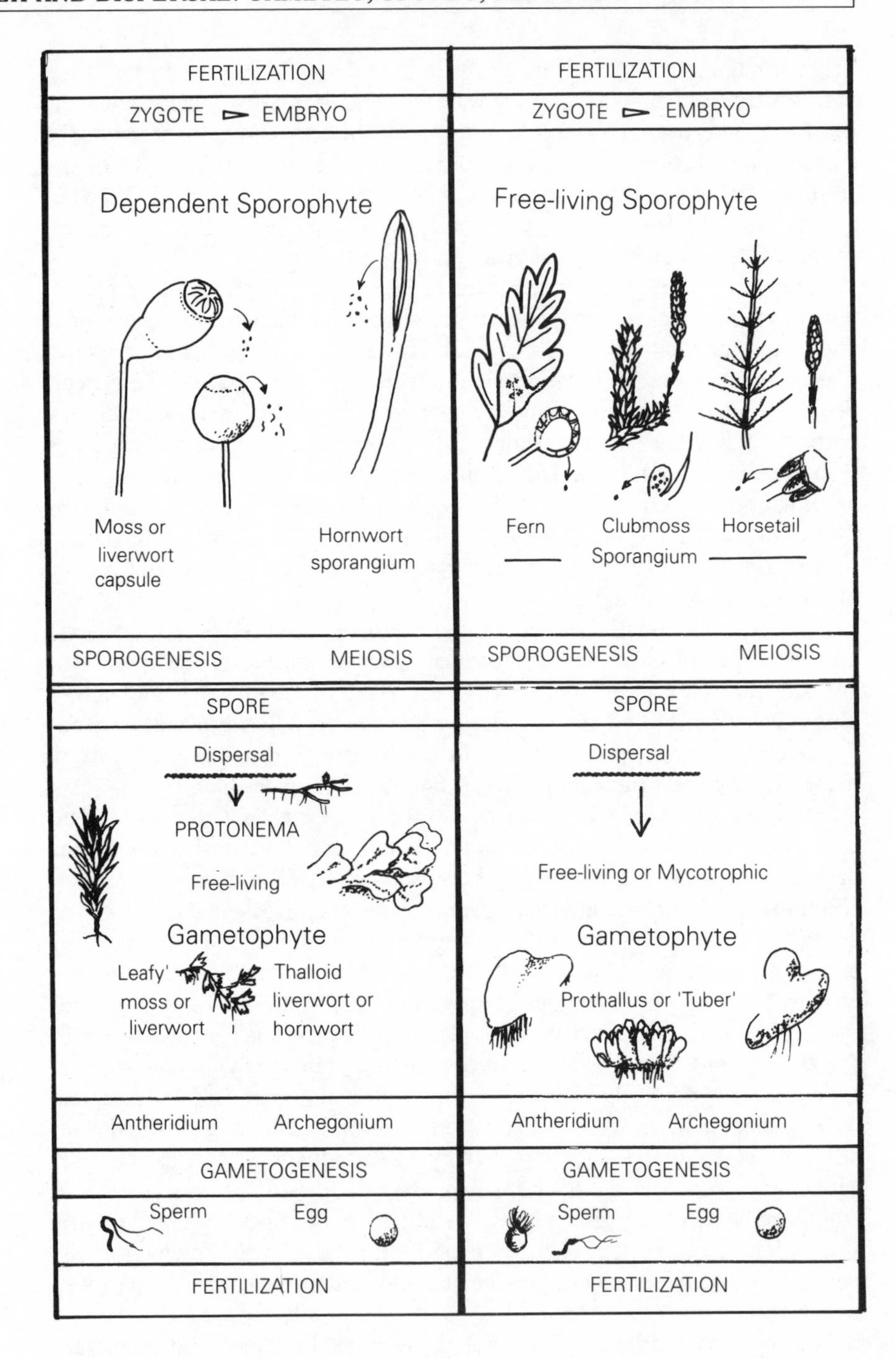

Fig. 4.17. The four main patterns of reproduction in land plants showing the sporophytic and gametophytic part of the alternation of generations: in bryophytes, homosporous pteridophytes, heterosporous pteridophytes and seed plants.

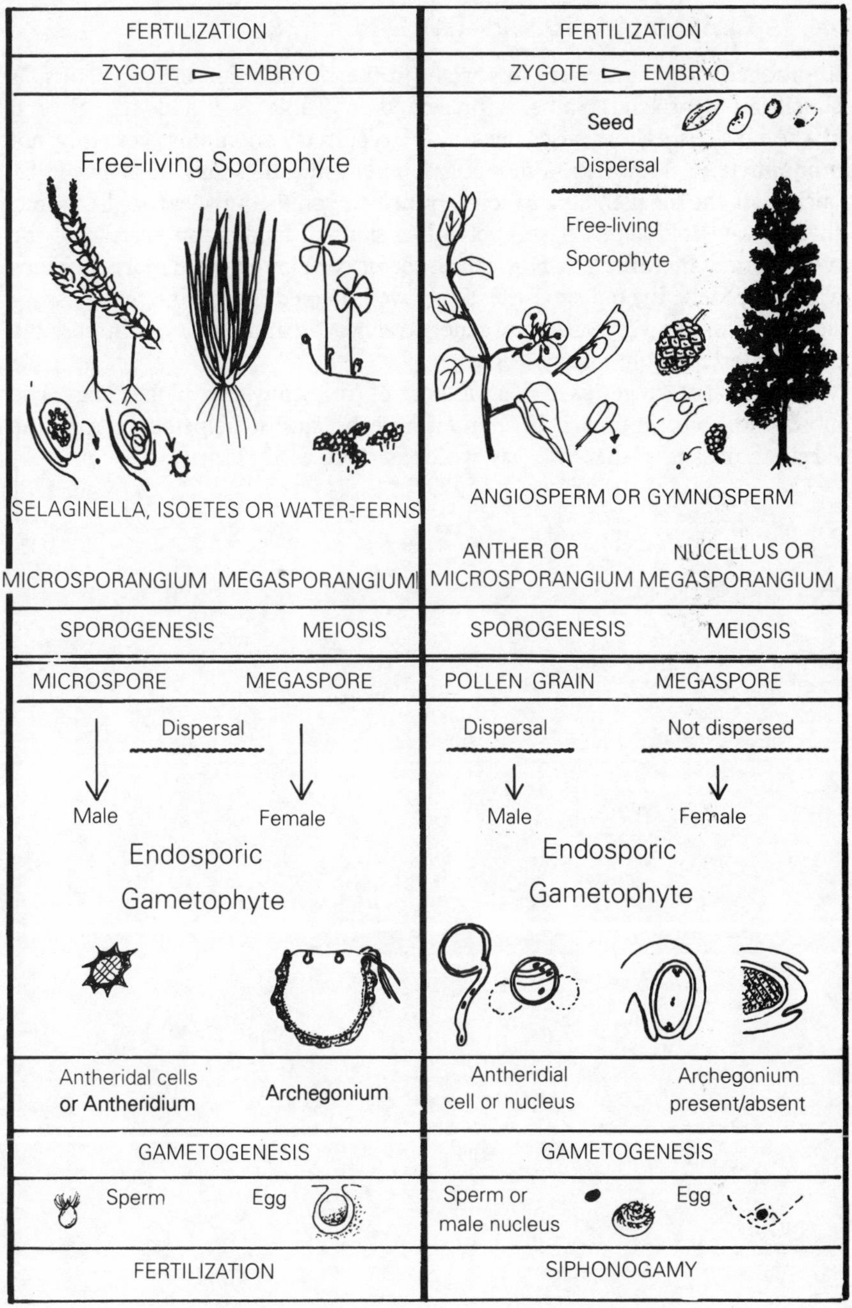

FERTILIZATION	FERTILIZATION
ZYGOTE ▻ EMBRYO	ZYGOTE ▻ EMBRYO
	Seed
	Dispersal
Free-living Sporophyte	Free-living Sporophyte
SELAGINELLA, ISOETES OR WATER-FERNS	ANGIOSPERM OR GYMNOSPERM
MICROSPORANGIUM MEGASPORANGIUM	ANTHER OR MICROSPORANGIUM NUCELLUS OR MEGASPORANGIUM
SPOROGENESIS MEIOSIS	SPOROGENESIS MEIOSIS
MICROSPORE MEGASPORE	POLLEN GRAIN MEGASPORE
Dispersal	Dispersal / Not dispersed
Male Female	Male Female
Endosporic Gametophyte	Endosporic Gametophyte
Antheridal cells or Antheridium / Archegonium	Antheridial cell or nucleus / Archegonium present/absent
GAMETOGENESIS	GAMETOGENESIS
Sperm / Egg	Sperm or male nucleus / Egg
FERTILIZATION	SIPHONOGAMY

4.6 A CLASSIFICATION OF LAND PLANTS

Reproductive characters are important in the classification of land plants. A classification of land plants is presented in Tables 4.1 and 4.2. Several of the taxa, especially large ones like Bryophyta, are almost certainly not **monophyletic**. However, a monophyletic classification may be achieved by raising all the smaller taxa at lower ranks to some equivalent higher rank. In this way Bold *et.al.*, have twelve Divisions of land plants instead of the two recorded in Table 4.1. However it seems certain that even more divisions will be needed. Even a small relatively well defined group like the Glossopsida contains the very different genera *Selaginella* and *Isoetes*, with possibly quite different evolutionary origins.

We may have to get used to a plethora of 'monophyletic' higher taxa. The more traditional approach is adopted here because it helps summarize the variation of land plants in a way which is more easily learnt.

Table 4.1 A classification of major living land plant groups (vernacular equivalents are recorded in inverted commas. Only a few distinguishing features are listed. They are mainly reproductive characters

EMBRYOBIONTA – 'embryophytes', 'the land plants'
- plants with a well developed alternation of generations with the young sporophyte, the embryo, parasitic on the gametophyte
- having multicellular antheridia and archegonia

BRYOPHYTA – 'bryophytes'
- with a sporophyte not connecting directly to the ground and dependent on the more conspicuous gametophyte for nutrition
- usually lacking a vascular system

HEPATICOPHYTINA
- 'liverworts'
- spherical capsule with elaters
- thallose or leafy with leaves in 3s

ANTHOCEROTOPHYTINA
- 'hornworts'
- sporophyte a horn
- thallose

BRYOPHYTINA
- 'mosses'
- 'protonema produced
- 3 classes

SPHAGNOPSIDA
- capsule explosive
- bog mosses, *Sphagnum*

ANDREAEOPSIDA
- capsule lantern-like

BRYOPSIDA
- peristome present

TRACHEOPHYTA – 'vascular plants'
- having an inconspicuous gametophyte and a large conspicuous sporophyte
- having a well developed vascular system, with xylem and phloem

PSILOPHYTINA – 'whisk ferns'
- bearing synangia at ends of side shoots
- no roots, rhizoids arising from rhizomes
- 2 genera *Psilotum*, *Tmesipteris*

SPHENOPHYTINA – 'horsetails'
- sporangia borne on peltate sporangiophores
- leaves and branches in whorls, arising at alternate nodes
- one genus, *Equisetum*

LYCOPHYTINA 'lycopods'
- sporangia in strobili
- microphyllous
- 2 classes

AGLOSSOPSIDA 'clubmosses'
- homosporous
- 2 genera *Lycopodium*, *Phyllog lossum*

GLOSSOPSIDA
- heterosporous
- ligulate
- 2 genera *Selaginella*, *Isoetes*

FILICOPHYTINA – 'ferns'
- megaphyllous, with (usually) large compound leaves
- 3 classes on basis of sporangium type

OPHIOGLOSSOPSIDA
- eusporangiate sporangia on fertile spikes

MARATTIOPSIDA
- eusporangiate sporangia on leaves

FILICOPSIDA
- leptosporangiate
- includes heterosporous forms

SPERMATOPHYTINA – 'seed plants', 'phanerogams'
- producing seeds
- megaspore retained within the megasporangium, the nucellus
- by far the most important and diverse group
- including *Gingko*, conifers, cycads, flowering plants.

Table 4.2 Characteristics of seed plants, SPERMATOPHYTINA, in more detail

CYCADOPSIDA -cycads, 106 spp.	GINKGOPSIDA -1 species, *Ginkgo*	PINOPSIDA -conifers, 628 species	GNETOPSIDA -3 genera *Gnetum* (28 spp.), *Ephedra* (40 spp.), *Welwitschia* (1 sp.)	MAGNOLIOPSIDA -angiosperms, 240 000 species
Motile sperm	Motile sperm	Non-motile sperm	Non-motile sperm	Non-motile sperm
Pollination drop	Pollination drop	Pollination drop	Pollination drop	Stigma and style
Pollen tube haustorial	Pollen tube haustorial	Siphonogamy	Siphonogamy	Siphonogamy
Female gametophyte very large (1–3000) nuclei	Female gametophyte very large (8000 nuclei)	Female gametophyte large (–›2000 nuclei)	Female gametophyte in *Ephedra* 256–1000	Female gametophyte 4–16 nuclei
Archegonia	Archegonia	Archegonia	Archegonia absent in *Gnetum* and *Welwitschia*	Archegonia absent
1 integument to ovule	1 integument to ovule	1 integument to ovule	2–3 ‘integuments’ around ovule	Usually 2 integuments around ovule
Perisperm made from female gametophyte (haploid)	Perisperm made from female gametophyte (haploid)	Perisperm from female gametophyte (haploid)	Perisperm from female gametophyte (haploid)	Endosperm from endosperm nucleus (triploid)
Embryo development begins with free-nuclear phase (64–1000 nuclei)	Embryo begins development with free-nuclear divisions (256 nuclei)	Embryo development begins with short free-nuclear phase, polyembryony	Direct development of embryo in *Welwitschia* and *Gnetum*, free-nuclear phase (8 nuclei) in *Ephedra*	Direct development of embryo
Pachycauls, weakly branched, woody but relatively inactive cambium	Profusely branched woody tree, long and short shoots, active cambium	Profusely branched woody trees often with long and short shoots, active cambium	Profusely branched or unbranched, woody, active cambium	Variously branched or unbranched, woody or herbaceous, cambium present or absent
Vessels absent	Vessels absent	Vessels absent	Vessels present	Vessels usually present
Leaves large and compound	Leaves simple, broad	Leaves usually small simple, needle or scale-like	Leaves various	Leaves various

Flowers: evolution and diversity 5

5.1 THE EVOLUTION OF THE ANGIOSPERMS
5.2 CROSS POLLINATION
5.3 BREEDING SYSTEMS
5.4 EVOLUTIONARY TRENDS – THE PERIANTH
5.5 A REVIEW OF ANGIOSPERMS

SUMMARY

The angiosperms, Magnoliopsida (Angiospermopsida), are by far the most diverse group of plants that has ever existed with more than 240 000 different species. Flowers are extraordinarily diverse but a few evolutionary trends can be recognized. Many trends are related to the greater and greater specialization of insect pollination mechanisms: towards bilateral symmetry, fusion and elaboration of floral parts and the compound inflorescence. Others are related to changes in the number of seeds produced and their protection and dispersal: towards different kinds of placentation, the inferior ovary and different fruit types. Some taxa have become adapted for wind pollination, and highly specialized for this mode of reproduction (e.g. the grasses).

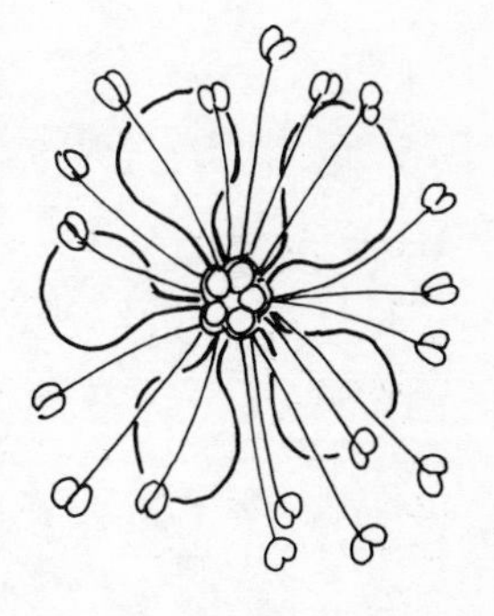

5.1 THE EVOLUTION OF THE ANGIOSPERMS

A flower is fundamentally a hermaphrodite reproductive axis in contrast to the unisexual cone or strobilus of most gymnosperms. A flower consists of a number of whorls of parts. The outer part is the perianth which may be differentiated into a green outer calyx made up of sepals and an inner showy corolla of petals. Inside the perianth is the androecium, the male part of the flower, made up of stamens. The innermost or uppermost part of the flower is the gynoecium with a pistil or pistils.

Below are some of the most important characteristics distinguishing angiosperms from the gymnosperms according to Stebbins (1974).

1. Leaves have a finely divided venation with both secondary and tertiary veinlets, the veinlets ending blindly.
2. The fertile shoot has both male and female reproductive structures with the male parts (stamens) situated below the female parts (carpels).
3. The ovules are protected within an enclosed structure, formed by one or more carpels.
4. The male gametophyte is very reduced producing only three haploid nuclei.

5. The female gametophyte (embryosac) is very reduced with only 4–8 nuclei.
6. There is a double fertilization to produce a diploid zygote and a triploid endosperm nucleus.
7. There is 'direct' development of the embryo from the zygote, i.e. with no intervening free-nuclear pro-embryo phase.

Primitive angiosperms are either early fossils or living taxa with a high proportion of primitive characters. Primitive characters are those which appeared early in the fossil record. Unspecialized characters which show great potential for evolutionary modification and diversification are also usually regarded as being primitive. Piecing together the nature of the first flowers is made difficult by the paucity and poor quality of early fossils and by the rapid diversification of early flowers.

There have been a number of theories about the origin of flowers, homologizing in different ways floral structures with the reproductive structures of other plants. For example two contrasting theories are the **euanthium** theory, which derives a flower from a uniaxial cone bearing both micro and megasporophylls and the **pseudanthium** theory which derives the hermaphrodite flower from a complex inflorescence of unisexual male and female flowers. It has been suggested that a flower-like structure could have arisen by a change of sex of some of the microsporophylls of a male cone, a process called gamoheterotropy (Meyen, 1988). This process is not hard to imagine in plants which are hermaphrodite and where the determination of sex is a developmental phenomenon which is only rarely associated with sex chromosomes. A similar transfer of sex has been proposed in the evolution of the maize cob. Hermaphrodite inflorescences have evolved in the Bennettites and the living gnetophytes where they are associated with insect pollination.

5.1.1 The environment of early angiosperms

The angiosperms probably originated in the tropics. Extant primitive families are tropical or subtropical and early fossils show no adaptations to temperate conditions. There is a consensus that the flowering plants evolved in West Gondwana, an area equivalent to modern South America plus Africa, from where they spread north into Laurasia and south and east into the rest of Gondwana.

In the list of angiosperm characteristics above, four of them allow the speeding up of the reproductive process. Stebbins (1974) has suggested that many of the innovations of the flower were selected in a seasonal environment where the period for reproduction was restricted by drought. An environment with seasonal drought may not just have encouraged reduction of reproductive structures but may also have encouraged the protection of the ovule and developing seed from drought by the development of the ovary. In a seasonal environment insect pollination may also have been encouraged because of seasonal cycles in pollinator abundance. A trend for the reduction of the reproductive structures has continued to be important in the subsequent evolution of some flowering plants. The reduction of the gametophytic phase of the life-cycle, like insect pollination, has increased efficiency of the reproductive process.

There is no fossil evidence to suggest that the angiosperms evolved in seasonal environments because fossil flower pollens appear in equable areas as well as arid areas. However, the sudden appearance of the angiosperms in the fossil record may only mean that an early diversification occurred in upland areas where seasonal heavy rains and summer drought would not provide the conditions for the preservation of fossil material. Fossils angiosperms would only be preserved on migration to tropical lowlands with its many sedimentary basins. A mountainous region would have favoured rapid evolution into the many different niches it provides. Such regions have today some of the highest levels of diversity of all communities. The highlands of Ethiopia, the Cape region of South Africa, and south central Mexico provide good examples (Stebbins, 1974).

Alternatively it has been suggested that the flowering plants evolved in coastal areas as early successional plants of moist habitats (Retallack and Dilcher, 1981). The earliest angiosperm fossil with inflorescence attached was a small, rhizomatous perennial (with secondary growth) with similarities to primitive herbaceous magnolioid dicots and 'basal' monocots (Taylor and Hickey, 1990) which agrees with an origin in a seasonally wet marshland. The divergence of monocots and dicots may date right back to the origin of angiosperms. Whatever the conditions of their origin by the Mid Cretaceous, 98 million years ago, the flowering plants had rapidly diverged and spread even into high latitudes in Laurasia (Crane, 1987). In northern Gondwana and southern Laurasia there was a new diversity of pollen types. Palms became abundant. By the end of the Cretaceous flowering plants, were dominant nearly everywhere.

5.1.2 The form of the ancestral flower

The earliest undoubted angiosperm fossils are pollen grains from 130 million years ago in the Early Cretaceous from England and West Africa. Angiosperm type leaves appear about 125 million years ago from eastern North America. The first fossil inflorescence comes from about 120 million years ago from Australia (Taylor and Hickey, 1990). The fossil has a mosaic of characteristics which are found in a range of living families of the super order Magnolianae. There is general agreement that living families with a large proportion of primitive characteristics, both vegetative and reproductive, are to be found in this group (Friis and Crepet, 1986). In particular the fossil shares several characteristics with the family Chloranthaceae possibly related to the more familiar Piperaceae (including *Peperomia*). The fossil has a female inflorescence, with a bract and two bracteoles at the base of female flowers which lacked a perianth (**achlamydeous** flowers). The flowers are effectively just tiny pistils, less than 1 mm in diameter.

By the Mid Cretaceous, from about 110 million years ago, these very simple flowers, some hermaphrodite, were coexisting with others with a more complex structure, with a spiral arrangement of parts and including a perianth. The archetypal flower may not have been hermaphrodite but it is from this later more complex type of magnolioid flower that most others can be derived. Floral parts were arranged spirally and varied in number between different flowers of the same plant. The perianth was a continuation of the leafy axis which supported the flower. It was not differentiated into a distinct calyx and corolla. There were several or numerous stamens, with pollen sacs separate (tetrasporangiate), and with short filaments. The pollen was monosulcate (having a single furrow). There were several separate carpels (**apocarpous**), each containing several seeds. Carpels were not completely closed and stigmas were sessile. The pistil developed into a fruit, the **follicle**. The derivation of carpels from a folded fertile leaf is obvious in follicles, though it may more truly have arisen as a tubular (peltate-ascidiate) rather than a folded leaf.

It is widely agreed that some of the most primitive flowers of this magnolioid sort are to be found in the family Winteraceae which grows in lands bordering the South Pacific, but like all living families it has some advanced features. Its pollen is different from the most primitive sort and some members have indehiscent berry-like fruits. In the family Winteraceae, there is a spectrum of variation from those species with an unsealed carpel, with the stigma running along the adpressed margins of the carpel like some species of *Drimys*, to those with a sealed ovary and a terminal style and stigma. Primitive carpels/follicles are retained in several taxa, outside the Winteraceae, including *Paeonia* (Paeonaceae) and many species in the Ranunculaceae like *Helleborus*.

Possibly three early strategies evolved in the ancestral group. One major strategy was the evolution of wind pollination which can easily be imagined from unisexual chloranthoid type inflorescences. Another strategy was the evolution of more and more specialized pollination mechanisms which promoted efficient pollination. A third strategy may have been paradoxically towards increased efficiency of being pollinated by the rather unspecialized insects such as beetles, flies, sawflies and micropterigid moths, which were the early pollinators i.e. being specialized for unspecialized pollinators. These plants retained many primitive characteristics but there was an increase in size of the flower and the number of floral parts, leading to the extant families Magnoliaceae and Annonaceae. The perianth became arranged as a cup surrounding a spiral of numerous stamens. The receptacle became elongated to bear the many carpels. These adaptations evolved early and were well developed by the late mid Cretaceous 95 million years ago (Crepet and Friis, 1987).

Similar flower like axes had evolved before the angiosperms in the fossil gymnosperm group, the Bennettitales (Fig 5.1). Pollination by beetles and other boring insects has also been suggested for these. Diptera flies as pollinators have been suggested for one called *Williamsoniella*. Some 'flowers' of Bennetttitales are strikingly similar if only superficially to *Magnolia* (Fig. 5.1). The Bennettitales, like angiosperms, have a pollen with a sculptured

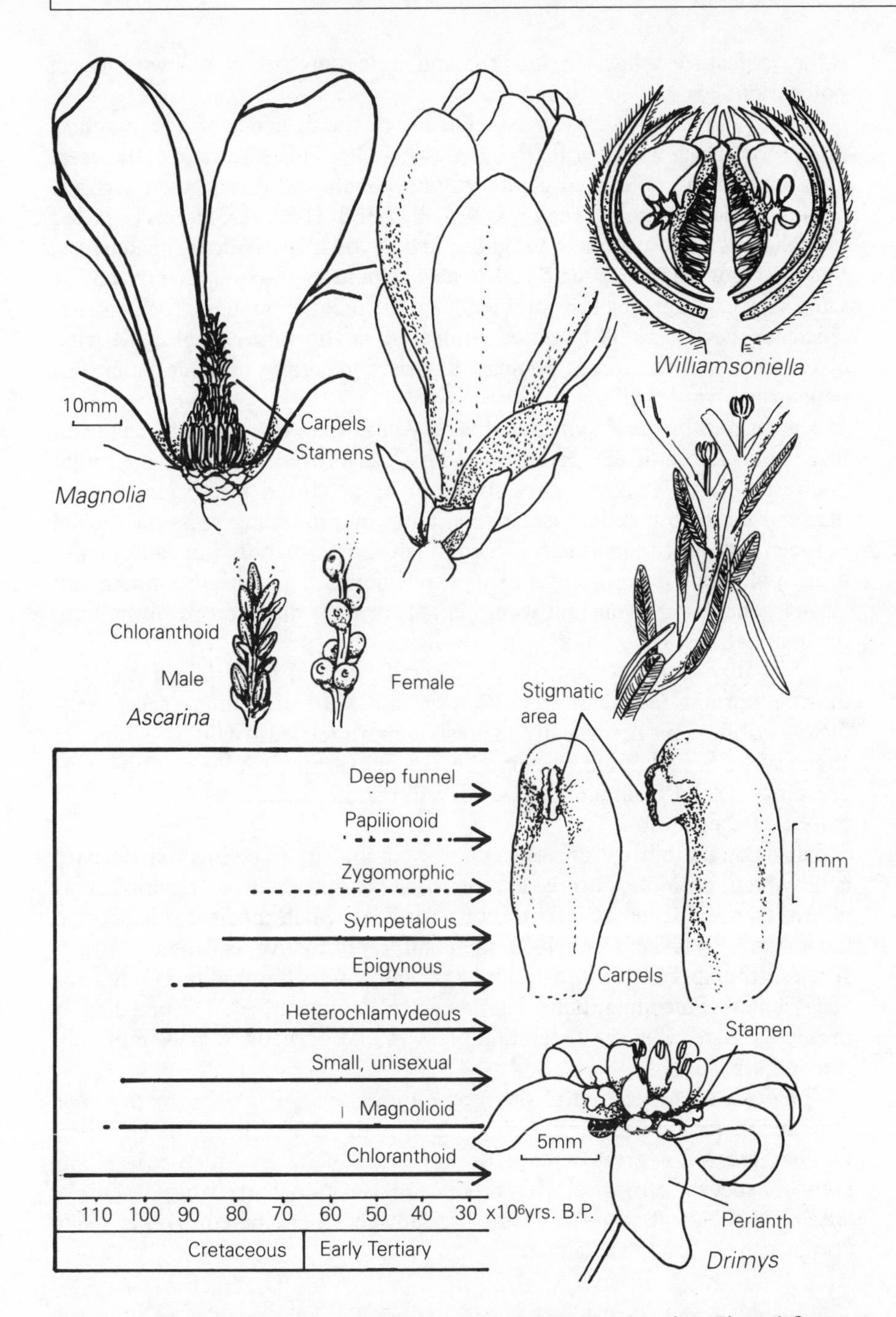

Fig. 5.1. Primitive flowers: *Williamsoniella* (after Thomas (1915) and Crane (1985)) is a fossil cycadeoid (not an angiosperm) which had a flower-like bisexual reproductive axis bearing naked ovules which probably functioned like *Magnolia*. Chloranthoid inflorescences have very simple flowers without a perianth. The time of the appearance of floral features is from Friis and Crepet (1986).

exine, a feature which in living plants is closely associated with insect pollination.

Insect pollination is closely associated with the presence of a hermaphrodite reproductive axis, because pollen was likely to have been an early insect attractant. Wind pollinated plants may have male and female parts separate because wind, as a pollen vector, is sexually blind. However, insects attracted by pollen as food are likely to ignore female areas in favour of male areas, but the combination of male and female structures in close vicinity on the same axis counteracts this problem. Insect pollination has much greater efficiency because it is targetted compared to the rather haphazard wind dispersal of pollen. Insect pollinated flowers generally produce much less pollen than wind pollinated flowers.

A great part of the astonishing diversity of flowers represents mechanisms which promote efficient pollination. This has proved especially important not just because it allows less pollen to be produced but because many mechanisms help maintain genetic variation by promoting cross-pollination between different individuals. Even though the hermaphrodite condition of insect-pollinated flowers seems to allow self-pollination, with the subsequent loss of genetic variation that would entail, in fact, many mechanisms limit self-pollination. The evolution of mechanisms to prevent self-fertilization, was probably a turning point in the diversification of flowering plants. The most important mechanism which prevents self-fertilization, called self-incompatibility, is not usually expressed morphologically. It is a kind of chemical self-recognition system. Self-incompatibility, by enforcing outbreeding, helped maintain genetic variation within species which is the raw material for evolution.

Self-incompatibility is one aspect of the complexity of pollination mechanisms which promote outbreeding in angiosperms. Self-incompatibility is clearly related to the self-recognition processes which mediate interspecific incompatibility reactions which maintain reproductive isolation between species. Interspecific incompatibility has been observed in conifers (Hagman, 1975) but self-incompatibility has not been demonstrated. Outbreeding is promoted instead by the separation of sexes and perhaps by polyembryony (section 4.5.7).

The great sophistication of the reproductive process in angiosperms, not only maintains the genetic diversity which is the raw material of evolution, but has created a greater range of ways in which reproductive isolation between species can arise. If reproductive isolation is a critical event in speciation the great numbers of angiosperm species compared to other plant groups can be partly explained.

5.2 CROSS POLLINATION

5.2.1 Primary attractants

The pollen vector must be attracted by the provision of food, either pollen or nectar, or brood places for its young. The provision of these primary attractants must be advertised by colour, shape and scent. The advertisements have the secondary advantage of encouraging fidelity of the pollinator. Pollinator fidelity improves efficiency and helps the reproductive isolation of species, and thereby species differences to be maintained. The flower must be shaped and behave in ways to suit the behaviour of particular pollinators. There has been extensive coevolution of flowers and pollinators.

Pollen was probably the attractant in early flowers. Many flowers produce an excess of pollen as a food for the pollinator. Up to 30% of pollen is protein. The earliest recognizable flowering plant pollen has a complex **exine** differentiated into two layers, an inner **endexine** and an outer **ectexine** (Doyle, 1978) which is associated with insect pollination. The pollen is **tectate** with a continuous ectexine through which beetles chew to open the grain. Vectors attracted by the provision of pollen include hover flies, bees, beetles and bats and other small mammals. Pollen is the attractant in many 'primitive' magnolioid flowers but it is the only primary attractant even in some fairly advanced flowers, such as the rose, tulip and poppy. These flowers have numerous stamens. Some pollen flowers, like *Melastoma* and *Cassia*, may produce food pollen in different stamens to the normal ones. Pollen is coloured by carotenoids and flavonoids and has its own odours different from the rest of the flower, which direct the pollinators attention to it.

Many flowers produce nectar. Nectar is an alternative food to pollen and its provision lessens the conflict between pollen surviving for pollination and being supplied as food. In those gymnosperms, especially cycads where insect pollination has been claimed, it is possible that the pollination drop, which usually helps capture the wind dispersed pollen, acts in some cases as a kind of nectar. This is certainly the case in the peculiar gymnosperm *Welwitschia*. The male cones contain a single sterile ovule whose sole function is to provide a sweet pollination drop to attract insects (Fig. 3.10). *Ephedra* also provides a kind of nectar.

Nectar is basically a concentrated sugar solution. The sugars vary from 25–75% by weight (Faegri and van der Pijl, 1979) depending on the kind of pollinator. Sucrose, glucose and fructose are the main constituents. There is some evidence that more specialized flowering plants are more likely to provide the sugars as glucose or fructose, i.e. in a more readily utilized form (Harborne, 1988). In a few advanced families, like the Scrophulariaceae, energy is also provided in a more concentrated form as lipids. Nectar also contains amino acids. Some butterflies are almost completely dependent on nectar nutritionally and so the amino acids, though in low concentration, provide all their nitrogen requirement. Flowers which are pollinated mainly by bees, which get their nitrogen from pollen, have lower quantities of amino acids in the nectar.

Nectar is provided by nectaries which may have a special 'glandular' structure. The nectaries are located either within the flower, floral nectaries, or are extra-floral. Extra-floral nectaries may be present on parts of the inflorescence, as on the margins of the cyathia in Euphorbiaceae (Fig. 5.4) but frequently they are elsewhere on the plant and not associated directly with a pollinator. They are found on petioles or the margins of leaves and in many cases encourage ant associations (section 8.2.2). Floral nectaries may be present on almost any surface of the different parts of the flower. A common pattern is for nectaries to be on the receptacle, for example forming a ring-shaped nectary.

In a few plants it is the pistil which provides the primary attractant. A mutual relationship has arisen where the plant suffers the loss of some ovules for the sake of constant pollination. It is a kind of tax on female reproduction. In the globe flower, *Trollius* (Fig. 5.10) a moth lays an egg inside a pistil at the base of one of the styles. The larva of the moth eats some but not all of the ovules. In *Yucca* (Fig. 5.15) the yucca moth collects pollen from the stamen of a flower. It then flies to another flower, finds a mature pistil and pollinates it by pushing the pollen in the style tube. Then it lays an egg in one of the locules. The developing seeds provide food for the larva but not all are eaten. Two remarkable kinds of brood inflorescences, the figs and arums, are described in section 5.5.7.

5.2.2 Advertising (secondary attractants)

Pollinators are initially attracted to flowers, and guided within them, by various kinds of advertising and sign posting. We can thank the secondary attractants of colour, shape and scent for the wonderful beauty and diversity of flowers.

The silhouette of a flower or inflorescence is important. Dissected, star shaped flowers attract bees more effectively than those with smooth outlines Usually the corolla determines the shape of a flower but sometimes the calyx is showy. In many species the stamens are part of the attractive structure. Some may be sterile (staminodes). The shape of a flower may be emphasized by movement in the breeze. Both insect and vertebrate eyes are attuned to detecting movement. The location of a flower on a flexible peduncle may exaggerate the motion.

Different colours are effective for different pollinators. Red attracts birds. Pale colours are effective for night pollinators like moths and bats, and in deeply shaded forests. Insects see in a different spectrum of wavelengths to humans. Bees cannot see red but see yellow, blue and ultra-violet. Red poppies are ultra-violet to bees. Patterns of colour can emphasize the centre of a flower like a target or provide guide lines directing the insect to the source of nectar.

Cyanidin, which is also found in gymnosperms, is probably the basic pigment. There are three basic anthocyanidins (*anthos* is Greek for flower); cyanidin (magenta), pelargonidin (orange–red) and delphinidin (purple). Different patterns of hydroxylation gives pigments like apigenidin (yellow) and petunidin (mauve). Other factors which modify colours are the pigment concentration, the presence of flavone or flavonol co-pigments, a chelating metal, carotenoids, an aromatic acyl or sugar substitution, or a methylation (Harborne, 1988). Taxonomic differences in flower pigments exist at various levels in the hierarchy. A well-known example is the presence of a unique set of pigments, the betalains, in 10 families of the Caryophyllales (Centrospermae) a characteristic which among others links such diverse families as the Cactaceae, Aizoaceae, Chenopodiaceae and Amaranthaceae.

Petals have textural patterning, mattness or shininess or even directional brightness (Kay, 1988). A common pattern is the possession of conical-papillate epidermal cells, like blunt projecting cones which maximize the reception of low angle light and evenly spread light reflected back from the mesophyll. Changes in pattern can indicate guide lines. Surface microsculpturing adds to the potential variations.

Scents are usually produced from the surface of the petals, but the whole plant may be scented as in the Lamiaceae which includes many herbs like the mints, *Mentha*. Such whole-plant scents are not necessarily attractive, but rather may be repellant to herbivores. Both insects and bats, but not birds, are attracted by scents. Scents are ethereal oils, mainly mono- or sesquiterpenes or many other compounds. The scent is composed of several or many components which reinforce each other. Some smell sweet, but aminoid odours, which are unpleasant to us, attract flies and other insects by mimicking the smell of dung or decaying meat. There are monoamines like methylamine or diamines like putrescine and cadaverine whose names are testament to their pungency.

Insects may be attracted by the warmth of a flower. The spadix of the arums produces heat by the respiration of its packed mitochondria. In arctic and alpine regions flowers are heliotropic so that the flower follows the sun, focusing its rays within the corolla. Mosquitoes bask at the focal point of the flower. Patterns of heat distribution within the flower may help direct the pollinator after arrival at the flower.

5.2.3 The shape of flowers

There is a bewildering diversity of floral architecture. They are designs for pollinators of different sorts. They may be divided into dish, bell-funnel, brush, flag, gullet, trumpet and tube kinds (Faegri and van der Pijl, 1979). The three main ways in which the architecture of a flower is commonly recorded are illustrated in Fig. 5.2. They record the number of parts, their relative position, their fusion, and their shape.

5.3 BREEDING SYSTEMS

Over 70% of angiosperms species are hermaphrodite and cross-pollinating. Mechanisms to prevent self-fertilization are very important in the evolution of flowering plants. Some flowers deposit pollen on an insect in a way which prevents self-pollination. However even where flowers have a pollination mechanism which prevents self-pollination, the pollinator may be able to carry self pollen to other flowers on the same plant. Methods which prevent self-pollination are of two sorts; those which separate the sexes, and self-incompatibility mechanisms. In some circumstances inbreeding through self-pollination and self-fertilization is advantageous.

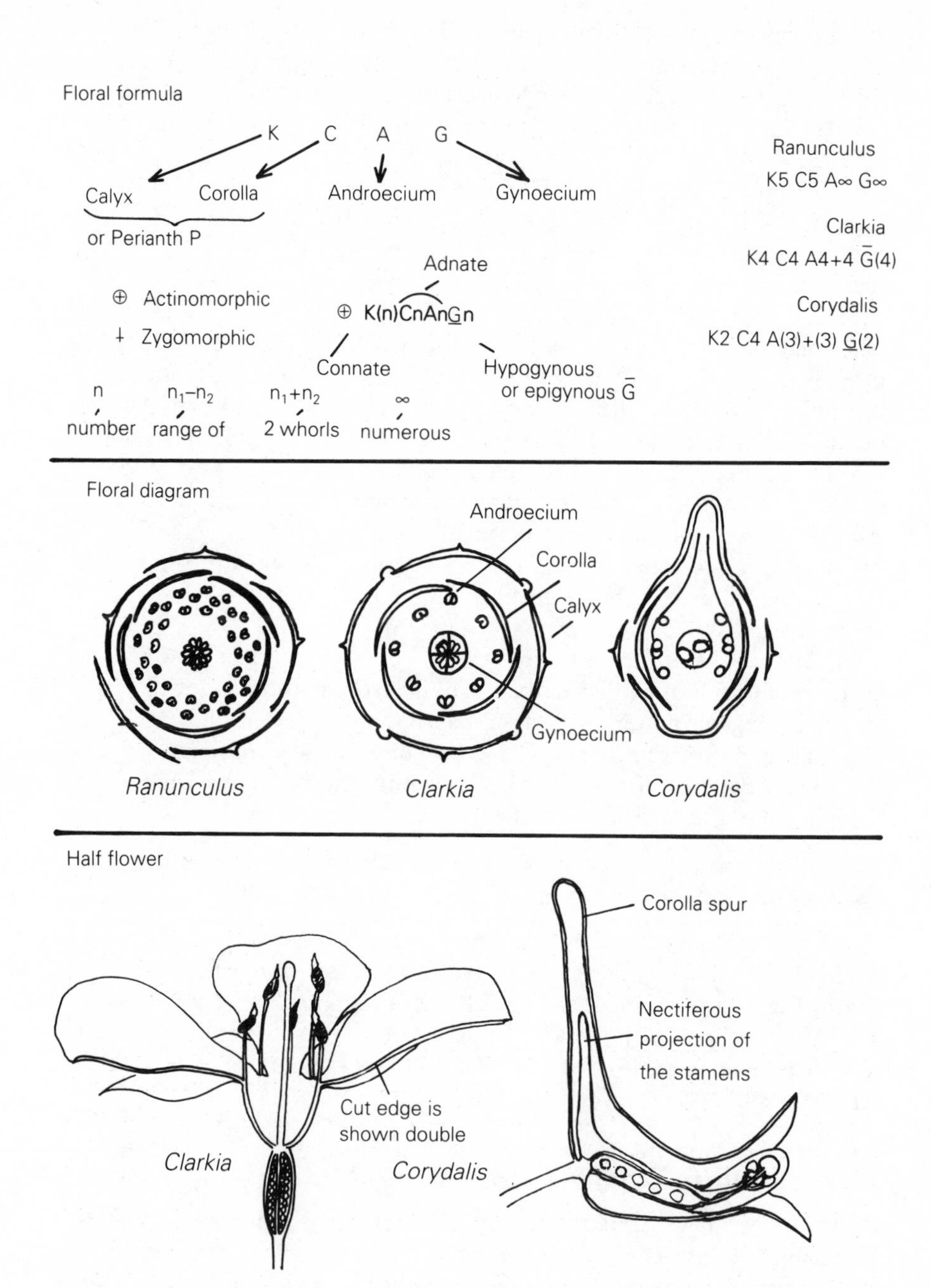

Fig. 5.2. Ways of recording flower structure: floral formulae, floral diagrams and half-flower drawings.

5.3.1 Self-incompatibility

Self-incompatibility is very widespread. It is present in at least half of all angiosperm species which have been tested. Self-incompatibility is a chemical/genetic self recognition between the pollen and stigma so that own pollen will not germinate on own stigma or if it germinates the pollen tube will not successfully grow down the style. In some cases the incompatibility reaction is late-acting, even after fertilization. The closure of the carpel and the evolution of a differentiated style and stigma has favoured the evolution of self-incompatibility by providing tissues where the incompatibility can act. Nevertheless similar incompatibility processes, but occurring interspecifically, have been demonstrated in some gymnosperms acting between the pollen tube and the nucellus (Pettitt, 1985).

A key step in the evolution of the angiosperms is likely to have been the evolution of self-incompatibility. Unfortunately the evidence for early self-incompatibility is somewhat equivocal. Self-incompatibility is present in most primitive orders (Thien, 1980) but many of the living primitive angiosperms such as the *Magnolia*, are self-compatible (East, 1940). However, it may be that these species have become secondarily self-compatible as an adaptation to pollination by relatively undiscerning insects such as beetles. One very interesting primitive genus with self-incompatibility is *Illicium*, in a family of its own the Illiaceae but in the same super order as the Magnoliaceae. It has a large number of primitive floral and vegetative characters. It is advanced in one feature, the possession of a tricolpate pollen. It is pollinated by numerous different kinds of insects, especially dipteran flies, which emerge from the leaf litter at the beginning of the rainy season (Thien *et al*, 1983). Where pollinators are seasonally abundant the penalties of not being able to self-pollinate are minimized. A similar pattern is present in some other primitive genera (Godley and Smith, 1981).

Several kinds of self-incompatibility are particularly interesting because they are associated with morphological differences between flowers. In these heteromorphic systems there are either two or three flowers morphs with pollinations only possible between different morphs.

In *Primula* there are two morphs (Fig. 5.3). The pin morph has a long style, anthers located low in the corolla tube, and produces small pollen. The thrum morph has a short style, anthers at the top of the corolla tube and produces large pollen. The incompatibility reaction either prevents the pollen tubes of self-pollen penetrating the stigma or inhibits their growth a short way down the style. The heterostylous condition helps to promote efficient cross-pollination between the morphs but does not mediate the incompatibility. The reciprocal positioning of the anthers and stigmas helps to prevent the stigmas becoming coated with pollen from their own flower. The pollen in pin flowers is less effectively dispersed from the hidden anthers, which is compensated for by a greater production of smaller grains (Richards, 1986). In *Lythrum salicaria* there are three morphs.

In family Plumbaginaceae the different morphs may actually play a part in mediating the incompatibility reaction. There are two morphs with different

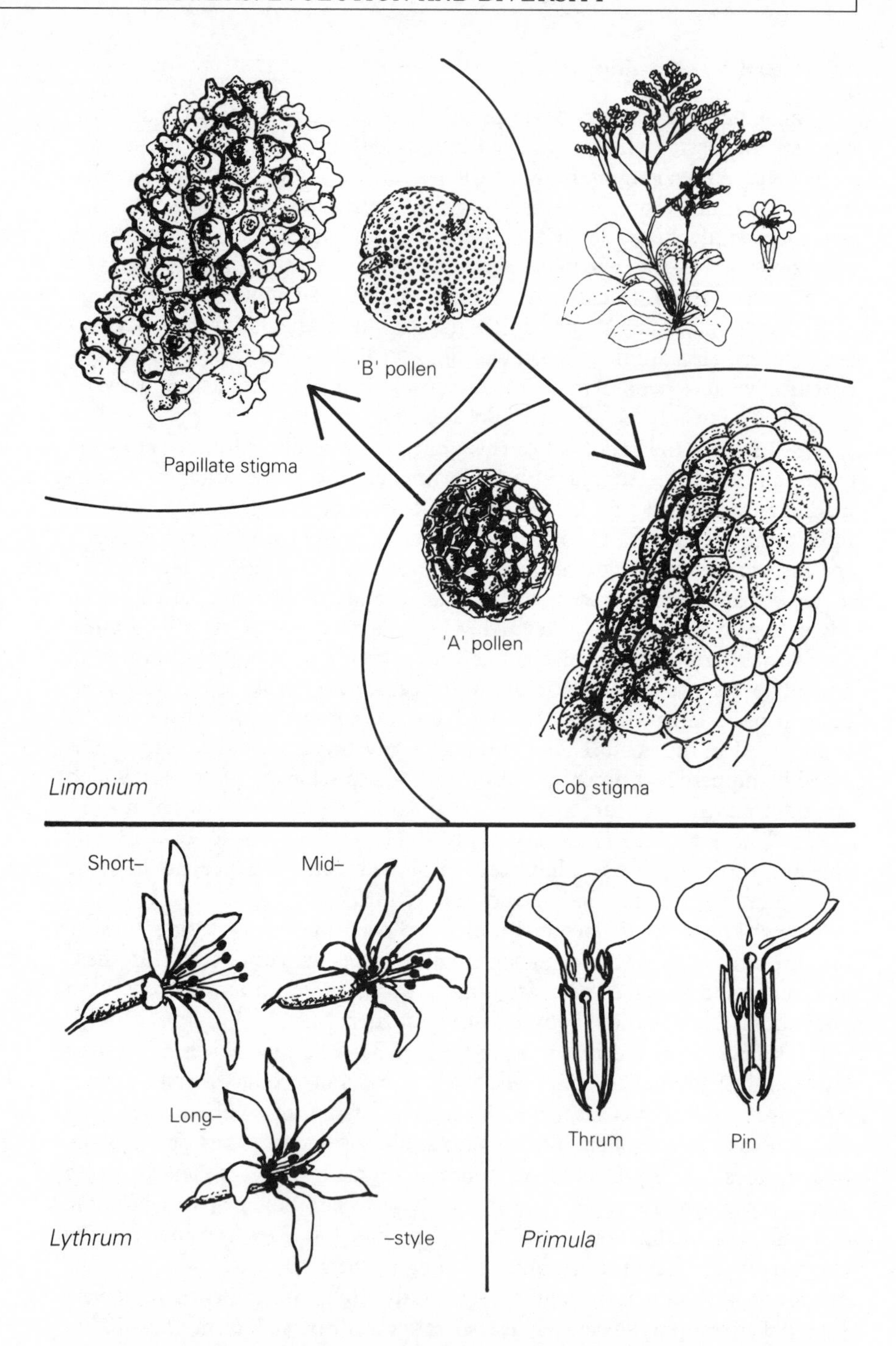

Fig. 5.3. Heteromorphic sexual incompatibility systems help to promote cross-pollination. Chemical self-recognition prevents self-pollination. In *Limonium* the morphology of pollen and stigmas varies. *Lythrum* and *Primula* are heterostylous.

stigma morphologies and different patterns of reticulation on the pollen (Fig. 5.3). 'A' pollen which is produced in flowers with 'Cob' stigmas germinates only on the 'Papillate' stigmas of the alternative morph. The 'Papillate' morph produces 'B' pollen which will germinate only on 'Cob' stigmas. The 'A' pollen has a pattern of reticulations which allows it fit closely onto the 'Papillate' stigma. 'B' pollen is relatively smooth like the surface of the Cob stigma. Since close connection of pollen and stigmatic cells is necessary for rehydration of the pollen grain prior to germination this may be an important part of the incompatibility reaction. One species of *Limonium*, has evolved self-compatibility by a cross-over within the self-incompatibility gene so that it is monomorphic with the self-compatible combination of 'A' pollen and Papillate stigma. A similar breakdown of incompatibility has been observed in many other groups.

Heteromorphic incompatibility systems are rare. Most systems are homomorphic. A distinction is made between those in which the haploid pollen grain itself determines its incompatibility, called gametophytic systems, and those in which the diploid/polyploid tapetum determines the incompatibility, called sporophytic. The former is more common and is probably primitive. *Illicium* has gametophytic self-incompatibility. This seems to be the pattern in the primitive orders such as the Magnoliales, Winterales, Nympheales, and Hamamelidales. Pollen with a granular wall is associated with gametophytic self-incompatibility but many self-compatible species also have this kind of wall.

There is a strong correlation between sporophytic self-incompatibility and pollen wall structure where the ectexine has three layers; the foot layer, a columellar layer and a tectum. In spaces within the ectexine, in the columellar layer, the proteins from the tapetum are held which play a key role in the recognition process (Zavada, 1984). These leak out through micropores or gaps in the ectexine onto the stigma where they signal the genetic make-up of the male parent sporophyte. Heteromorphic incompatibility systems are sporophytic. Two large important families the Astercaceae and Brassicaceae have homomorphic sporophytic systems.

5.3.2 The separation of the sexes

The separation of the sexes is called dicliny. Separate male and female flowers may be found together on the same plant (monoecy) or on separate male and female plants (dioecy). Dioecy is rare in the angiosperms. For example only 2% of the British flora is dioecious and 12% of the New Zealand flora. Dioecy is common in the gymnosperms which are mostly wind pollinated. In temperate regions wind pollinated angiosperms tend to be dioecious. The unisexual flowers have a very reduced perianth and flowers are aggregated into cones and catkins.

Where dicliny is associated with insect pollination nectar is usually the attractant but *Decaspermium parviflorum* is a dioecious Indonesian species which provides sterile pollen in the female. It is important that the male and female flowers look identical to the insect so that it does not discriminate between them. Sterile stamens are found in female *Silene dioica*, which also compensates for its lack of pollen by the production of more nectar. In the Cucurbitaceae, in which there are closely related monoecious and dioecious species, the stigma is lobed and somewhat remiscent of the three stamens of the male flower (Fig. 5.4).

Dioecious species in the tropics are more often insect pollinated than those in temperate regions. They tend to have large animal dispersed seeds and it is possible that dioecy in them is the result of competition between the sexes within plants. In dioecious species male plants sometimes produce more flowers and grow vegetatively more vigorously perhaps reflecting the extra energy cost of producing pistils and seed.

The important family Euphorbiaceae are entirely diclinous. They illustrate several interesting aspects of sexual separation. Niche separation, with each sex growing in slightly different environments has been observed in dioecious *Mercurialis perennis*. In monoecious *Ricinus* the male flowers have large numbers of branched stamens while the female have below them numerous conical nectaries. In the genus *Euphorbia* effective hermaphrodity has evolved. They are monoecious but the inflorescence mimicking a flower, called a **cyathium**, has one female flower at the centre of a cup which bears on its inside a number of male flowers, so reduced that they each look like a single stamen. Around the margin of the cyathium there are large nectiferous glands like a perianth.

In those hermaphrodite species without self-incompatibility, self-pollination is frequently a standby if cross-pollination fails. This option is not open to dioecious species but greater vegetative reproduction than in related hermaphrodite species can compensate for the greater hazards of achieving fertilization. Dioecy may have evolved via gynodioecy (separate female and hermaphrodite plants), that is via male sterility. Gynodioecy is much commoner than the alternative androdioecy (separate male and hermaphrodite plants). In gynodioecious species the female plants frequently have smaller flowers. Stable gynodioecy is common in those families in which dioecy is rare. For example it is associated with the specialization of different flowers within a single inflorescence in the daisy family, the Asteraceae.

The separation of male and female flowers in monoecious species promotes cross-pollination but self-pollination can occur. The degree of self-pollination depends on the density of flowers within plants, the distribution of plants in a site, and the behaviour and availability of pollinators.

Over 70% of all angiosperms have only hermaphrodite flowers. Nevertheless the sexes may be functionally separated by different timing of development. Either the stamens mature first (protandry) or the stigmas are receptive before pollen is released (protogyny). In the protandrous *Campanula*, pollen is shed onto a hairy style while in bud (Fig. 5.4). The hairs first trap the pollen but later they invaginate like the fingers of a glove to release it. The style elongates as the flower opens and the stamens wither. The lobed stigma opens to cross-pollination but later self-pollination may take place as the stigma lobes bend back to touch the style. In *Prunus* the mature stigma is poked out of the bud to allow cross-pollination. The anthers shed their

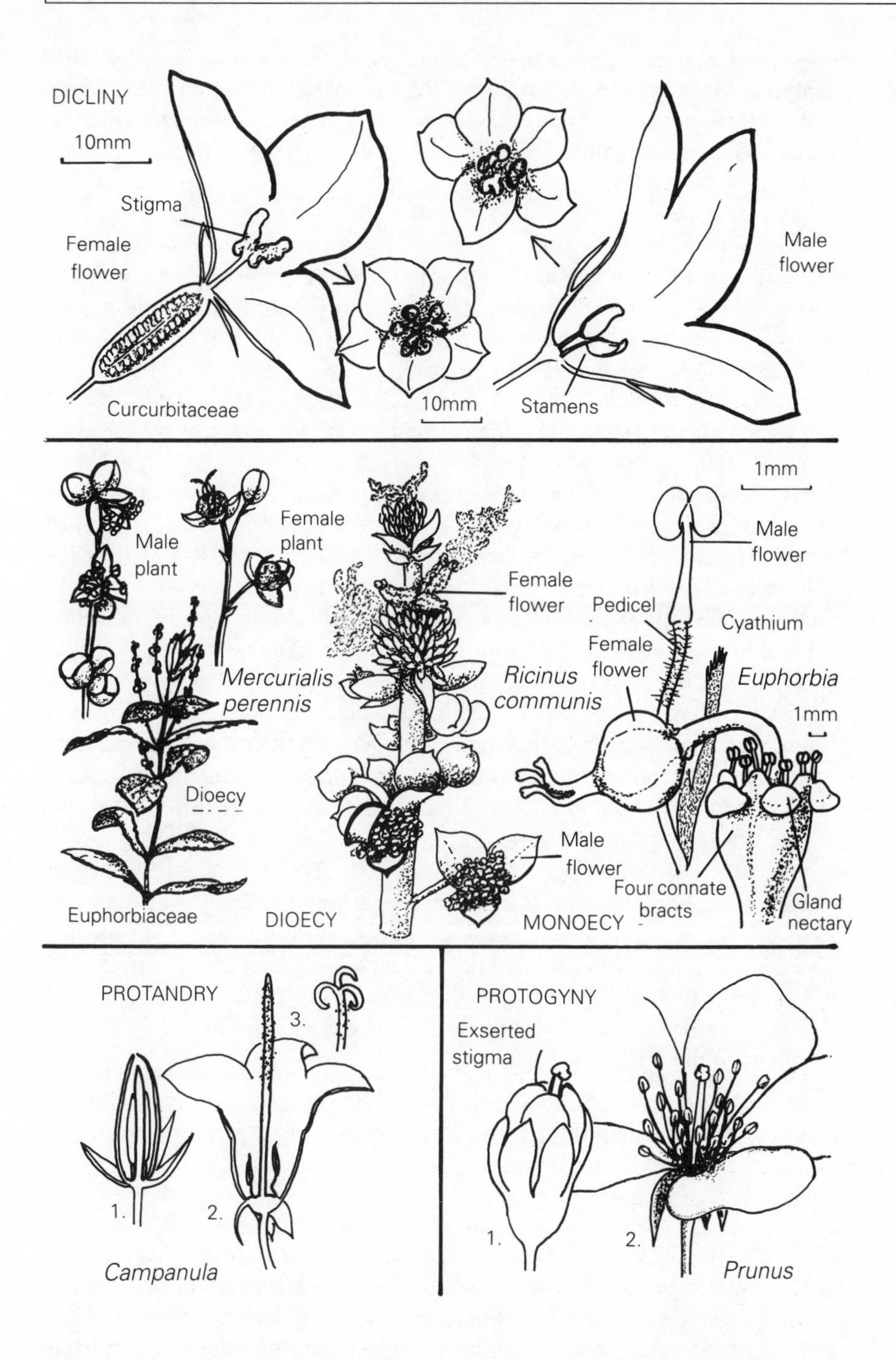

Fig. 5.4. The separation of sexes in the angiosperms: in different flowers on the same plant (monoecy) or different plants (dioecy) or in time of development (protandry and protogyny). The cyathium of *Euphorbia* is a flower-like group of separate male and female flowers.

pollen later when the flower opens properly. In this case, as in many other examples of protogyny and protandry, this is a mechanism to ensure efficient pollen transfer not prevent self-pollination, since a self-incompatibility system is also in operation.

5.3.3 Inbreeding

Adaptations for inbreeding at their simplest consist of the subversion of any mechanism for cross pollination. Many primarily self-incompatible groups have given rise to self-compatible species. Inbreeding species often have small flowers. Most have no need to attract insects. They can also save in the production of pollen. Pollen–ovule ratios for obligate inbreeding species are two orders of magnitude smaller than those for obligate outbreeding species (Queller, 1984).

The ability to inbreed has evolved many times in flowering plants. Many species are facultative inbreeders. A few inbreed so regularly that they must be regarded as obligate inbreeders. In regions where there is a shortage of pollinators or environmental uncertainty inbreeding may be an advantage since it ensures the successful production of seed. It may also be important in a weedy species where the rapid production of large quantities of seed is required. In those species which grow in small isolated populations, especially colonizing species, mechanisms which promote outcrossing may be disadvantageous because they reduce the chances of any seed production.

Cleistogamous flowers never open and pollination is therefore normally by selfing. Pollen–ovule ratios for cleistogamous species are lower even than non-cleistogamous but obligate inbreeding species. Total **cleistogamy** is impossible to prove and perhaps unlikely. The rate of cleistogamy may be affected by the environmental conditions. Several species like ground ivy, *Glechoma hederacea* produce cleistogamous flowers at the end of the season. *Commelina forskalaei* produces subterranean cleistogamous flowers in the dry season. Several aquatic plants are regularly cleistogamous but at least one of these, *Subularia*, has been observed to produce open flowers on a dried up lake margin (Richards, 1986).

5.4 EVOLUTIONARY TRENDS

5.4.1 The perianth

Where the **perianth** is not easily designated as being a **calyx** or **corolla**, its parts are called **tepals**. Primitive flowers either lack a perianth or have one in which there is a single whorl of tepals. In most flowers the perianth is differentiated into separate whorls, the calyx of sepals and the corolla of petals. Each of these whorls may have been derived from tepals, but it is likely that in some groups, such as the buttercups, the petals actually originated from stamens, to which they are anatomically similar. In the peony, *Paeonia*, there is a gradual transition from leaves, through modified leaves on the flowering stem called bracts, into the perianth with parts at first sepal-like and then petal-like.

The calyx has a protective function. It is usually green with the sepals tightly abutting each other (**valvate**), or overlapping (**imbricate**), in bud. The calyx protects against drought, temperature shock and predatory insects. Protection is increased by **synsepaly**, the fusion of the sepals together. In some extreme cases, as in *Eucalyptus*, the calyx has been modified into a kind of cap, a **calyptra** (cf. moss capsules) which abscises at the base as the bud opens pushing off the cap. The shedding of the calyx after the bud has opened, having **caducous** sepals, is widespread and is seen for example in the poppies, *Papaver*.

In the monocotyledons there is often no clear separation of the calyx and corolla. The outer whorl of tepals forms part of the showy perianth. In many of these cases the flower or inflorescence is protected instead by a **bract** or **spathe**. Bracts are very important in the flowers of many other groups. They may be showy as in poinsettia or may form an **epicalyx** or many other modified structures.

In most taxa the number of sepals and petals in each whorl has become fixed for each species and the petals or sepals are arranged precisely, usually alternating with each other, and with the stamens, but other patterns are not unusual. In the monocotyledons tepals are arranged in threes. Patterns of fours or fives are very common in the dicotyledons.

The way the petals and sepals overlap in the bud, their **aestivation**, is a characteristic which is useful in classification (Fig. 5.5). A very important trend in the corolla has been from separate petals (polypetaly) to having the petals fused (**sympetaly**). The fusion of similar structures is said to be **connation**. In many flowers this is associated with fusion of the stamens to the petals, called **adnation** because the structures joined are dissimilar. In this case the stamens are said to be **epipetalous**. Both sympetaly and epipetaly increase the precision of the floral structure for particular kinds of pollination.

Another important trend has been away from radial symmetry (**actinomorphy**) towards bilateral symmetry (**zygomorphy**). This is associated with particular kinds of pollination and is associated with the development of compound inflorescences where the flowers are lateral on the floral axis rather than terminal. The lower part of the corolla is expanded providing a landing platform.

5.4.2 The stamens (the androecium)

Stamens are very diverse (Fig. 5.6). Normally the stamen has a **filament** and an **anther**. The anther consists of two thecae joined by a continuation of the filament, the **connective**. Each **theca** has two **pollen sacs** fused in a kind of **synangium**. Primitive stamens are found in the Lauraceae. The anthers have two (*Lindera*) or four (*Laurus*) locules which are not arranged in the characteristic way of most other angiosperms (Fig. 5.6). In *Drimys* (Winteraceae) the stamens are more similar to those of advanced angio-

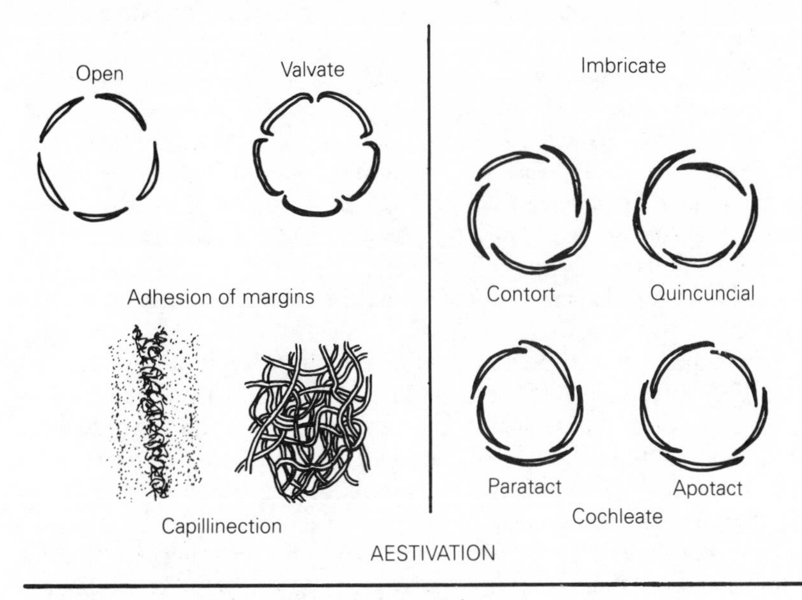

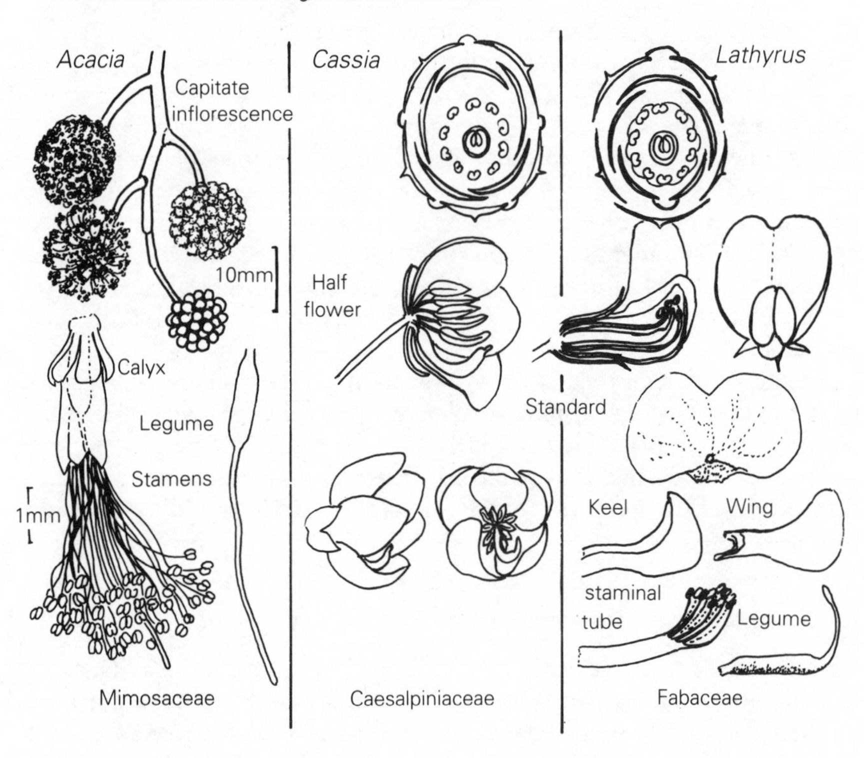

Fig. 5.5. The perianth: aestivation following Weberling (1989). The three families of legumes have characteristically different inflorescences.

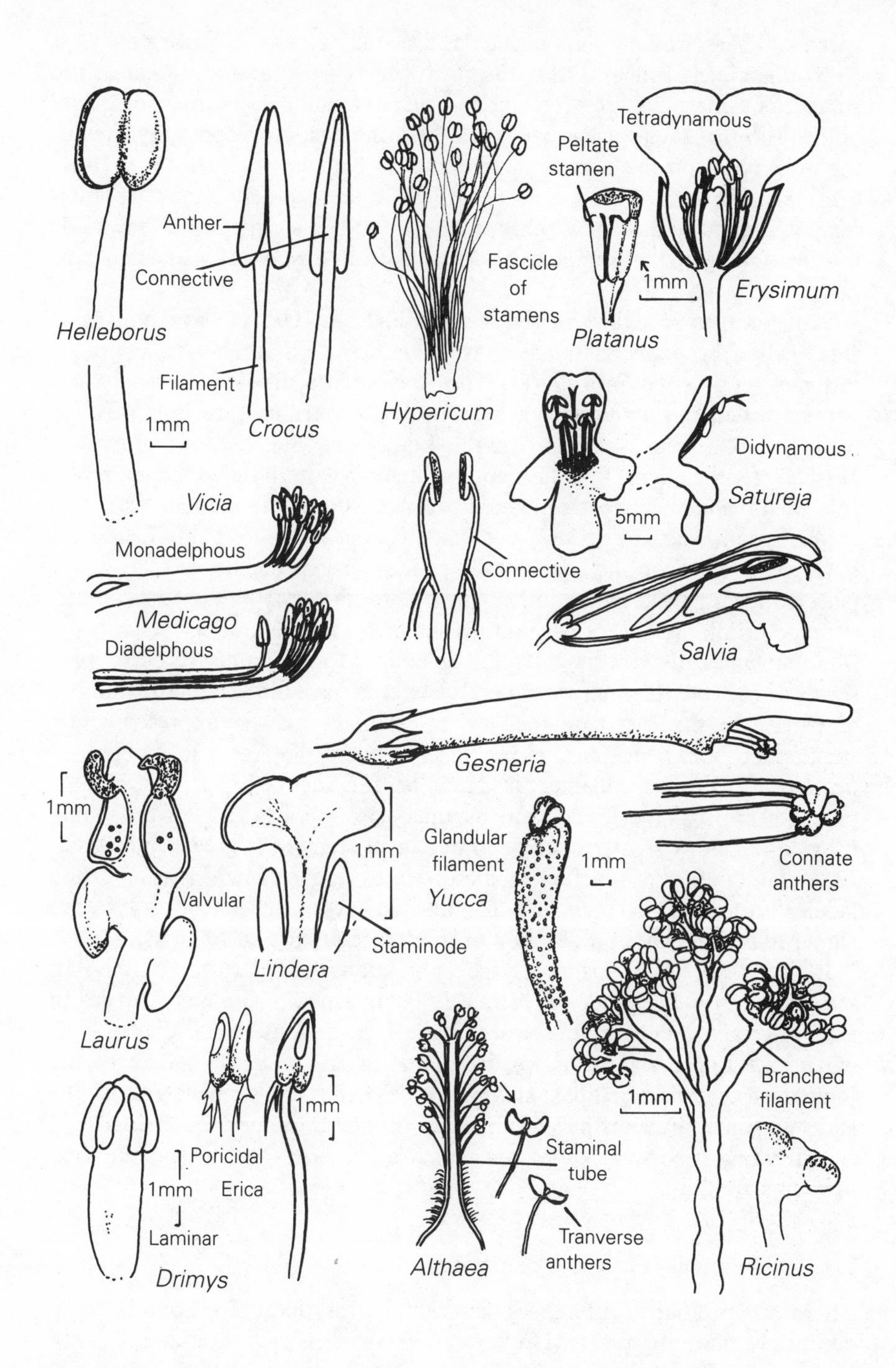

Fig. 5.6. Variation in the form and arrangement of stamens. *Lindera, Laurus* and *Drimys* have primitive kinds. The branched stamens of *Ricinus* are unusual.

sperms. There are two pairs of locules situated more or less marginally on a broad laminar filament. Other 'primitive' sorts are known. In *Platanus* the stamen is almost peltate recalling the peltate sporangia of the horsetails. Some primitive angiosperms have tepal-like stamens, microsporophylls, with the pollen sacs on the adaxial surface and no filament or connective. However, tepaloid stamens are not necessarily primitive but may be an adaptation for pollination by insects which clamber around in a cup shaped flower. It is easy to imagine the origin of the perianth from a whorl of stamens of this kind.

Stamens usually dehisce by a longitudinal slit. Others have transverse slits, valves or pores. Stamens may be either all of the same length or different lengths within a flower. The Brassicaceae, like *Erysimum*, have a pattern called **tetradynamous** where two outer stamens are short and an inner whorl of four are long. In the Lamiaceae, like *Satureja*, there are normally four stamens, two short and two long, a pattern called **didynamous**. In *Salvia* the connective is elongated bearing a sterile theca at one end. The whole structure acts like a lever with the filament as the fulcrum. The insect pushing its way into the corolla tube pushes against the sterile theca in its path, so levering the fertile anther down onto its back.

Some primitive flowers have numerous or a variable number of stamens. More advanced flowers have a number equal to the number of petals plus sepals or equal to the number of petals alone, or have fewer than the number of petals. In this last case the 'absent' stamens may be represented by staminodes, sterile stamens, as in *Lindera*, where they seem to function as nectaries. In *Yucca* the filament is glandular. It is very common for zygomorphic flowers to have fewer stamens than corolla lobes. In flowers which provide pollen as an attractant the number of stamens is greatly increased. In *Ricinus* pollen production is increased by each stamen bearing many locules on a multiply branched filament. In *Hypericum* it looks as if each whole stamen has been multiplied to produce a fascicle of stamens.

In many species the stamens are **epipetalous**, fused to the corolla. In some plants the filaments are **connate**. There are two common patterns in the Fabaceae: **monadelphous** where there is a single tube and diadelphous where the upper stamen is free. In *Althaea* and other members of the Malvaceae there is a complex staminal tube with many stamens around the style. In some liliaceous flowers, like *Pancratium* the stamens are fused in a ring forming a **corona**. In *Gesneria* the anthers are connate and the long filaments are free.

5.4.3 The pistil or carpel (gynoecium)

There are profound differences in the form of the **gynoecium** between families of flowering plants (Hodgson, 1989) as well as within families. The gynoecium of primitive angiosperms is made up of one **pistil** or a group of separate pistils (apocarpy). Primitive pistils have an ovary and stigma but only a short style if it is present. The style has become longer. Separate **carpels** have become fused to form a compound pistil with a **syncarpous** ovary. The pistil may have separate styles with one style per original carpel or have a single style. In the latter case the stigma may be lobed or form a single knob. In a syncarpous ovary the number of original carpels may be recognized by the number of compartments containing ovules, each called a **locule**. In many groups the inter-locule wall has broken down to give a unilocular ovary. Nevertheless it may still be possible to recognize the ancestral condition and so it is possible to describe an ovary, for example, as having three carpels but being unilocular. The alternative condition of having more locules than carpels is also possible but rare. Extra walls divide each carpel.

The evolution of a syncarpous ovary has brought greater protection to the ovules, while minimizing the amount of ovary wall material required. Ovules in separate carpels are much more exposed to drought, temperature shock or predation than they are in a compound pistil. A compound ovary has a single exterior wall which can be made strong. It also has great potential for adaption to encourage dispersal either of the fruit or the seed. Species with a syncarpous pistil have an additional advantage when they have a common style or stigma, that a single pollen load on the stigma may effect fertilization of ovules in all the locules of the compound ovary.

In primitive flowers and many others, the whorls of the calyx, corolla and stamens originate on the receptacle around the base of the gynoecium. The gynoecium is said to be superior and the flower is **hypogynous**. One of the most important trends in the evolution of flowers has been the evolution of **epigynous** flowers where the ovary is inferior. In epigynous flowers the calyx, corolla and stamens have the appearance of originating above the ovary around the base of the style. In fact the floral axis bearing the outer whorls of floral parts has extended up as a cup and is fused to the ovary wall. In some groups an intermediate condition between epigyny and hypogyny, called hypoepigyny can be seen. The whole sequence can be seen in different members of the genus *Saxifraga* (Fig. 5.7). Some of the same evolutionary pressures which lead to the development of a **syncarpous** ovary may have encouraged the evolution of **epigyny**.

A pattern called perigyny, in some respects intermediate between hypogyny and epigyny but without the fusion of the floral axis to the ovary wall, is seen frequently in the Rosaceae (Fig. 5.11) as well as other taxa. **Perigynous** flowers have a bowl-shaped floral tube, the **hypanthium** around the pistils. Separate whorls of calyx, corolla and stamens originate around the rim of the hypanthium.

Epigyny has evolved many times over. In some groups it seems to have happened quite easily without an intervening perigynous stage, by delays in the separation of the meristems which give rise to the various floral whorls, what Stebbins (1974) calls intercalary concrescence.

The stigma arose along the sealed margins of the carpel, clearly seen in the follicular stigmas of *Platanus* and *Helleborus* (Fig. 5.7) and *Drimys* (Fig. 5.1). A variety of other types have evolved. They may be long and filiform, often covered with papillae as in *Limonium* (Fig. 5.3) or capitate. In *Iris* they are petaloid (Fig. 5.15). In *Papaver* the stigmatic crest has lobes made

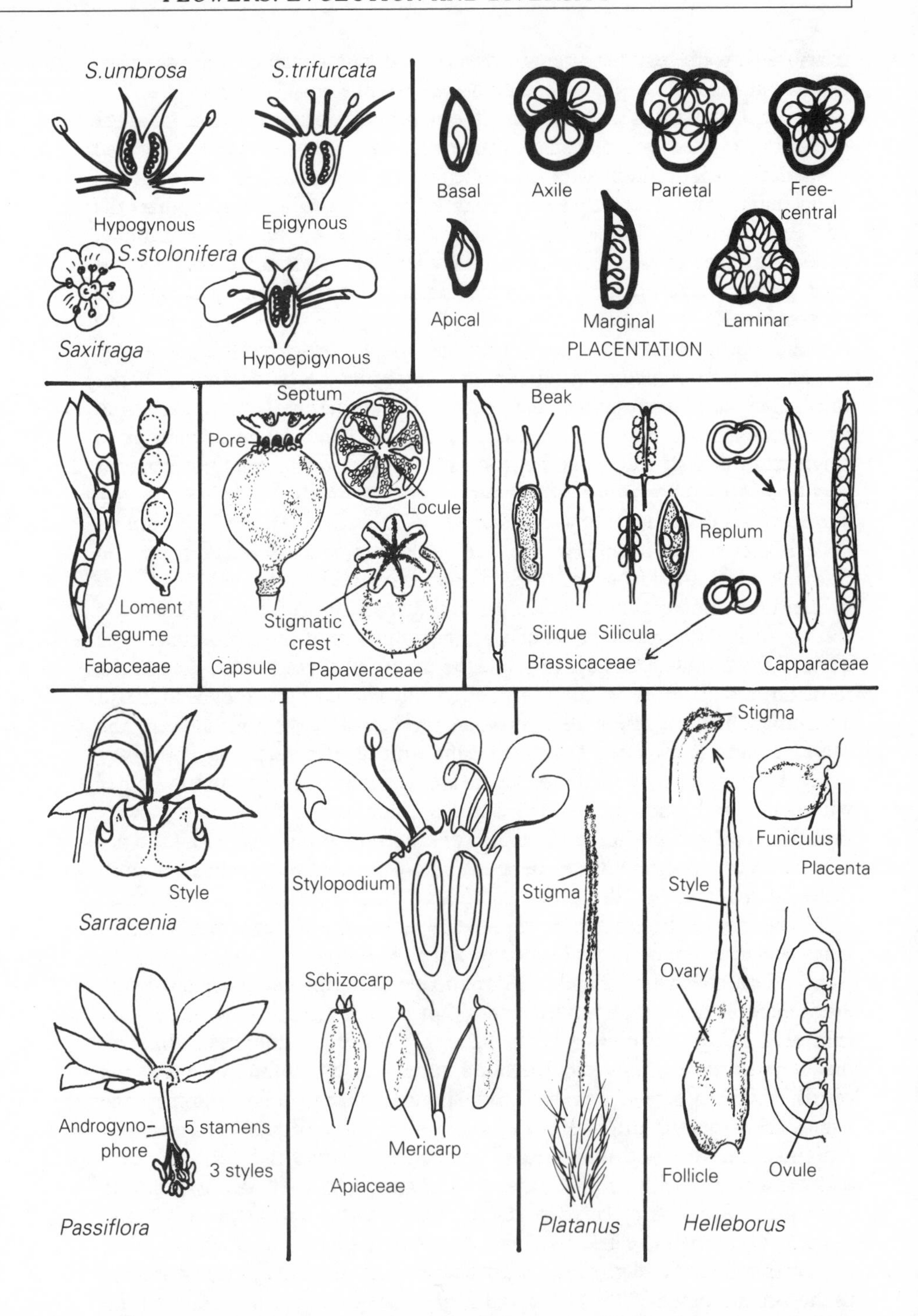

Fig. 5.7. Variation in the pistil and fruit. Some families have a characteristic pistil and fruit type though this can be variable even within a genus like *Saxifraga*.

from the two halves of stigma of adjoining carpels (Fig. 5.7). In *Sarracenia* the stigmas are hidden below an extraordinary umbrella-like style.

The arrangement of ovules within the ovary is called placentation. Changes in placentation have accompanied changes in the number and size of seed produced. The number of ovules produced in a pistil varies from one to many thousands. The primitive placentation is the arrangement of ovules on the adaxial (i.e. inner) surface of the carpel with the adaxial surfaces exterior to the ovules pressed together. The ovules therefore appear to be marginal along the broad suture of the carpel. In a syncarpous ovary, where the sutures of each carpel are in the centre of the ovary, this primitive placentation gives rise to axile placentation. An increase in ovule number is associated with a reduction of the number of locules, as if, having fused, the carpel walls have now shrunk back. More space is created by losing the unnecessary inner walls. In orchids and some parasitic plants, which have thousands of tiny ovules in the ovary, the ovules are not restricted to particular areas of the ovary wall but are spread over the whole wall, giving laminar placentation.

A reduction in ovule number is often associated with increased seed size. It has been achieved by basal or apical placentation in different groups. This has led, in some groups, to a pistil with a solitary ovule giving a single seeded fruit such as an **achene** or **caryopsis**, with the fruit adapted for dispersal without dehiscence. Development of single seeded fruits may be more rapid than those containing many ovules. Solitary ovules are found in a very wide range of taxa. Single seeded fruits may also have the advantage of restricting the loss of ovules following predation. In the syncarpous ovary with separate locules predation may be restricted to individual locules. In a few groups like the primroses, Primulaceae, there may have been evolutionary pressure for increased seed number after a period of selection for reduced seed number. This may explain the change to free central placentation, with ovules arranged on a central column, from a basal placentation.

Differences in the fruit have been used to distinguish many taxa. Ovaries are sometimes dehiscent. They open by slits, either between the carpels (**septicidal**) or down a locule wall (**loculicidal**) or by pores (**poricidal**). The order Fabales are characterized by having legumes or pods, fruits which split open along both the ventral suture as in follicles, and a dorsal line. In the two related families the Brassicaceae and Capparaceae the pistil has carpels. Where they join there is a frame called the replum to which the ovules are attached. In the Brassicaceae the replum has running across it a septum so that the ovary is bilocular. In the Capparaceae there is no septum and the ovary is unilocular. The fruit is characteristic; an elongate **siliqua** or a short **silicula**. At maturity the outer walls, valves, fall away leaving the replum. Differences in the fruit, such as the presence or absence of a beak and the relative length to breadth of the ovary, distinguish different genera in the Brassicaceae.

Syncarpous dry dehiscent fruits are called **capsules**. Single-seeded indehiscent dry fruits are **achene**, **nut** (with a sclerenchymatous pericarp), **cypsela** (from an inferior ovary), **samara** (winged) and **caryopsis** (with testa fused to pericarp). Multiple seeded dry indehiscent fruits called **loments** can split into single seeded segments, or in a **schizocarp** each segment (called a **mericarp**) is a single seeded carpel, as in the Apiaceae.

The **pericarp** may be dry or fleshy. Anatomically three layers may be detected; an outer **exocarp**, an inner **endocarp** and the **mesocarp** between. In fleshy fruits it is the mesocarp which becomes fleshy. Fleshy fruits are usually indehiscent. They are simple (e.g. **berry**, **hesperidium**, **drupe**) or compound (e.g. **pome**, **drupecetum**, **hep**, **pseudocarp**) or multiple (**syconium**, **sorosis**, **coenocarpium**). Some are described in greater detail elsewhere, as in the section on Rosaceae (section 5.5.2).

There is a good deal of variation in the the structure of the embryo within the developing seed. These variations are important in classifying some groups of flowering plants. A classic example are the so called Centrosperms (order Caryophyllales) which include 12 families and 10 000 species. They include diverse families like the cacti (Cactaceae), pinks (Caryophyllaceae), amaranths (Amaranthaceae), goosefoots (Chenopodiaceae) and fig-marigolds (Aizoaceae). Most have a **campylotropous** ovule in which the inner integument protrudes, and a peripheral embryo surrounding a nutritive central perisperm tissue plus a number of other technical characters. Other kinds of ovule are **anatropous**, **orthotropous** and **amphitropous**.

There seem to be few limits to variations in the ovule and seed. In the mistletoes, families Loranthaceae and Viscaceae, the ovary is sometimes solid without locules. Embedded in a massive placenta-like body called the mamelon, there are a number of embryo-sacs without a clearly defined nucellus or integuments. Following fertilization embryos with 2–6 cotyledons, or the two cotyledons fused, develop inside an aril or are capped by it. The fusion of the cotyledons is functionally analogous to the situation in monocots. It aids the elongation of the embryo to plant the radicle on the host plant. Being pedantic these are not true seeds since they lack a testa and frequently contain more than one embryo. The 'seeds' are enclosed in a fleshy berry which is eaten by birds. The birds smear the seeds out onto a branch or if the seeds are ingested they are rapidly regurgitated or defecated. The sticky aril glues the rapidly germinating embryos to a branch. The necessity of rapid germination may also explain the replacement of the testa by an aril.

5.4.4 The inflorescence

A number of evolutionary trends have been recognized in inflorescences. Generally flowers have become smaller and aggregated into a compound structure which functions as a single unit. Within the inflorescence the timing of flowering is controlled and different florets may be specialized for different functions. An example of an advanced inflorescence is the capitulum of the the daisy family, Asteraceae, where the inflorescence superficially looks like a single solitary flower though it is made up of many florets, some functionally and structurally differentiated.

Flowering shoots can be **determinate** giving a cymose inflorescence or

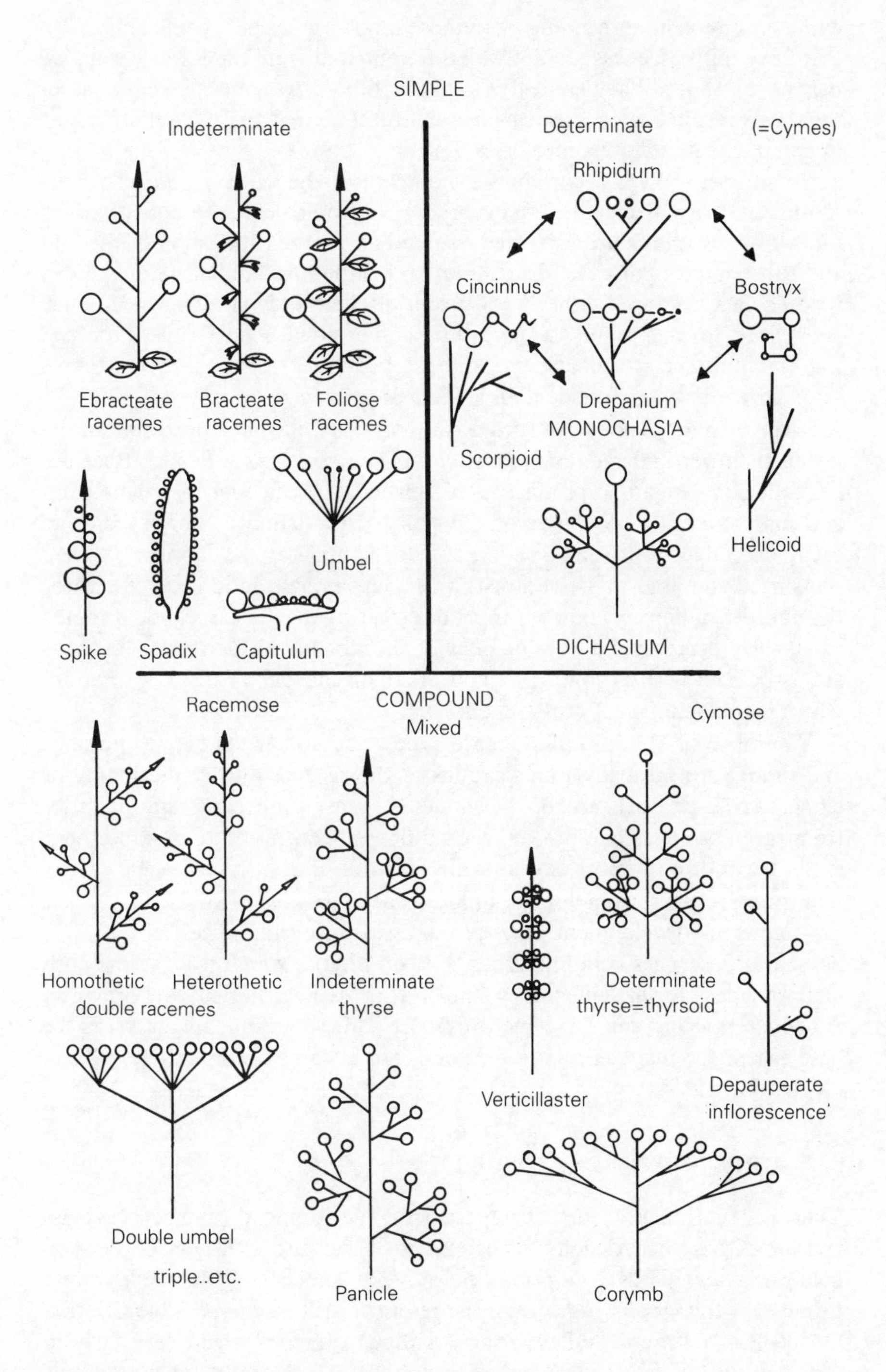

Fig. 5.8. Inflorescence types following Weberling (1989).

cyme or **indeterminate** giving a racemose inflorescence or **raceme** (Fig. 5.8). Primitive inflorescences are not well differentiated from the lower vegetative part of the shoot. They are leafy (foliose) cymes or racemes. A reduction of the leaves resulted in bracteate cymes and racemes. Further reduction results in an ebracteate inflorescence or even solitary flowers.

In a cyme growth occurs by development of the axillary bud below the terminal flower. If a single lateral branch is produced at the node, various kinds of monochasia are produced, depending on the orientation of the bud and subsequent buds. The **drapanium** and **rhipidium** are flat inflorescences with flowers produced either to one side, or alternately to either side. They are related to each other via more three-dimensional inflorescences the zig-zag **cincinnus** or scorpioid cyme and the **bostryx** or helicoid cyme. If two buds grow out at each node a **dichasium** is produced.

Racemes may have evolved from compound cymes by suppression of the terminal flower and reduction of lateral cymes to a single flower. Racemes have the advantage that production of flowers is spread over the whole plant and not localized at the apex of the stems. Production of flowers is more easily extended over a long period and for as long as the flowering season remains favourable. This contrasts with determinate inflorescences where the number of flowers and the timing of flowering may be largely determined at an early stage in the development of the first floral meristem. Racemes are associated with **zygomorphy** and other mechanisms such as **protandry** which promote efficient cross-pollination.

A number of special inflorescence types, the **spike**, **spadix**, **capitulum**, and **umbel** are normally modifications of the raceme by the shortening or elongation of **pedicels** and/or internodes. Cymose umbels or capitula differ from racemose ones because the central flower matures first. In a raceme it is the lower or outermost flowers which mature first.

In practice the terminology of inflorescence types is very difficult because the timing of development may be modified obscuring ancestral patterns. Some inflorescences combine types, like the **thyrse** which is a raceme with lateral cymes. A **verticillaster** is similar with dense whorled lateral cymes. A **panicle** is a compound raceme with more and more branching towards the base except the main axis and the laterals are in some examples determinate (Weberling 1989).

5.5 A REVIEW OF ANGIOSPERMS

There are two major groups of angiosperms; the monocots (monocotyledons) and the dicots (dicotyledons) recognized by Cronquist (1981) as Class Magnoliopsida and Class Liliopsida respectively, subclasses Magnoliidae and Liliidae in this book (Fig. 5.9). At present it still seems possible that the monocots are a monophyletic group. Although they diverged very early in the evolution of flowering plants perhaps from a chloranthoid type plant, the Piperaceae have been suggested, this may not have been the earliest major diversification of the flowering plants. If there was a previous diversification

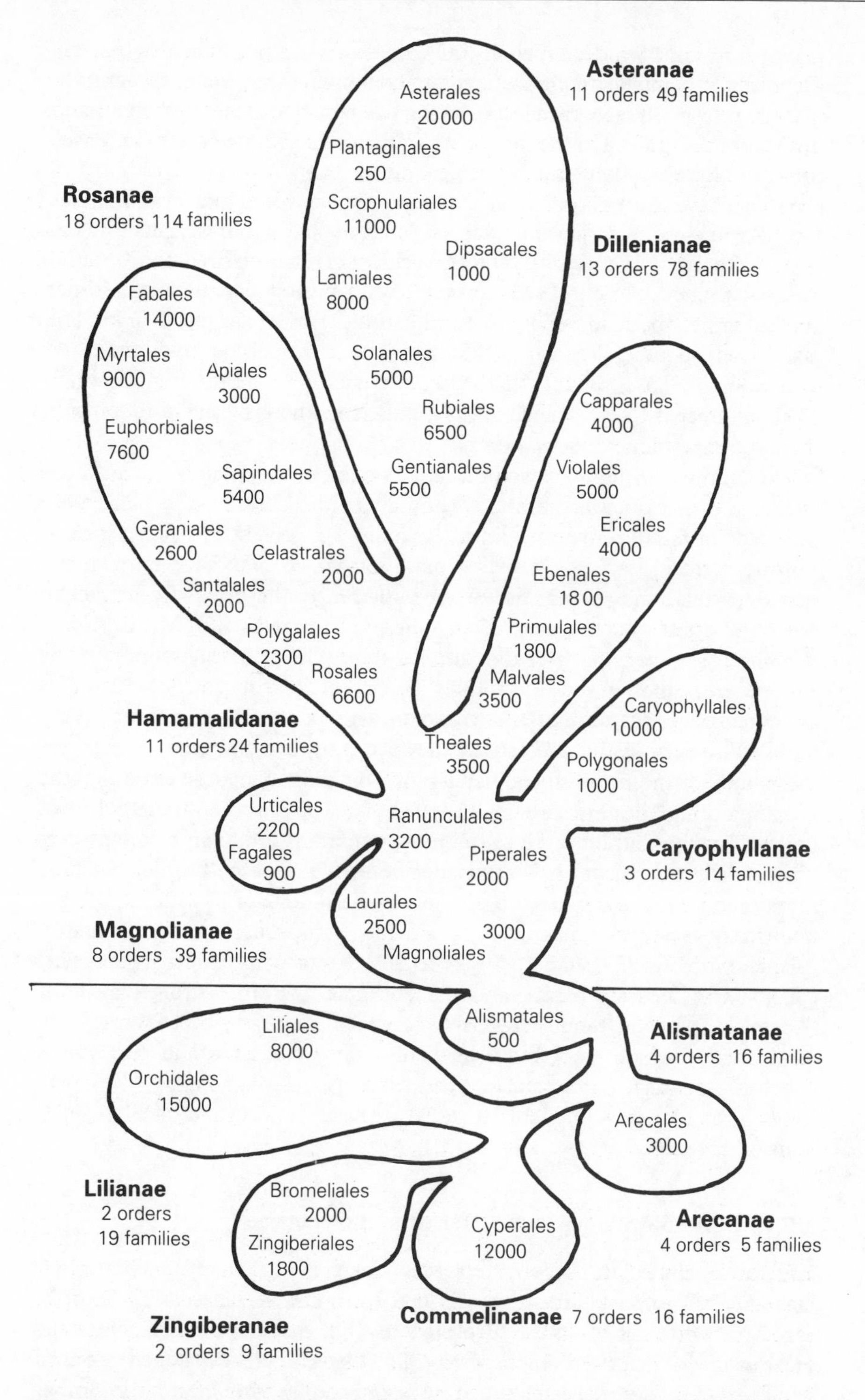

Fig. 5.9. Relationships between the main orders of angiosperms (after Cronquist (1981)). The number of species is shown under each order. The dicots are above the line, the monocots below; the common origin of the two lies in the Magnoliales/Alismatales group.

perhaps giving a wind-pollinated line, the dicots are not monophyletic and according to some taxononomists not a taxonomic group worth recognizing.

Currently six major groups are recognized in the dicots and five major groups are recognized in the monocots. When in the 18th century De Jussieu produced his very influential classification of plants he created 100 'natural orders'. These 'natural orders' are mostly what we would today call families. De Jussieu's classification included cryptogams and gymnosperms but that there were 100 'families' was not an accident. It summarized the variation in a comprehensible number of groups. With greater botanical exploration and attempts to define more natural families there has been a kind of taxonomic inflation. Cronquist (1981) in his *Integrated System of Classification of Flowering Plants* divides the flowering plants into 383 different families. Their number is likely to increase as cladistic methods of classification split them to make monophyletic groups.

This number of families would seem to place an impossible block in the way of the new student getting to grips with the classification. There are hundreds of families to get to know. Even in a relatively small area such as Europe, with its recent history of glaciation which may have made many species extinct, and present narrow climatic range, there are 173 families of flowering plants. There are 237 families in Australia and 355 in North America. However, many of the families are small and rarely seen. Outside tropical areas the number of families seen is much reduced. Some families are much larger than the others, commonly met and so easy to learn. Many families are very distinctive so that they are easily remembered.

An understanding of the diversity of flowering plants is eased by the grouping of families into orders. In Cronquist's classification there are 83 of these. The distribution of 40 of the most important orders into superorders (Cronquist's subclasses) is illustrated in Fig. 5.9. These 40 orders include more than 90 % of angiosperm species. Some older but now defunct taxonomic names are still useful. They are the 'Sympetalae' or 'Gamopetalae' (plants with fused petals), 'Polypetalae', 'Choripetaly' or 'Apopetalae' (plants with separate petals) and 'Amentiferae' (catkin bearing plants) or 'Apetalae' (plants without petals) which record some very broad syndromes of characters which are present in dicots. They correspond in part to the subclass Asteranae (Sympetalae), the superorders Magnolianae, Rosanae, Dillenianae and Caryophyllanae together (the Polypetalae) and the Hamamelidanae (Amentiferae).

5.5.1 Flowers with separate petals, the 'Polypetalae'

Here are included the superorders Magnolianae (12 000 spp.), Caryophyllanae (11 000 spp.), Dillenianae (25 000 spp.) and Rosanae (58 000 spp.), together two thirds of all the dicotyledons. The Magnoliales are one of the most primitive orders of angiosperms. They have a well developed perianth which is poorly differentiated into a calyx and corolla, with a variable number of tepals spirally arranged or in whorls of three. Stamens and carpels are numerous and varying in number. The carpels are sometimes incompletely joined and have a sessile stigma.

The Ranunculales were probably derived from the more primitive Magnoliales via something like the Illiciales (Cronquist, 1981). Evolutionary trends illustrated in the order include changes in the symmetry and complexity of the flower, providing a great diversity of garden flowers, and in the structure of the pistil (Fig. 5.10). The largest family is the Ranunculaceae. The calyx is often petaloid or falls off when the flower opens (caducous). 'Honey leaves' are showy staminoidal petals with nectaries towards the base. The flower is **apocarpous** with superior pistils. Some genera like *Ranunculus* have simple, actinomorphic flowers. In *Aquilegia* the petals are long spurred with nectaries at the base. *Aconitum* and *Delphinium* have complex zygomorphic flowers.

The evolution of achenes (single seeded indehiscent fruits) from many seeded follicles has been associated with the evolution of different seed dispersal mechanisms. The achenes are often hooked or tuberculate and in *Clematis* the style is persistent in fruit, forming a long feathery aid to dispersal. *Myosurus* has a peculiar elongated receptacle on which the achenes are situated. It is as if there has been selection in a buttercup like ancestor for the increased production of achenes.

Species with follicles or achenes are sometimes separated into two distinct tribes the Helleboreae and the Anemoneae, or even into subfamilies, but chromosomal evidence divides the genera differently and even suggests that the peculiar wind pollinated *Thalictrum* with achenes may be quite closely related to the showy *Aquilegia* with follicles. Some species in the alternative genera have strikingly similar foliage, e.g. *Aquilegia thalictrifolia*, and *Thalictrum aquilegifolium*.

The second largest family in the order, the Berberidaceae is more advanced. It is distinguished from the Ranunculaceae by having a single pistil which becomes a berry.

The genus *Paeonia* was at one time placed in the Ranunculaceae but it has persistent sepals, sepaloid petals and stamens which mature from the centre outwards, centrifugally, whereas in the Ranunculaceae and indeed in the Magnolianae as a whole they develop centripetally. It is now placed in superorder Dillenianae in its own family. One genus *Glaucidium* has characters which link it to the Ranunculaceae, Berberidaceae and Paeonaceae.

The poppies, Papaverales, are derived from the buttercups from which they differ by having a syncarpous gynoecium and only 2–3 sepals. There are two families, the actinomorphic Papaveraceae and the strongly zygomorphic Fumariaceae.

5.5.2 The roses and their relatives, the Rosanae

The Rosaceae is a large family which is central to the evolution of many other families in the Rosanae. It has about 100 genera and 3000 species. The ovary may be hypogynous, perigynous or epigynous. Perhaps the only widespread feature is the possession of a **hypanthium** (Fig. 5.11) (Kalkman,

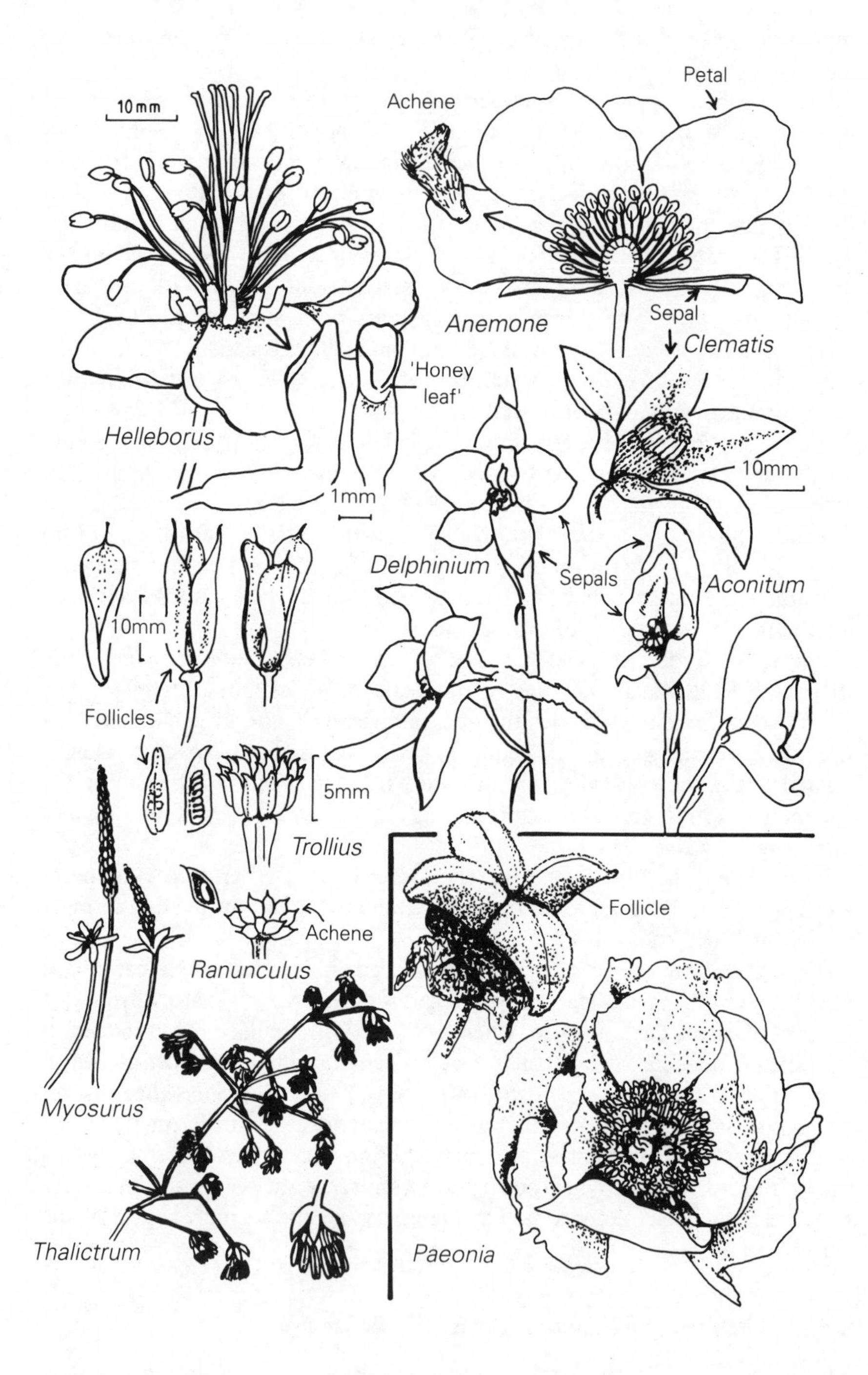

Fig. 5.10. Polypetalous flowers in the Ranunculaceae and Paeoniaceae. There are differences in the symmetry of flowers and in the presence of follicles or achenes.

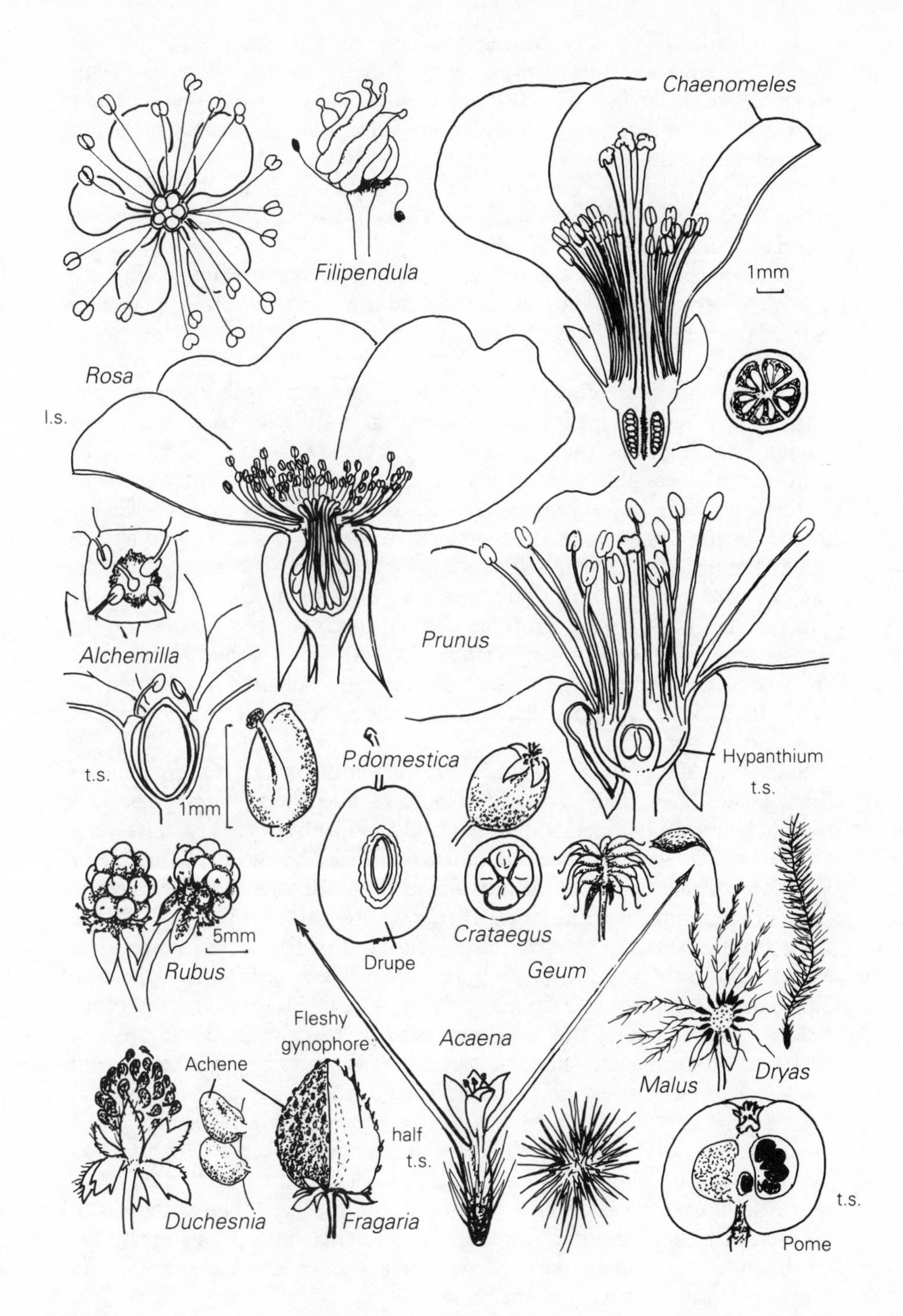

Fig. 5.11. Variation in the flower and fruit of the Rosaceae: showing states of hypo-, peri- and epigyny.

1988). This floral cup is primitively small and saucer- or cup-shaped. It has evolved to large and ultimately connate with the carpels. As a result the Rosaceae are particularly diverse in their fruits, many of which we eat. There are usually several carpels to each flower. Primitively these are arranged in a single whorl with each carpel containing several ovules.

The Spiroideae is a primitive but variable subfamily from which the other subfamilies may be derived (Cronquist, 1981). It has one too many dehiscent follicles. Some species have winged seeds.

The cherries, peaches, almonds and plums (subfamily Prunoideae) have a single superior carpel with one ovule producing a fleshy drupe. The inner part of the pericarp is sclerenchymatous giving the familiar stone of these fruits.

The subfamily Rosoideae is very variable. They are apocarpous and have single-seeded indehiscent fruits. In *Acaena* these are achenes with the calyx modified as long hook-tipped spines. In *Alchemilla* the style is basal. The fruits are often combined with other fleshy tissue derived in different ways. In the rose, *Rosa*, and the strawberry, *Fragaria*, the fruits are hard achenes though in the former they are contained within the fleshy swollen hypanthium forming the hep, and in the latter the achenes are situated on the exterior of a swollen torus. In *Rubus*, the blackberry and raspberry, the fruits are fleshy themslves, drupelets (little drupes) grouped together on the torus, to give the 'berry' (drupecetum). In both *Fragaria* and *Rubus* there are a large number of carpels which are arranged spirally on the extended torus. In *Rosa* and *Fragaria* there is one ovule per carpel, in *Rubus* two but only one develops.

Subfamily Maloideae have an inferior ovary with 2–5 carpels, producing a fleshy pome. They exhibit a condition called pseudosyncarpy. The carpels are free on their ventral (inner) surface but held together by the fleshy fruit tissue.

The Fabales are sometimes recognized as a family, the Leguminosae or Fabaceae, within a broadly defined order of the Rosales. More frequently three distinct families within their own order are recognized (Fig. 5.5). The three families share the characteristic of a kind of fruit, the legume. The Mimosaceae have small regular flowers in dense spikes or heads, brush blossoms with numerous exserted stamens. The family, includes the genus *Acacia* which is one of the most important in arid and semi-arid areas of Africa and Australia and one of the largest of all genera of angiosperms with 1200 species. The Caesalpiniaceae have large showy more or less regular flowers. The Fabaceae *sensu stricto* has many agriculturally important species providing peas and beans of all sorts. Clovers are especially important in pasture. They have the zygomorphic 'papilionaceous' flower. The flag blossom is pollinated by large bees which land on the keel. The nectary is at the base of a staminal tube. In forcing its proboscis into the staminal tube the heavy insect pushes the keel petals (alae) down and the stamens and stigma rub against its ventral surface.

The Proteales and Myrtales are two orders of the Rosanae which are well represented in the southern hemisphere. The family Proteaceae in particular shows a distribution which records the old Gondwanan super continent.

Both orders have families in which brush blossoms have evolved. Brush blossoms have conspicuous stamens and flowers crowded together to increase visual attractiveness. In Australia there has been a pattern of coevolution between the bird pollinators, especially honeyeaters and the evolution of *Banksia* and related genera in the Proteaceae. The pollinators, honeyeaters, lorikeets and honey possums have brush tipped tongues to help collect nectar. Bats are also visitors. The honeyeaters are also partly insectivorous, and obtain amino acids from their insect prey. However, the nectar of different species of *Banksia* also contains different amounts of amino acids, as a kind of dietary supplement for the pollinator. Highest concentrations of amino acids are found in those species with the least insectivorous honeyeater pollinators.

In the Myrtaceae, which includes the important genus *Eucalyptus*, pollination by birds and bats, has been accompanied by an increase in the number of stamens (Briggs and Johnson, 1979). In *Eucalyptus* the corolla has been lost and the calyx forms a calyptra which falls off as the flower opens.

The Apiales is one of the most adavanced order in the Rosanae. There are two families the Apiaceae or Umbelliferae and the Araliaceae. The Apiaceae is a family which has many species cultivated for food (carrot) or more frequently for spice (coriander, cumin etc.). Diversification has been accompanied by chemical evolution (section 8.2.3). This compound umbel is so characteristic of the Apiaceae that it was one of the first angiosperm families to be clearly recognized and gave it its other name, Umbelliferae. The compound umbel is not however a perfect diagnostic because it is confined to subfamily Apoideae. In some genera like *Hydrocotyle*, in subfamily Hydrocotyloideae, the compound umbel is reduced to a single floret and in *Eryngium*, subfamily Saniculoideae, the inflorescence is globose.

The umbel is a relatively unspecialized inflorescence and is normally visited by a range of different insects, which move freely over the platform of the umbel. In each floret, the ovary has a disc at the base of the paired styles. The styles are swollen, together forming a nectiferous stylopodium (Fig. 5.7). Flowers in different parts of the umbel may be specialized. Those in lateral umbels are sometimes female sterile with abortive ovaries and shorter stylopodia. The ovary has two carpels and matures into a dry fruit, a schizocarp, which splits into two mericarps joined by a carpophore. Some species have fruits with thick corky walls so that they float, have wings for wind dispersal, or have hooks for animal dispersal. The Apiaceae share many similarities with the tropical family, the Araliaceae. However, the latter only rarely forms regular compound umbels and also differs in producing a fleshy fruit.

5.5.3 Flowers with a tubular corolla, the 'Sympetalae'

The 'Sympetalae' is an old taxonomic group which contained families which had a tubular corolla of connate (fused) petals. Today several sympetalous families are thought not to be related to the others and are put in different groups. However, the bulk of the Sympetalae remain in the superorder

Asteranae (Cronquist's Asteridae). As well as their pentamerous sympetalous corolla most have an equal number of epipetalous stamens, alternating with the five corolla lobes. Superorder Asteranae (60 000 spp.) contains the most advanced members of the dicotyledons, and the most recently evolved. They have diversified especially in having specialized pollinators. Floral architecture and behaviour show many individual adaptations to particular kinds of pollinator. In many different lines the corolla has become zygomorphic, stamens have been lost or converted into staminodes, and compound inflorescences of different sorts have evolved. There are possibly two main lines of evolution. One lineage leads to the daisy order Asterales with their capitulate inflorescence. This line is epigynous. The other lineage leads to the orders Lamiales (mints, dead-nettles), and Scrophulariales (figworts, snapdragon, foxglove) which are mainly hypogynous. An early offshoot of this line are the epigynous Campanulales (bellflowers).

The order Gentianales has the most generalized flowers with five epipetalous stamens close to the origin of both lines in the Asteranae, but in the asclepiads they also exhibit one of the most peculiar floral morphologies of any angiosperms (Fig. 5.12).

The Gentianaceae, although they are relatively recently evolved, are closely related to the more primitive Loganiaceae. They are regular, sympetalous, actinomorphic with five normal epipetalous stamens.

The asclepiads, Asclepiadaceae, have two carpels only united by the head of the style. The stigma consists of five lateral grooves alternating with the anthers. The anthers are adnate to the style to form a structure called the gynostegium. The pollen in each theca is a compact mass, the pollinium. Pollinia from adjacent thecae are united by an acellular yoke called the translator and the whole structure is released as a pollinarium. Evolution of the pollinium has been accompanied by a merging of the pollen sacs of each theca so that each anther is bisporangiate. In a primitive subfamily the mechanism is not perfected: each theca has two pollen sacs and so produces two small pollinia. The translator clips the pollinaria to the pollinator and they are later pulled off by being caught in a groove in the stigma of another flower.

Intermediate between the Gentianaceae and the Asclepiadaceae are the Apocynaceae. Simple ones like *Vinca* have completely united carpels, and anthers which are distinct and fully fertile. More complex ones like *Nerium* have carpels separated up from the base and only united by their style and stigma (c.f. the asclepiad gynostegium). The anthers, of which only the top part produces pollen, are grouped closely together (connivant) in depressions around the top of the expanded style. The evolution of separate locules of an ovary from the bottom up is very peculiar.

The Solanales show many similarities to unspecialized members of the Gentianales. There are two large families, the Solanaceae (potato, tomato, tobacco, petunia) and the Convolvulaceae (morning glory, bindweed). Morphological tendencies are present which are fully realized in the orders Scrophulariales and Lamiales. These include evolution from actinomorphy to zygomorphy and trends in the gynoecium.

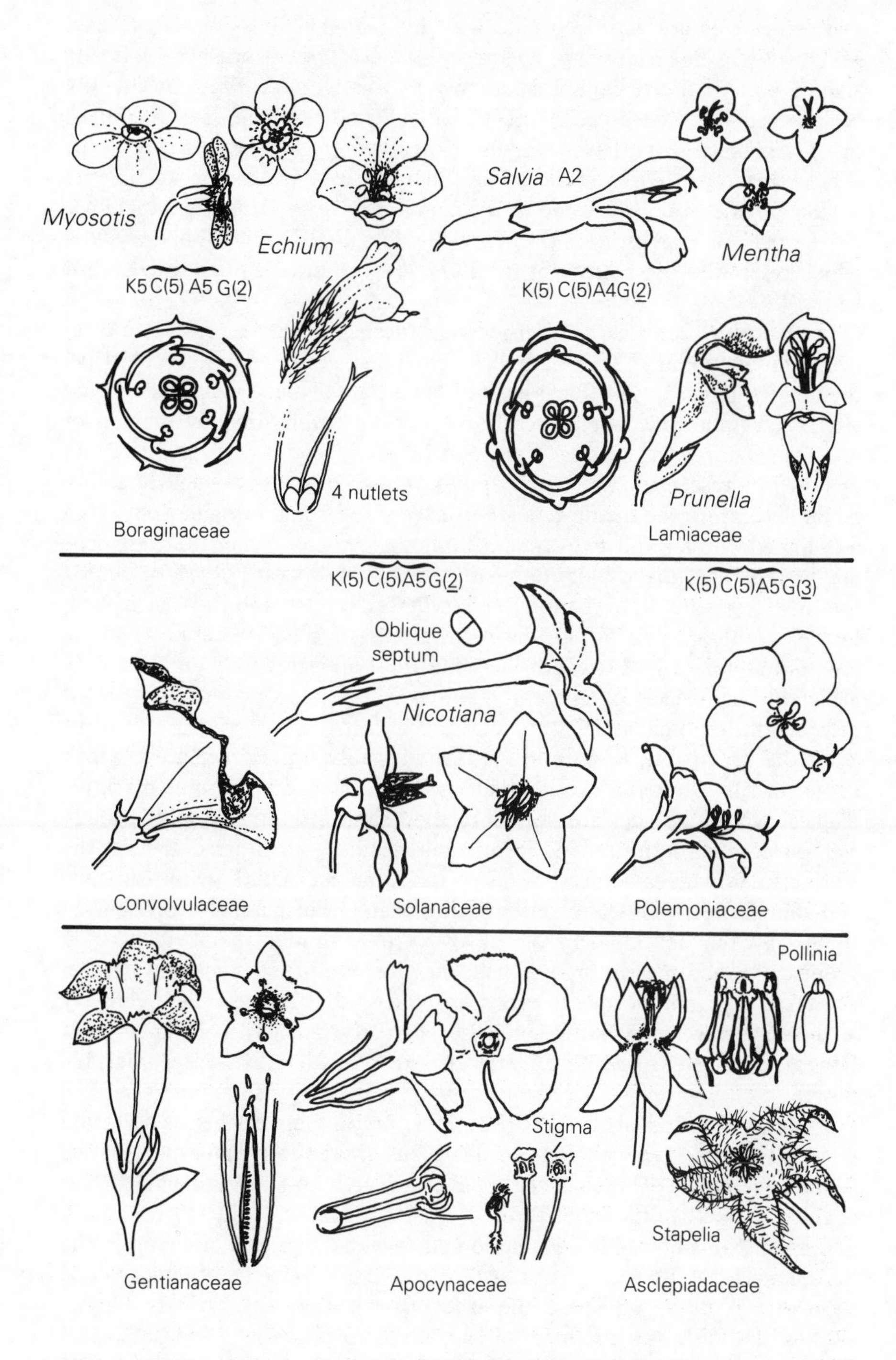

Fig. 5.12. Some related families with a sympetalous corolla.

Most species are actinomorphic with five equal corolla lobes but some genera of the Solanaceae are zygomorphic. *Schizanthus* has even lost one stamen and of the remaining four only two are fertile. The small family Nolanaceae has five carpels, the primitive condition, but other families in the Solanales have fewer carpels, either two (e.g. Solanaceae) or three (Polemoniaceae). The Solanaceae are characterized by having an oblique septum in the ovary. In some of the Convolvulaceae (*Dichondra*) there are two carpels and the ovary is deeply two or four lobed with gynobasic style, which resembles the pattern of the Boraginaceae and Lamiaceae described below.

In the order Lamiales, the Verbenaceae (verbena, teak) represent an early stage of evolution. The 2–carpellate pistil has four locules, each carpel divided by an extra wall. Each locule has a single ovule. The corolla is only slightly zygomorphic but bilateral symmetry is emphasized by the loss of one stamen in most species. Zygomorphy is associated with a shift from cymose inflorescences to racemose spikes where the flowers are held laterally. In family Lamiaceae the flowers are usually very strongly zygomorphic with a 2–lipped corolla and two (*Salvia*) or four stamens. When four stamens are present they are usually didynamous. The 4–loculate ovary is divided into four segments from the top. The style reaches down to the base of each locule (gynobasic). Each segment is dispersed as a separate nutlet when the seed is mature. The distinction between the Lamiaceae and Verbenacae is arbitrary; the former contains several intermediate genera with an undivided ovary and a terminal style.

Parallel trends can be seen in the related family, the Boraginaceae which differs in having alternate rather than opposite leaves. In fact some primitive tropical woody borages are very similar to some primitive tropical woody Verbenaceae (Cronquist, 1981). Some borages have an entire ovary like the Verbenaceae whereas others are like the Lamiaceae with a gynobasic style and four nutlets. This is a very striking example of parallel evolution. In some way the development of the ovary of primitive members of both groups allowed, or even preordained, the later evolution of nutlets. Parallel evolution did not occur in every character so that even the relatively advanced borages are mostly actinomorphic. *Myosotis* has five stamens and flowers arranged in cymes. *Echium* is exceptional with its zygomorphic flower.

In some advanced labiates like the mints *Mentha* there has been a reversion to radial symmetry, with four corolla lobes alternating regularly with the four stamens. The lower lobe, repesenting two connate corolla lobes is slightly larger than the other three.

Similar kinds of parallel evolution and reversions can be observed in the Scrophulariales. The Scrophulariaceae are a large diverse family from which most of the other families in the order can be derived (Fig. 5.13). Some species like *Verbascum* are more or less actinomorphic with five stamens. In *Penstemon* flowers are zygomorphic with four fertile stamens and the fifth converted into a staminode. *Antirrhinum* has four stamens and *Calceolaria* has two. Spikes or similar kinds of inflorescences are common. Many species

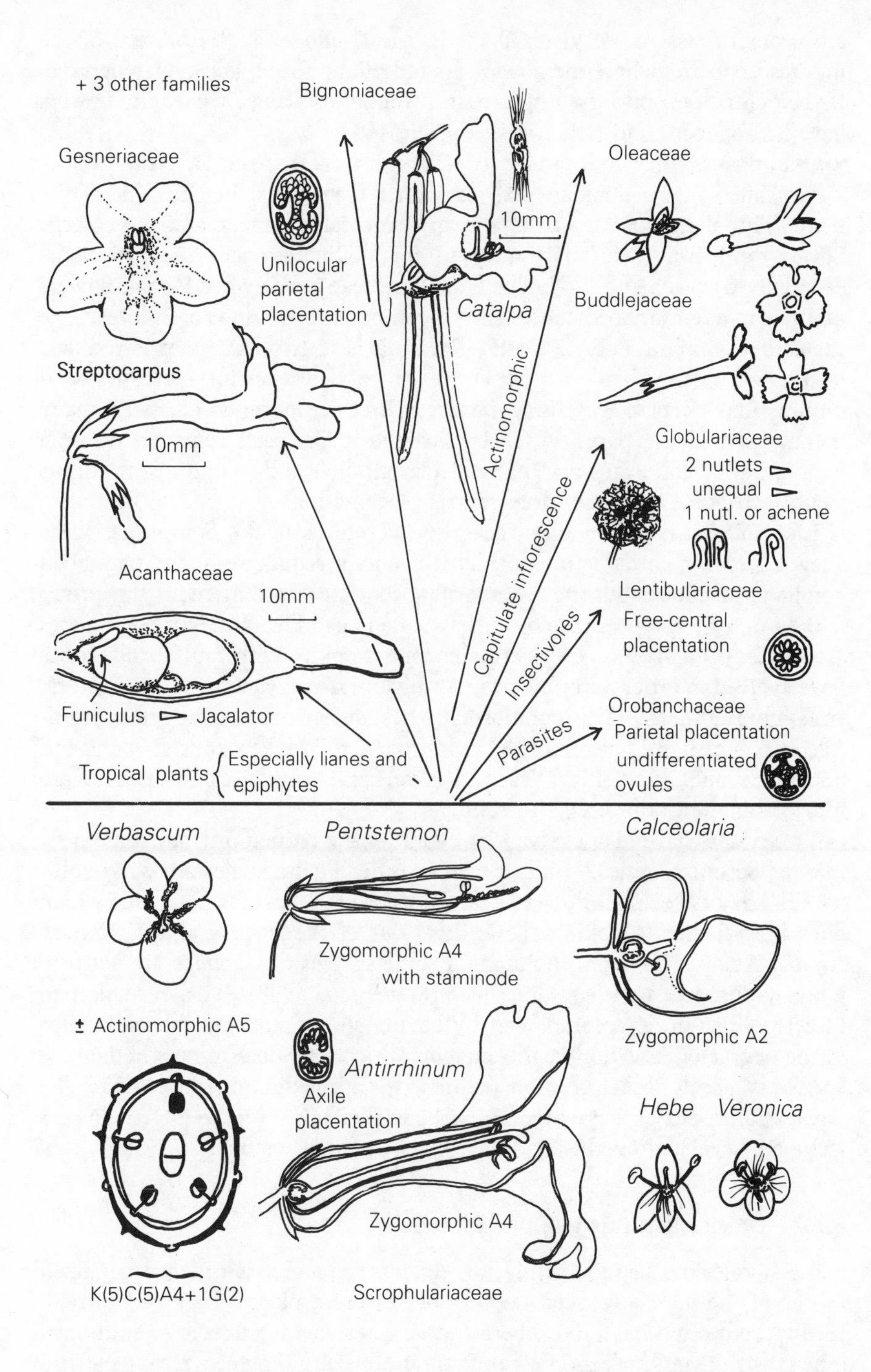

Fig. 5.13. Evolutionary trends in the Scrophulariales: reduction in the number of stamens, increase in the number of ovules in the ovary, specializations for dispersal and the secondary evolution of radial symmetry.

are protandrous. A bee visits *Digitalis*, the foxglove, at the bottom of the inflorescence first where the flowers are oldest and the style is now receptive. It then climbs up into the upper part of the inflorescence where the flowers have just opened, and pollen is being released.

One, possibly derived, family, the tropical Acanthaceae, have the funiculus modified into a hook-shaped jaculator which flings the seed out of the fruit. The Bignoniaceae include many tropical climbers and have large flowers and fruits variously adapted for pollination and seed dispersal especially by birds and bats. The Globulariaceae have evolved small flowers grouped in a tight inflorescence. As is common in these kinds of inflorescence the carpels contain a single ovule. The fruit is of two nutlets or is reduced to a single nutlet shed in a persistent calyx. Selection for increased seed number has occurred in three families. In the Gesneriaceae, with many epiphytes, and the parasitic Orobanchaceae it has been achieved through having unilocular ovary and parietal placentation and in the insectivorous Lentibulariaceae by having free central placentation.

Like the mints in the Lamiaceae, the veronicas in the Scrophulariaceae show a trend towards actinomorphy, through a reduction of the five-lobed corolla to four lobes but the zygomorphic condition, primitive for this group is indicated by the presence of only two stamens. The closely related hebes are more zygomorphic. Two actinomorphic families with four corolla lobes have evolved in order Scrophulariales probably from zygomorphic ancestors. In family Buddlejaceae, which normally has four stamens, *Sanango* is primitive in being slightly zygomorphic with five corolla lobes, four stamens and one staminode. In family Oleaceae (olive, ash, privet, jasmine) there are four corolla lobes but only two stamens.

Two orders have also evolved by evolutionary reduction from something like the Scrophulariales. Some species in order Plantaginales are wind pollinated. They have small flowers with four corolla lobes, four stamens which have long filaments poking the anthers out of the flower, and a feathery stigma. Adapted for wind pollination some species of *Plantago* are actually insect pollinated, at least sometimes (Stelleman, 1978). This reversion to insect pollination is accompanied by having showy stamens. Species differ in the degree of exsertion of the stamens which may determine whether the species is largely insect or wind pollinated, outbreeding or inbreeding. The Order Callitrichales has three aquatic families with very reduced flowers lacking a corolla altogether and having only a single stamen.

5.5.4 Daisies and thistles, the Asterales

The Asterales has a single family, the Asteraceae, which is widely recognized as one of the most advanced families of flowering plants. Its origin is relatively recent but it has 1100 genera and 20 000 species. All have a capitulum, a head with many florets on a common and usually flattened receptacle and surrounded by an involucre of bracts. The involucre is particularly obvious in 'everlasting' flowers like *Helichrysum* where it is showy. Florets are epigynous with a single ovule. They are protandrous and mature centripetally (i.e.

the capitulum is racemose) (Fig. 5.14). Florets are of different sorts, either actinomorphic tubular or zygomorphic with three corolla lobes connate and greatly expanded to form a strap. Florets may be pistillate, hermaphrodite or functionally staminate. Heads are made of a single kind of floret or a combination of kinds, commonly with a central disk of hermaphrodite actinomorphic florets and a margin of showy zygomorphic florets, with the latter pistillate or sterile, as in the daisy and sunflower. Other kinds are those with just tubular florets with all the florets the same or with the marginal florets modified or with bilabiate florets. One group which includes lettuce and dandelion have all ligulate florets which differ from normal marginal ray florets in having all five corolla lobes forming the strap.

Many of the structures found in the Asteraceae are paralleled in other families, either distantly or closely related. Capitulate inflorescences are found in other families of order Asterales including the Campanulaceae (*Jasione*), and the more closely related Dipsacaceae (*Scabiosa*, *Knautia*). However in these cases either the head is not bracteate or it is basically cymose because the florets mature centrifugally. The order of maturation of florets has apparently shifted in a small family, the Calyceraceae, so that they have centripetally maturing florets, paralleling the Asteraceae. However, they are probably an anomalous offshoot of the Dipsacaceae (Cronquist, 1981). More distantly related groups have superficially similar **corymbs** and other kinds of inflorescence. They also often show a similar specialization of florets in different parts of a head, with the outer floret more showy.

The mode of pollen presentation has parallels with that in the Campanulaceae described above (section 5.3.2). Florets are protandrous. Anthers are connate or connivant, forming a tube. Pollen is shed into the anther tube and then the immature style elongates and pushes the pollen onto the surface of the capitulum. Later the stigmatic lobes open. A similar kind of pollen presentation mechanism is found in the Australian family Goodeniaceae (order Campanulales). The one species of the monotypic Brunoniaceae, which is sometimes put in the Goodeniaceae, shows a further parallel in having an involucrate head, though the head is cymose and the florets are hypogynous.

The fruit of the Asteraceae is usually crowned by a pappus derived from the calyx. This has parallels in the persistent calyx of Dipsacaceae and Valerianaceae. The pappus in Asteraceae is very variable either, absent or cup-like, or with scales, bristles, simple or feathery hairs, which are barbed, or glandular. The fruit, called a **cypsela**, is a kind of achene of an inferior ovary. The dispersal adaptations of the fruit contribute to the success of many species as weeds.

It may not be an accident that the Asteraceae have an unusual multiallelic homomorphic sporophytic self-incompatibility. The role of this sexual incompatibility system in the evolution of reproductive isolation between incipient species is worthy of study. Alternatively many weedy species are secondarily self-compatible and self-pollinating. One example is the ubiquitous groundsel, *Senecio vulgaris*, which lacks the ray florets of its relatives, no longer needing to attract pollinators. Polyploidy, sometimes following

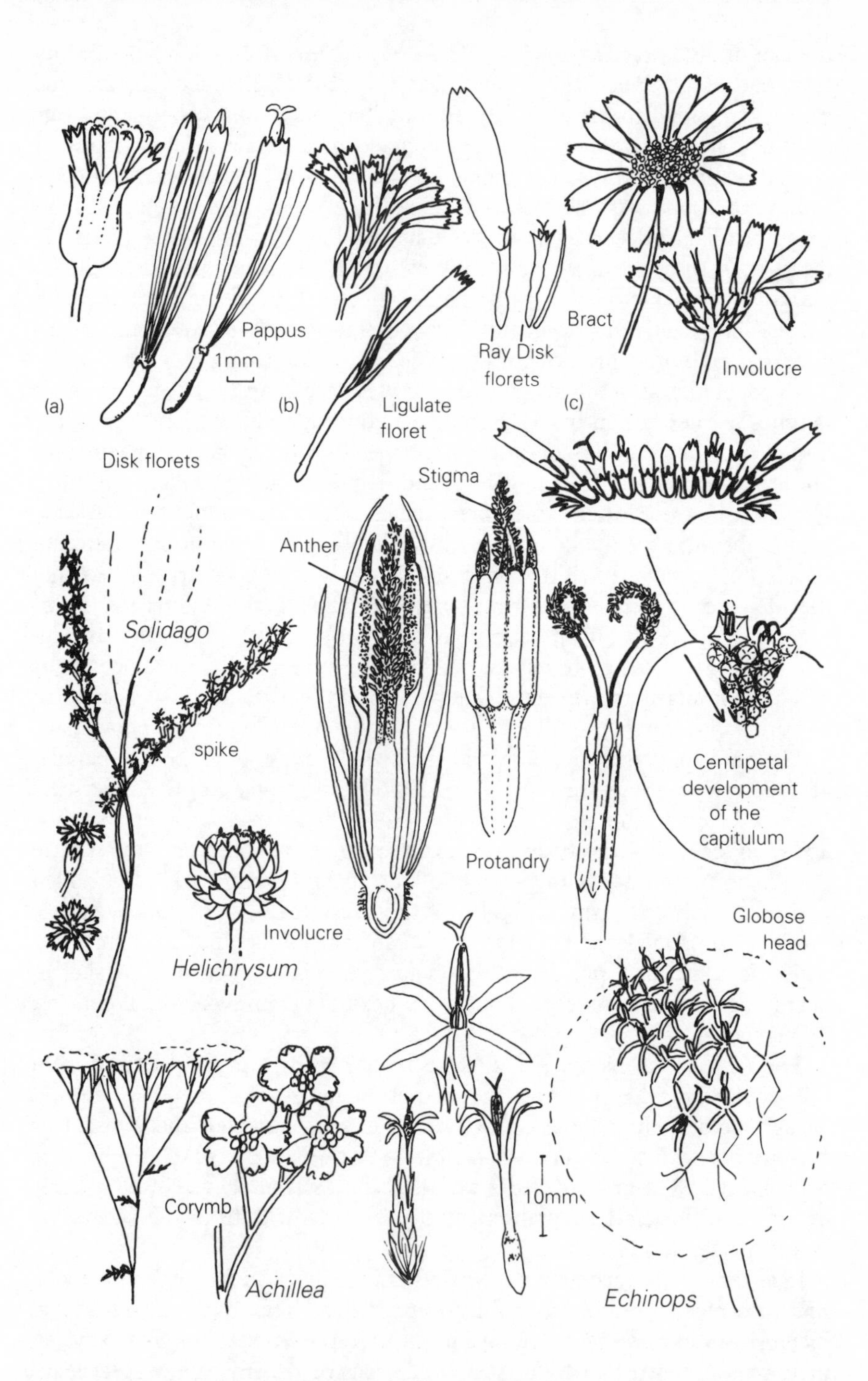

Fig. 5.14. Variation in the capitulum of the Asteraceae (Compositae). The three kinds are (a) all disk florets, (b) all ligulate florets and (c) central disk florets and marginal ray florets. Capitula are arranged in various ways.

hybridization, seems to play a significant role in the evolution of the weedy species, especially since it can destroy the self-incompatibility.

The features described above have evolved in many groups outside the Asteraceae. Their repeated evolution argues strongly that they are adaptive and yet because they are widespread they do not explain the particular evolutionary success of the Asteraceae. It is possible that, as in the Apiaceae, their diversification is closely tied to their chemical evolution (Cronquist, 1981). They have a particularly effective combination of antiherbivore chemical repellants: polyacetylenes and sesquiterpene lactones. Possibly chemical diversification in different niches and resisting different coevolving insect herbivores has resulted in much speciation. The genus *Senecio*, one of the largest of all seed plants with over 1500 species, has a peculiar type of alkaloids.

One advantage of having a capitulum is the protection given to the ovule and seed. Functionally it provides a large showy target for pollinators and yet each ovule is packaged separately, as a defence against predators and for dispersal. There is a lot of diversity in the size and number of florets which capitula contain. There are the familiar huge capitula of the sunflowers which have been selected by plant breeders. At the other end of the spectrum many species have capitula containing very few florets. Frequently species with small capitula have the capitula grouped in some way. In *Solidago* the capitula are arranged in spikes. In *Achillea* they are grouped in a corymb. In *Echinops* each capitulum only has a single floret and they are grouped in a globular head with its own involucre.

One of the most important evolutionary trends that can be observed in land plant reproduction has been the greater and greater protection of the developing female gametophyte, through endosporic development and then the evolution of the ovule protected by integuments. In the Asteraceae this has reached its greatest development. Like a treasure protected in a series of chinese boxes, the embryosac is inside the nucellus, inside an integument, inside an ovary, protected by being inferior and so surrounded by other 'receptacle' tissues, protected within a capitulum surrounded by an involucre which in some cases itself is part of a head with its own involucre.

5.5.5 The monocots, the Liliidae or Liliopsida

The dicots and monocots diverged very early in flowering plant evolution (Friis *et al.*, 1987). The monocots, containing only 50 000 species, might be considered less important than the dicots (188 000 spp.) except that they include in one of their families the Poaceae which provides by far the most important food crops for humanity: wheat, maize, rice, sorghum and millet. Proper names are Liliidae (-opsida) or Monocotyledonidae (-opsida).

The possession of a single cotyledon is found in several 'dicotyledons'. This actually emphasizes the value of using proper taxonomic names like the Magnoliidae and Liliidae. The differences between the monocots and dicots are shown in Table 5.1.

Table 5.1 Differences between monocots and dicots

	DICOT	*MONOCOT*
Mature root system	Derived from radicle and adventitious	Wholly adventitious
Stele	Vascular bundles open (with cambium) in a ring	Vascular bundles closed (lacking cambium) 'scattered'
Leaves	Net veined with petiole and a lamina	Parallel veined derived from petiole lamina vestigial
Flower parts	5s less commonly 4s or rarely 3s	3s
Cotyledon	2 rarely 1, 3, 4 or undifferentiated	1 or undifferentiated

A single cotyledon could have evolved by the loss of one cotyledon or the fusion of the cotyledons. It is the latter which is likely to have happened in the monocotyledons (Stebbins, 1974). As the seedling germinates the cotyledon elongates to push the radicle out of the seed and down into the mud. The possession of a single cotyledon makes this process more efficient. It is very interesting that some members of the dicot Nympheaceae, which grow in wet mud, show a trend for the fusion of the two cotyledons. This feature as well as several vegetative similarities like the lack of an active cambium, and the production of adventitious roots may be the result of convergent evolution in the aquatic habitat. Monocots are well able to undertake elongating growth but poor at thickening (secondary) growth. Elongating growth adapts them for the aquatic and epiphytic niche where they predominate.

Monocotyledonous flowers are characteristically trimerous, have a uniaperturate pollen, have apocarpous flowers and a relatively unspecialized perianth, all characteristics shared with some primitive dicotyledonous families such as the Annonaceae in the Magnolianae. The most primitive monocotyledons are the aquatic Alismatanae. The monocotyledons probably evolved from aquatic dicotyledons but they diverged rapidly and in the late Cretaceous the palms were well represented in many floras. Floral adaptations in the monocotyledons occurred in three directions: wind pollination (anemophily), water pollination and animal pollination (zoophily).

Many lilies and their relatives have adaptations for insect pollination but the trimerous condition usually remains recognizable. The perianth is usually not clearly differentiated into a calyx and a corolla. Very frequently the inflorescence is protected in bud by a bract, the **spathe**. Most flowers are actinomorphic. Many have a large tubular or trumpet like perianth with varying degrees of fusion of the **tepals** (Fig. 5.15). Stamens are often large with broad filaments and large anthers. In *Pancratium* the fused filaments form a rim, the **corona**. In the daffodils, *Narcissus*, a similar structure has arisen from the perianth. Many monocots have a trilocular syncarpous ovary. An important difference, which was used to separate two large families of

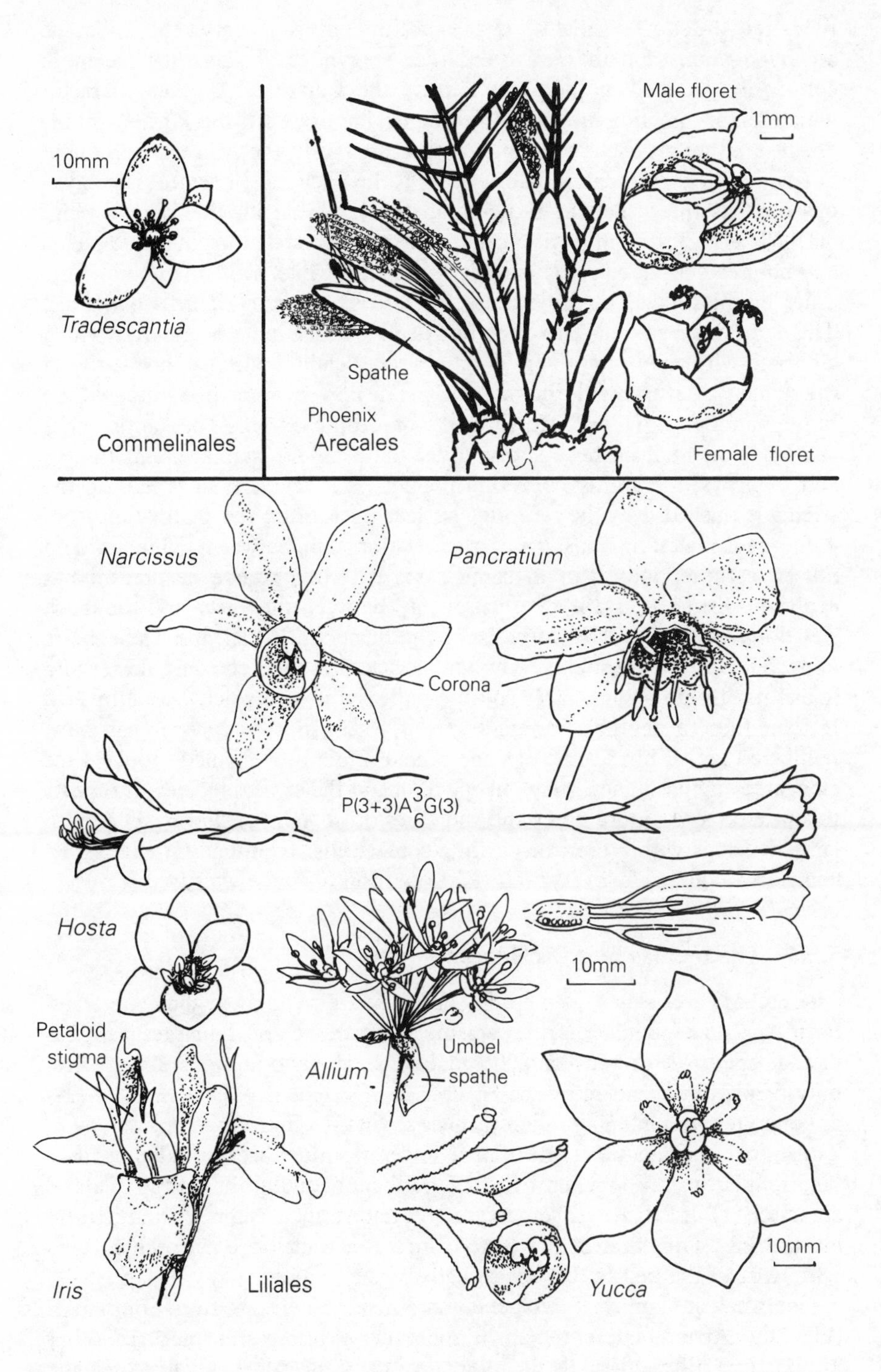

Fig. 5.15. Trimerous florwers in the three orders of the monocotyledons.

the large order the Liliales, is the position of the ovary: the Liliaceae are hypogynous and the Amaryllidaceae epigynous. Today they are more commonly grouped in the same family, the Liliaceae. Another character which has in practice proved of dubious value has been the kind of inflorescence; either a raceme or single or an umbel as in the Alliaceae (onions).

Another important family in the order is the Iridaceae (irises). It is epigynous but has three stamens rather than the six usual for the Liliales (Fig. 5.15). It also differs in having calcium oxalate crystals in some of its cells, and not needle-shaped crystals, called raphides, which some members of the other families have. In some irises the tepals are arranged in two whorls. The outer deflexed 'falls' are often bearded while the inner 'standards' which alternate with them are glabrous and erect. Petalloid stigmas hooded over the stamens alternate with the standards. The flower is radially symmetrically but functionally divided into three zygomorphic parts. A bee approaches and lands on a fall pushes its way under the petalloid stigma, pollinating it, and then past the anther collecting pollen. As it reverses out a flap on the stigma is pushed over the receptive surface preventing self pollination.

In the Liliales, in most cases the relationship, between pollinator and flower is rather loose but in some cases a rather bizarre mutualism has evolved. One such case is the relationship between the yucca and the moth *Tegeticula yuccasella*. The yucca has large flowers in which the stamens are kept well clear of the stigma. The female yucca moth lands on a flower and hooks itself onto a stamen. It collects pollen with two prickly mouthparts. It then flies to another flower. First, it establishes whether it has been pollinated yet. If not it pushes some pollen into a tube formed by the three elongated stigmas. Then it lays an egg in one of the ovaries locules. It repeats this process a few more times and then flies off to another flower. The moth larva develops within the ovary eating some seeds. Eventually it escapes by chewing a hole in the ovary wall. A proportion of the seeds survive.

5.5.6 The orchids, the Orchidaceae

The orchids are related to the lilies but they are strongly zygomorphic and have a peculiar pollination mechanism. They are one of the largest families of angiosperms with between 800 and 1000 genera and up to 20 000 species, only rivalled in numbers by the Asteraceae. Orchids may have evolved first as epiphytes with hanging inflorescences so that the flower was upside down. The upper median petal, held in a lower position became adapted as a landing platform, a labellum, for the pollinator, giving the flower bilateral symmetry (Fig. 5.16). In some cases the labellum developed callosities to aid landing. This kind of flower remained functional in erect orchids by a 180° twist of the pedicel, a process called resupination.

One small subfamily of two genera and about 20 species from South East Asia, the Apostasiodeae remained more or less actinomorphic. In other orchids two different kinds of pollination mechanism have evolved. In the Cypripedoideae, like *Paphiopedalum*, an outer whorl of three stamens was lost but the median one was converted into a **staminode** (Fig. 5.16). In the

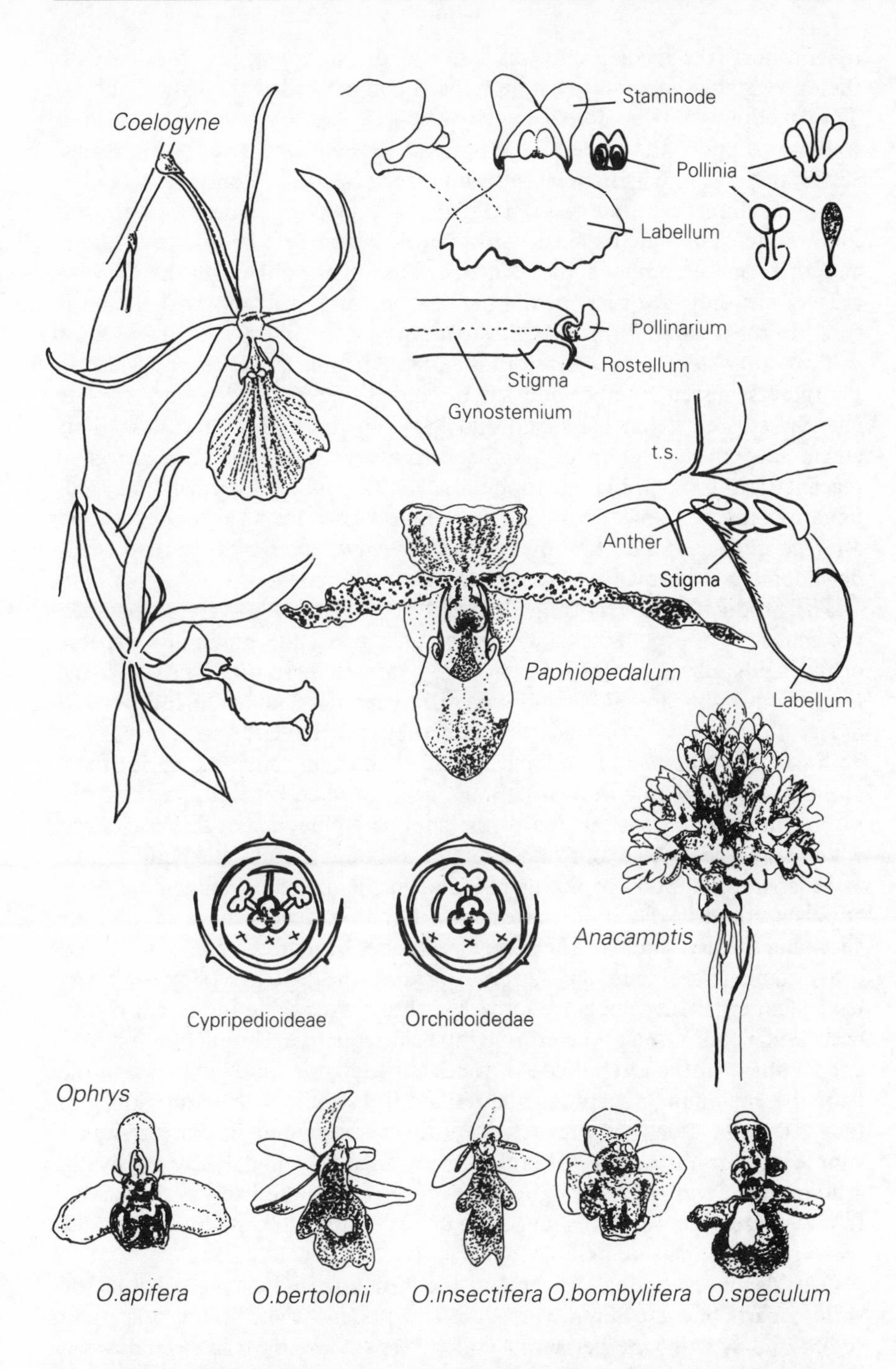

Fig. 5.16. Orchids (Orchidaceae) showing the structure of the two main kinds of orchid flower in the Cypripedioideae (*Paphiopedalum*) and the Orchidoideae (*Coelogyne, Anacamptis* and *Ophrys*). The variation of *Ophrys* illustrates the kind of specialization for particular pollinators that has taken place.

inner whorl, the median one was lost and the surviving pair formed with the outer staminode and the style a fused column called the gynostemium. The labellum petal became a pouch which insects fall into. The labellum walls are slippery and there is only one exit route. Aided by a track of hairs, and in the case of *Cypripedium calceolus* directed by thin translucent patches in the labellum wall, like the illuminated exit signs in a cinema, the insect follows the route out. Pushing its way up, its back first brushes the stigma and then one or other of the stamens. The other subfamily, the Orchidoideae, has only one stamen, the median one of the outer whorl. It too is fused to the style to form the gynostemium.

Two other steps have evolved in the epiphytic niche. There was selection for increased seed number and small wind dispersed seed to increase the chances of seed being blown up and lodge in the bark of the host. This encouraged the evolution of a unilocular ovary with parietal or marginal placentation (c.f. epiphytic Scrophulariales). Orchid seed is small and seedlings live mycotrophically relying on a mycorrhiza for an extended period after germination. The tiny ovules can develop very rapidly and so ovule development is normally delayed until after pollination.

The production of very large amounts of seed in each ovary encouraged the release of the pollen as a single compact mass, the **pollinium**, capable of fertilizing all the ovules from a single pollination. In the Cypripediodeae the pollen grains are sticky and cohere in irregular lumps. In the Orchidoideae they remain as tetrads but also they are regularly aggregated into pollinia. The number of pollinia produced in each pollen-sac varies from one to six but commonly two pollinia are connected by separate caudicles to a common stipe. At the end of the stipe is a gluey mass, the viscidium, which sticks the whole unit, the pollinarium, onto the insect's body.

An analogous monocot flower is found in the gingers (family Zingiberaceae) except that the style is free, and lies in a deep groove of the one surviving median stamen. The stigma protrudes beyond the pollen sacs. The other stamens are staminodes, some forming the labellum. Some gingers have a superficial resemblance to orchids but there is one important difference which seems to have determined the subsequent evolution of the different families. In the Orchidoideae, the median stigma lobe became a sterile flap, the **rostellum**, separating the anther from the fertile stigma lobes so that self-pollination was prevented. With the rostellum in place it was a simple step for it to play a role in the development of pollinaria by producing a fluid which would act as a glue sticking pollen to the back of the insect. In the gingers the filaments and style are long and the anthers and stigma are exposed so that the pollinator is not effectively 'controlled'.

Orchids are exceedingly diverse in the structure and arrangement of the various parts of the column, rostellum and pollinaria and in the shape and colour and scent of the perianth. This diversity has arisen in several ways. Orchids are adapted for pollination by wasps, moths and butterflies (Lepidoptera), flies (Diptera) and even humming birds, but about 60% are pollinated by bees. Non-social bees, like bumble bees in the northern hemisphere, and other solitary bees are more important pollinators than social bees because they are more effective in pollinating widely dispersed populations (Dressler, 1981). Cross pollination favours the maintenance of variation but an important factor in speciation has been pollination with pollinia. This very effectively multiplies and stabilizes new variants. Orchid pollination biology mirrors one of the most effective methods of plant breeding to create new cultivars; crossing two genetically different variants and then creating a diverse range of distinct cultivars by selection and inbreeding.

Few bee species visit only one or few species of orchid. Most will visit a whole range of flowers. In any one area in America the commonest and most widespread orchid is pollinated by the commonest and most widespread bee. However, the orchid shares the bee with other orchids. Reproductive isolation is maintained because each species places its pollinaria on a different part of the bee's body: 13 different places where the pollinaria can be placed have been scored.

Some species of orchid have encouraged faithful pollination by their pollinator. They do it by mimicking the female bee so that the male bee attempts to copulate with the flower. There is no food reward. The provision of nectar would encourage the bee to visit every flower in an inflorescence leading to self-pollination. A scent or wax is provided, which may act like an insect pheromone. Euglossine bees collect droplets of perfume from the surface of the flower and store it in their hollow hind legs. Pseudo-copulation is a common pollination mechanism, well known in the European flora because of the different species of *Ophrys* (bee-orchid). It has the advantage that male bees searching for a female to copulate with are likely to range widely and are more discerning of shape and scent than they would be if looking for food. Thereby cross-pollination is encouraged.

5.5.7 Spathe, spadix and syconium

The Arecanae is a superorder which is characterized by having small, usually unisexual, flowers crowded together in an inflorescence called a **spadix** which is subtended by a large bract called the **spathe**. It includes the Palms, Arecaceae, the screwpines, Pandanaceae, and the aroids, Araceae. A similiar inflorescence is also present in the bulrushes, the Typhaceae which are placed in the same superorder as the grasses (Fig. 5.17). Most plants with a spadix are insect pollinated by unspecialized insects like beetles but this kind of inflorescence seems to have been designed primitively for wind pollination. The Typhaceae are wind pollinated. They are monoecious with the male part of the spadix above the female. The Pandanaceae, which are dioecious, are wind pollinated. However, even within this family some animals take advantage of the abundant pollen and there has been a reversion to entomophily, and also especially pollination by bats and birds. Wind pollination is a rare and derived condition in the largest family, the palms, Arecaceae (Henderson, 1986).

The palms have a very variable inflorescence. It may be a branched panicle or a spike. Very often the inflorescence is very large making up in numbers of flowers for the relative smallness of the individual flowers. The talipot

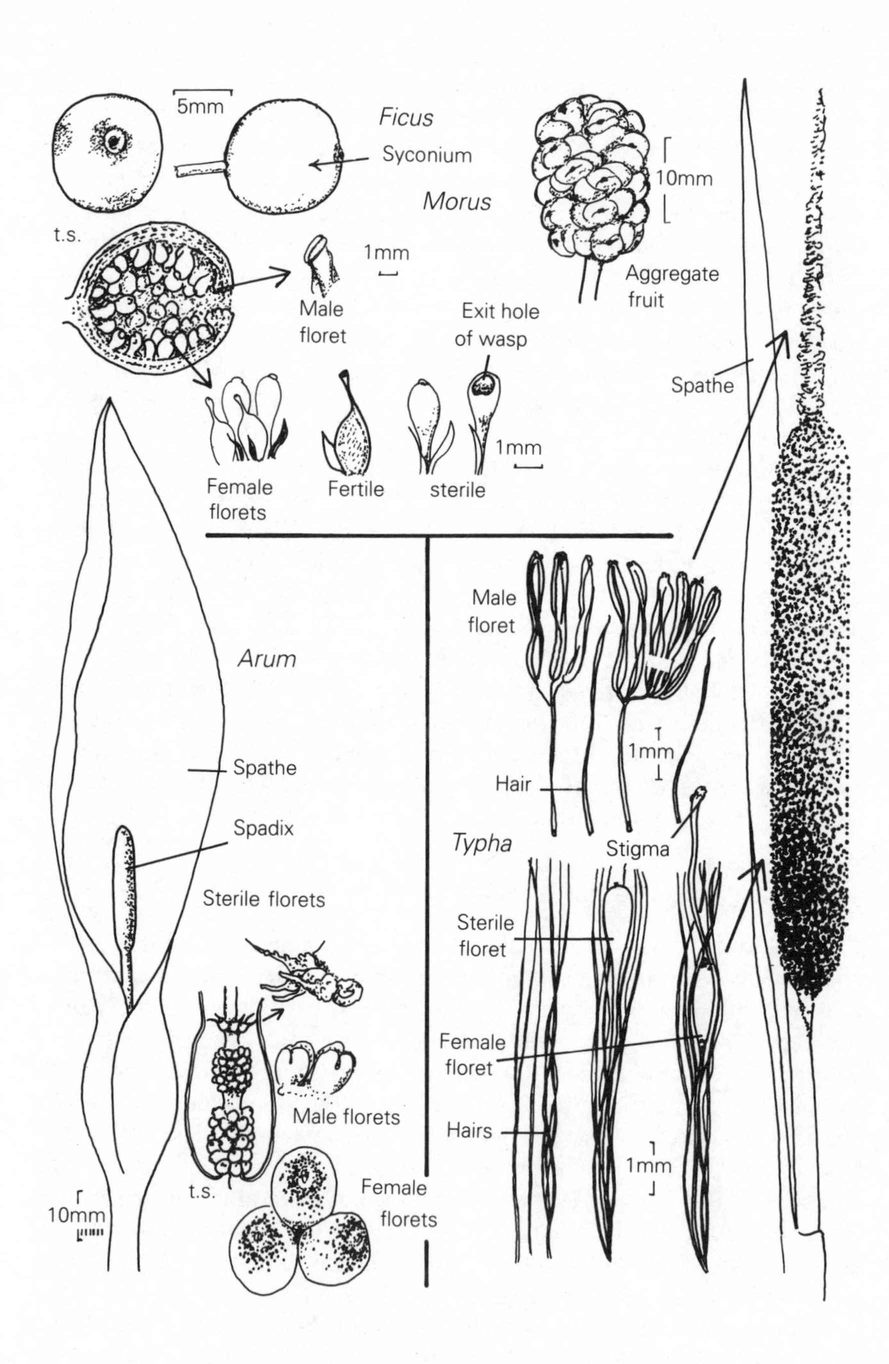

Fig. 5.17. Spathe, spadix and syconium, inflorescences with many tiny florets crowded on an axis. *Ficus* and *Morus* are in the same family.

palm, *Corypha umbraculifera* has 250 000 flowers in an inflorescence. The inflorescence is enclosed until nearly mature in large overlapping bracts, the last of which is the spathe. The flowers are very simple trimerous ones on a standard monocot pattern. The beetle pollinated forms are apocarpous and have persistent bracts (Henderson, 1986). Others are pollinated by flies. Those pollinated by bees lack bracts at anthesis and are syncarpous. There is also a trend towards sympetaly and synsepaly. Many palms are dioecious or monoecious. Artificial pollination of the dioecious date palm *Phoenix dactylifera*, by the throwing of a male inflorescence up into a female tree, is mentioned in cuneiform texts from Ur about 4000 years ago.

The spathe is often a conspicuous feature of the functioning inflorescence in the aroids, Araceae. Many produce scents which are disgusting to humans, of carrion, dung or urine, to attract the pollinators which are mostly flies. The spadix has cells which are packed with mitochondria. Respiration can significantly raise the temperature, up to 16°C above ambient in *Arum*. The raised temperature mimics that of the rotting animal products and also drives off the scents effectively. The spathe forms part of a trap in *Arum maculatum*. The lower part forms a chamber around the columnar spadix. The walls are slippery with oil and covered with downward pointing warts. The lower part of the spadix has regions of female, male and sterile florets. The sterile florets seal off the chamber. Once the insect enters the trap it cannot escape until the sterile florets wither. In the meantime it pollinates the female florets which mature first and then gets covered by pollen from the male florets. Not all aroids have a closed trap. Some aroids in the genus *Amorphophallus* have huge inflorescences. *A. brooksii* from Sumatra is alleged to have an inflorescence over 4 m high (Mabberley, 1987). Some species in the Araceae are wind pollinated.

The figs, *Ficus*, in the Moraceae, may represent the most extraordinary reversion to insect pollination in the angiosperms. Some primitive species in the family are wind pollinated. They have small flowers aggregated into a head common in wind pollinated taxa. In the 500 or more species of fig, *Ficus*, the head has invaginated to form a bottle shaped syconium with the florets on the inside.

Pollination is by species-specific gall wasps. There is a complex pattern of pollination. Inside the fig, the syconium, there are male, female and sterile flowers. The sterile flowers have short styles enabling the female wasp to reach the ovary to lay its egg with its ovipositor. These flowers are on a longer peduncle so that to the gall wasp all stigmas appear to be at the same level. The wasp larvae develop to maturity inside the fig. They give rise to wingless males and winged females. The males never leave the fig but fertilize the females before they depart. On leaving the fig the female gall wasps collect pollen from the male flowers. They find another fig at the right stage and pollinate all the female flowers inside, before laying eggs in the short-styled ones.

5.5.8 Wind pollination

The hermaphrodite flower which characterizes the Angiosperms evolved for insect pollination. There has been extensive coevolution of angiosperms and their pollinators. Some angiosperm groups have reverted to a much older kind of pollination, anemophily or wind pollination. This has occurred separately in many different groups and yet wind pollinated flowers show many convergent features: a dense, clustered inflorescence, especially in catkins (aments); small flowers with a reduced perianth or apetalous; lack of nectar; pollen abundant; anthers on long filaments or liberated explosively as in *Ricinus* and *Urtica*; expanded surface of the stigma or sometimes feathery; sexes separated (dioecious or monoecious); pollen dry, rounded and smooth and generally smaller than that of insect pollinated species, 20–30 μm in diameter (10–300 μm in entomophilous plants).

The origin of anemophily in most catkin-bearers is very ancient but some bear traces of their insect pollinated ancestors (Fig. 5.18). In many the evolutionary shift to wind pollination may have been the result of growing in regions where high winds are prevalent or where insects are scarce at the time of pollination, or where there is a short season for growth so that flowering early in the season is necessary to give adequate time for fruit to mature. There are extra costs in anemophily, notably in the amount of pollen which has to be produced to overcome the haphazard nature of the pollination.

There have been many reversions to insect pollination. If the perianth has been lost it is replaced functionally by showy, colourful bracts or stamens. Nectar is produced. The flower may be fragrant and instead of being pendulous the stamens are held erect. A remarkable example is in the Salicaceae which are dioecious with highly reduced flowers. The two genera of the family differ in type of pollination. *Populus*, has long dangling catkins and is probably wind pollinated. The willows, *Salix*, are entomophilous. They have rigid erect showy catkins and nectaries at the base of each flower.

Engler's great classification *Die natürlichen Pflanzenfamilien* (Engler and Prantl, 1887–1915) started with the willows, Salicaceae, because of their very simple apetalous flowers, placing them in a group called the Amentiferae. The Amentiferae included other catkin bearers like the Fagaceae. Although the concept of the Amentiferae was soon challenged because it included a mix of unrelated, derived taxa, similar only because they were anemophilous, the heart of it remains as the Hamamelidanae. The Salicaceae are now placed in the Dillenianae.

The Hamamelidanae have a lineage traceable to the mid-Cretaceous (95 million years ago) where there was a kind of pollen called *Normapolles*. It is triangular, triporate with a complex pore and wall structure. Its small size and smooth surface adapts it for wind pollination. However, at least 70 different types of *Normapolles* pollen have been distinguished and in some cases insect pollination is suspected where *Normapolles* pollen is associated with small hermaphrodite chlamydeous flowers. The pollen of living Hamamelidanae is clearly derived from *Normapolles* type pollen. It has three pores

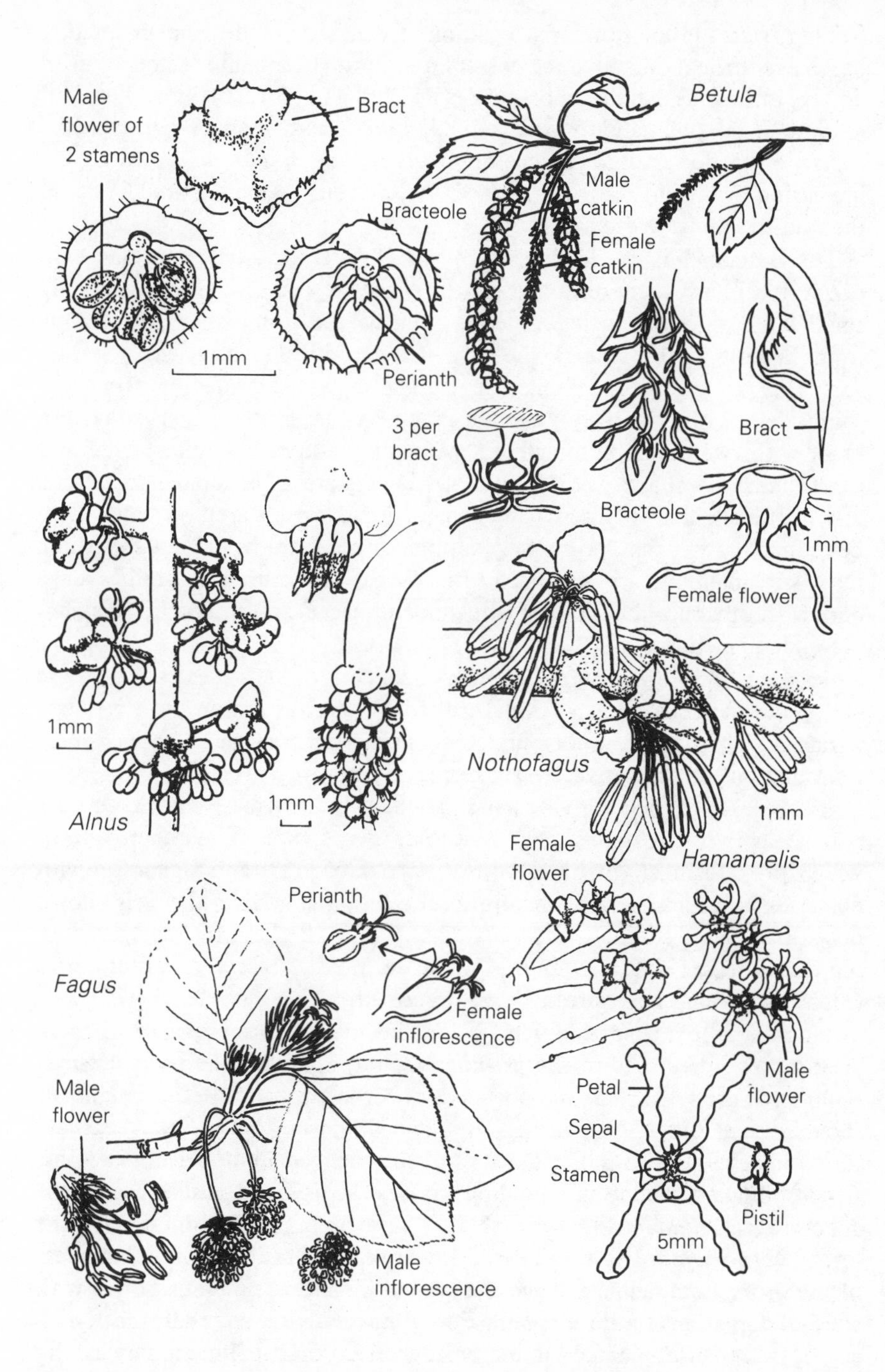

Fig. 5.18. The Hamamelidanae, showing various stages in the reduction of the floret in the evolution of wind pollination from flowers with a corolla to apetalous flowers crowded together in a catkin.

(triaperturate) in an equatorial position. Evolution in the Hamamelidanae has been through dicliny, the loss of the corolla, and the aggregation of flowers in various ways with bracteoles and bracts to form catkins. The shift to pure wind pollination, with the subsequent reduction of non-essential parts of the flower and the evolution of dicliny, has been linked to the increasing aridity in the rift system between South America and Africa in the mid to late Cretaceous.

The Hamamelidanae, includes many familiar temperate tree species: oak, beech, birch, sweet chestnut and plane. There are also some very interesting small families which include species such as the primitive *Trochodendron* from Japan and the strange she-oaks (*Casuarina*) of South-East Asia and Australia. In Engler's classification *Casuarina* was considered to be very primitive because of its very simple inflorescence: a flower consisting of a bract with two scale-like bracteoles with, in the male a single stamen and in the female a single pistil. The male flowers are aggregated into catkins and the female flowers into woody cones. These flowers are now believed to be reduced rather than primitively simple. Other features of *Casuarina*, like its wood anatomy, are advanced. In comparison primitive families have perfect (hermaphrodite) flowers (Trochodendraceae) and a perianth (Platanaceae, Hamamelidaceae).

An interesting order of the Hamamelidanae are the Urticales. Some are trees, like the elms, *Ulmus* (Ulmaceae) with a perfect flower and a reduced perianth. They are probably insect pollinated, but in some species most reproduction is vegetative by suckering. Some are herbs like the nettles (Urticaceae) with their stinging hairs. The family Cannabaceae includes only two genera both used by man: *Humulus*, two species of climber, one of which provides hop, giving the bitter flavour of beer; and *Cannabis*, with one species, which provides hemp fibre, seed oil as well as the well known drug (marijuana, pot, hashish, ganja, bhang, etc.). Perhaps the most interesting family in the Urticales are the Moraceae, the figs (*Ficus*) and mulberries (*Moraceae*) which have already been described.

There is a broad range of wind-pollinated monocots some of which have already been described in the previous section. The most important wind-pollinated groups in the monocots are herbaceous. Rather unmodified, though small but typical trimerous monocot flowers on a long scape are found in several wind-pollinated taxa of the monocots like *Triglochin*. Most wind-pollinated monocots including grasses, sedges and rushes are in the superorder Commelinanae (Fig. 5.19). The progressive reduction of the flower with anemophily can be seen in various different families. Members of the order Commelinales have the typical monocot trimerous flower with simple adaptations for insect pollination. One family in the order, the Xyridaceae, are insect-pollinated but are preadapted to wind-pollination by having an inflorescence held on a long leafless scape above the leafy rosettes. They have normal, perfect hermaphrodite flowers. Derived from the Xyridaceae are the mostly wind pollinated Eriocaulaceae, with similar dense heads of flowers. A few species in the Eriocaulaceae are insect-pollinated with

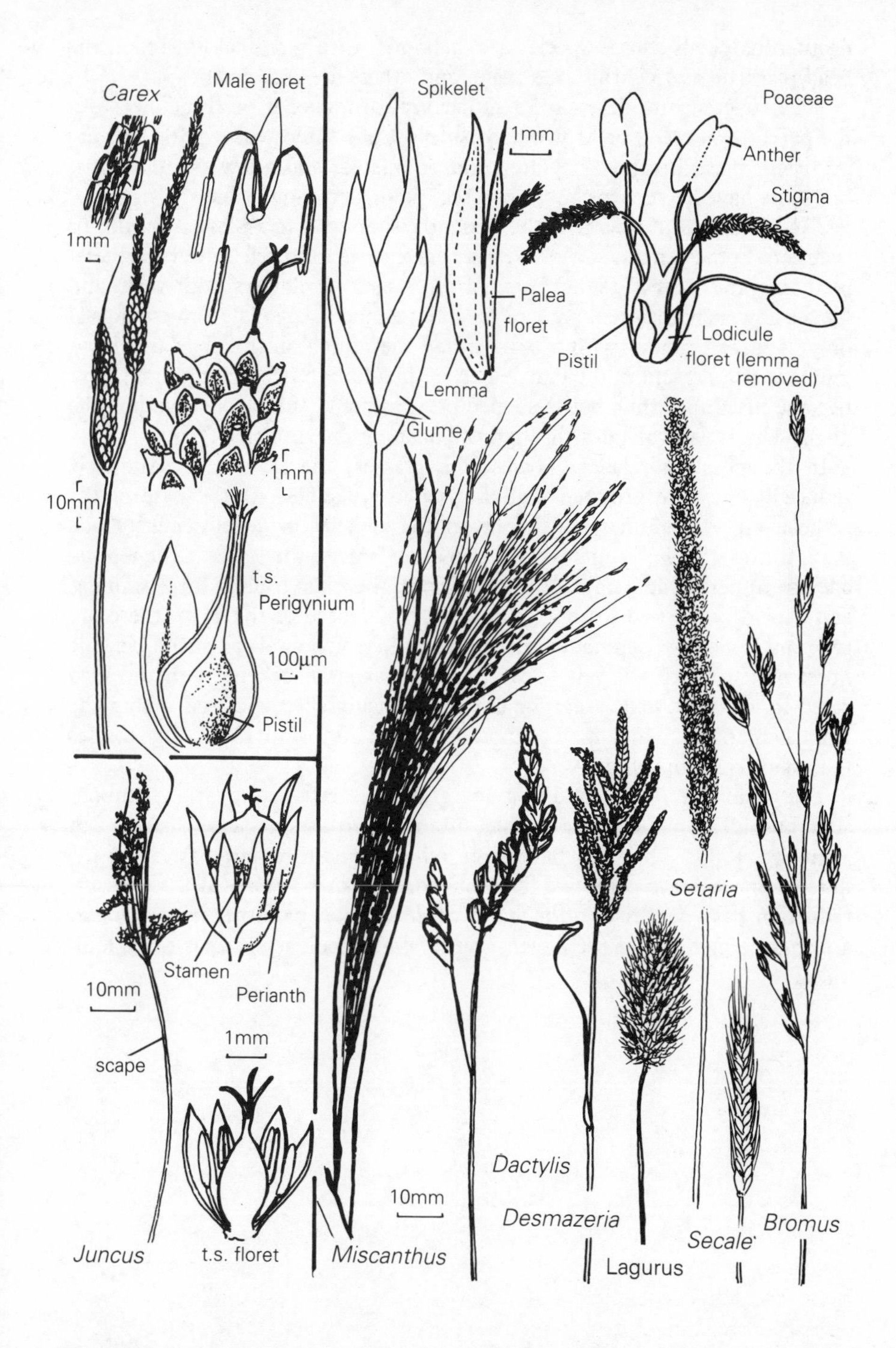

Fig. 5.19. Grasses (Poaceae) showing spike variation, spikelet and floret morphology; with sedges *Carex* (Cyperaceae) and rushes *Juncus* (Juncaceae) for comparison.

nectiferous petals. Some species are diclinous, with marginal flowers in the head pistillate and central ones male, and others are dioecious.

The rushes, Juncales, have a rather typical monocot perfect flower but the perianth is green or brown and papery. Associated with the flowers are persistent floral bracts. They are wind pollinated, and a few are dioecious, but some have become secondarily insect pollinated though they lack nectaries. In the sedges, Cyperales, the perianth is reduced to scales or bristles or is absent. Some genera, such as *Scirpus*, have hermaphrodite flowers. Others, including the largest genus *Carex*, have unisexual flowers with male and female flowers in different parts of the inflorescence. The male flowers consist only of three stamens with a bract called the glume on the abaxial side. A glume is also present in the female flower. In addition two inner glumes have become fused to form a bottle-shaped peryginium or utricle surrounding the pistil. The style protrudes through the opening in the utricle.

In the possibly related, Poales, the grasses, the flowers are similarly reduced. They are grouped **distichously** in **spikelets**. At the base of the spikelet are two glumes which protect the spikelet in development. Each floret in the spikelet is enclosed by two other bracts; the lower is the **lemma** and the upper is the **palea**. Within the floret there are usually three stamens and a pistil with two feathery stigmas. At the base of the pistil there are two tiny **lodicules**, remnants of the perianth, which swell to push open the floret or shrink to allow it to shut. The fruit is an achene with the seed fused to the fruit wall (a **caryopsis**). It is usually shed enclosed within the lemma and palea, which may be modified to aid dispersal. Threshing releases the grain from this chaff.

The primitive floral condition for grasses is retained by the bamboos (Bambusoideae). They have simple, often trimerous spikelets (Clayton and Renvoize, 1986). They may have three lodicules and three stigmas so that the florets approach the primitive liliaceous formula of P3 A6 G(3). Progressive reduction has given rise to the standard grass floret pattern. *Ampelodesmus*, a primitive member of the advanced subfamily Pooidae also has three lodicules.

Trees: adaptations in woods and forests

6

6.1 WOOD ANATOMY
6.2 TREE ARCHITECTURE
6.3 LEAVES
6.4 BARK AND PERIDERM
6.5 EPIPHYTES

SUMMARY

This chapter is about plants which grow in forests. One of the major trends in land plant evolution has been the evolution of trees. Most groups of land plants have given rise to tree species. There is a great diversity of both living and extinct forms. The evolution of secondary growth by a lateral meristem, enabling a tree to increase in girth as it grows in height, was a crucial early event in land plant evolution. The two groups with efficient secondary growth, the conifers and the dicotyledonous angiospems, dominate the forests of today. Trees are the naturally dominant form in all parts of the world except cold environments at high latitudes or high altitudes, and very dry environments. They have been removed from large areas by the activities of large herbivores and man. A large number of species of epiphytes have evolved to exploit trees as their habitat.

6.1 WOOD ANATOMY

There is such a great diversity in wood anatomy that the majority of tree species can be identified from characteristics of the form, range and arrangement of their cells (Fig. 6.1). The greatest diversity of wood anatomy is found in the trees of the tropical rainforest. Increasing seasonality in the climate, whether because of increasing latitude or from seasonal drought is correlated with a narrower range of tolerant anatomies. The greater diversity of types in the tropics may record the successive invasion by drought adapted taxa with more specialized xylem into tropical forests.

Wood consists of two interconnected systems; the axial system and the radial system. The axial system of **tracheids**, **vessels** and **fibres** is adapted for conduction and for strength. Many of its cells lack a protoplasm at maturity. The radial system of **parenchyma** and **ray tracheids** is also conductive but has an important metabolic and storage function.

Wood tissue may be composed of some or all of the following cell types: vessel elements, tracheids, fibre-tracheids, libriform fibres, parenchyma

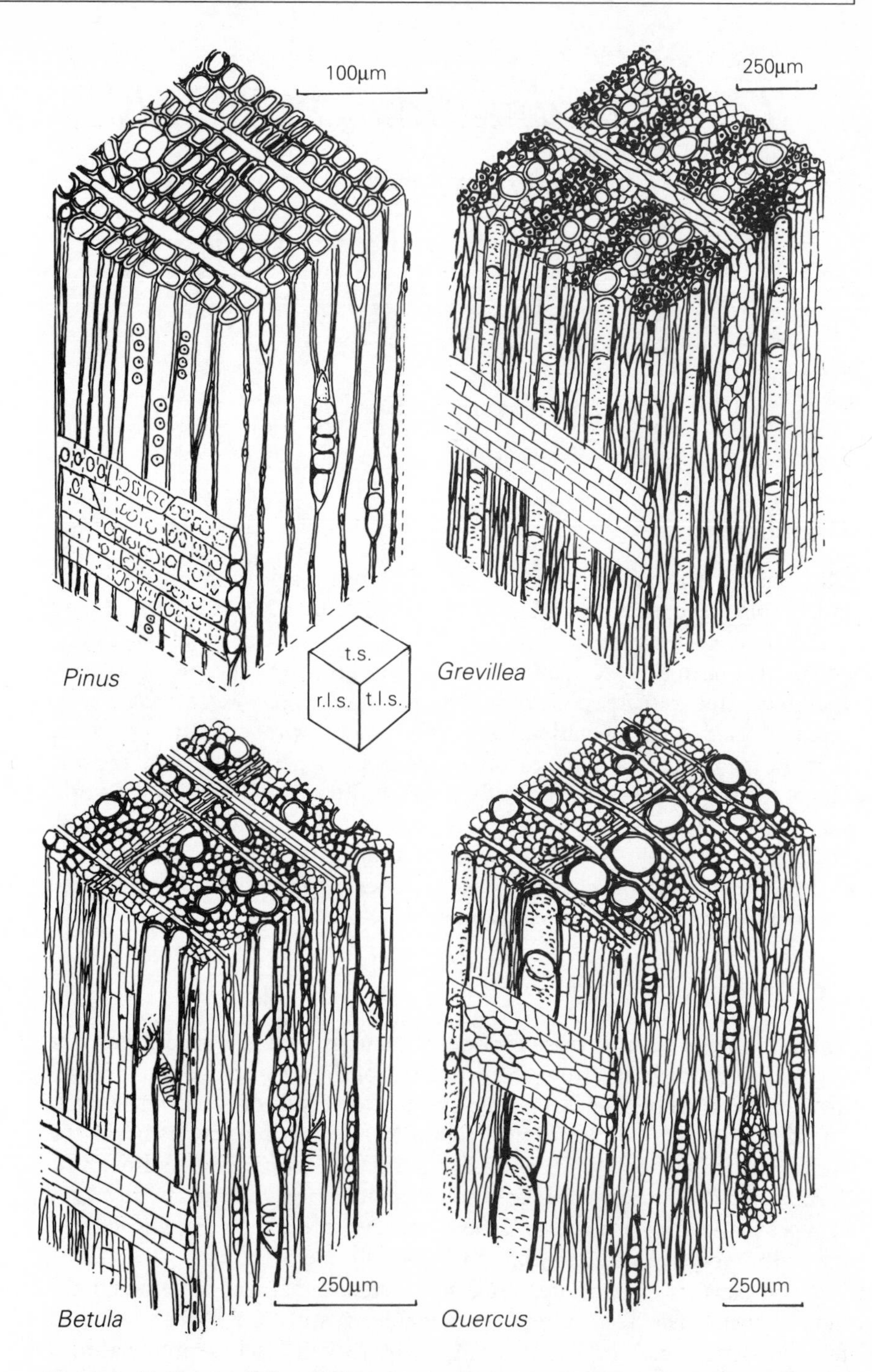

Fig. 6.1. Xylem variation: *Pinus* is a softwood with tracheids, *Grevillea* is a tropical timber with alternating bands of vessels and fibres, *Betula* is diffuse porous and *Quercus* is ring porous. t.s. = transverse section; r.l.s. = radial longitudinal section; t.l.s. = tangential longitudinal section.

cells, and secretory cells (Figs 6.2 and 6.4). The first three are collectively called tracheary elements because they conduct water. The cell contents of the tracheary elements and fibres usually break down at maturity after the secondary cell wall has been laid down and been lignified, though some mature fibres may retain a living protoplasm for several years. Vessel elements and fibres have evolved as specialized derivatives of tracheids.

Tracheary elements connect with each other by pits, areas where the secondary wall does not form. **Bordered pits** are present where the secondary cell wall overarches the margins of the pit. The pit membrane is flexible. In conifers the centre of the pit membrane is thickened to form a torus which can act as a plug sealing the pit pair if there is a sudden pressure difference in the tracheids on either side. Angiosperms with **vesturing** have warts or branched ornamentations on the walls of the tracheary elements or concentrated around the pits. Vesturing is commoner in taxa which predominate in dry areas. The increased surface area of the wall improves the bonding of water and thereby helps prevent the catastrophic formation of bubbles, embolisms, a process called cavitation, which then interrupts the tension causing flow in the water column.

Some structures are associated with the response to injury and deterring damage by animals. Resin canals or ducts are found in the wood of conifers. The canal is extracellular but is surrounded by epithelial cells. Laticifers are present in widely different groups and differ markedly in structure. Xylem cells may contain crystals, frequently of calcium oxalate, either rhomboidal or rod like, or in clusters called druses.

Unlike the secondary xylem the secondary phloem is usually a thin layer. When the bark is stripped from a tree it splits at the weak cambial layer and includes the thin phloem layer. Phloem sieve cells, sieve tubes and companion cells are usually accompanied by phloem fibres as well as parenchyma. As new phloem is produced when older phloem is crushed to the outside or converted into phloem fibres.

6.1.1 Patterns of secondary growth

Woody plants essentially have two phases of growth. The first phase is like that of herbaceous plants. After the development of the primary vascular tissue from the procambium, the secondary or wood tissue arises from the vascular cambium. Secondary growth allows the stem or root to increase in girth. In the shoot, a cambium arises in the procambium as a ring between the xylem and phloem of the vascular bundles (**fascicular cambium**). The arcs of fascicular cambium become connected into a continuous cylinder by an interfascicular cambium. In the root a cambium arises between the xylem and phloem, at first at the points of the xylem star and then connecting together to form a continuous ring.

The cambium produces xylem cells towards the inside (centripetally) and phloem cells towards the outside (centrifugally) by tangential divisions. This occurs both in shoots and roots. As more and more secondary tissue is produced the secondary stem and root grow more and more similar in appearance, because the primary root xylem star or the primary stem vascular bundles become insignificant in relation to the mass of secondary tissue. Radial division of cambial initials occurs, allowing the cambium to increase in circumference as the girth of the stem or root increases. Some species produce successive cambia with a new cambium arising in the cortex.

A growing tree undergoes different mechanical stresses. Horizontal branches suffer two kinds of stress because of gravity; tension on their upper sides and compression on the underside. Stressed trees produce reaction wood. Conifers produce compression wood with rounded, thick walled tracheids with intercellular spaces and cell walls with a higher lignin content (Zimmerman and Brown, 1971). About 60% of angiosperms produce tension wood (Hoster and Liese, 1966), which has gelatinous fibres. Gelatinous fibres have a cell wall in which the innermost layer has little or no lignin but is rich in hemicelluloses. In old wood neighbouring parenchyma cells may bulge into tracheary cells through the pits to produce tyloses.

Some fossil trees, and living pachycaul trees, have different patterns of growth to modern plants. *Lepidodendron* had a unifacial cambium producing nothing to the outside. Fusiform initials apparently only had a limited life span and new cambial initials were produced only occasionally as the diameter of the stem expanded. Increased girth was allowed by the cambial initials becoming wider. Secondary growth was limited by the maximum size of cambial cell allowable and by the absence of any secondary phloem. In the giant horsetail *Sphenophyllum* there was usually a bifacial cambium but like *Lepidodendron* it did not undergo radial divisions. Instead the cambial initials grew longer, growing intrusively between the initials above and below. In this way the number of initials at any one level of the stem was increased. This pattern of expansion is present in some living single stemmed trees. In evolutionary history, modern anatomy was soon represented by the evolution of *Archaeopteris* whose wood is very like that of modern gymnosperms.

Even tropical trees exhibit cycles of growth though in the tropical forest they are not as well synchronized as they are in temperate latitudes. The cambium ceases activity for a period and then later begins growth again. In temperate trees this results in well marked annual rings because the first formed cells of the 'spring' renewed growth are larger than the last-formed cells of the previous late summer.

6.1.2 Specialization in tracheary elements

Many of the differences in the kind of wood cells observed in different species can be related to an evolutionary trend towards increased specialization; improved efficiency of conduction and improved mechanical strength. The primitive multipurpose tracheid has been replaced by a xylem vessel element for conduction and a fibre for strength. This has been accompanied by a number of other changes.

Tracheids are imperforate, that is they have a continuous primary cell wall (Fig. 6.2). The secondary cell wall also covers the surface except for the pits. In ferns and other lower plants the pitting is typically scalariform. Greater mechanical strength is conferred where the pits are circular. In higher plants the pits are packed closely together either opposite or alternate to each other. The latter, in which the pitting has a hexagonal pattern, provides the closest possible packing. The tracheids of vesselless flowering plants have scalariform pitting on their oblique end-walls. The pit membrane, which is primary cell wall, is perforated by many tiny micropores. The oblique orientation of the end wall allows a greater concentration of micropores in the important area of contact between two successive cells. The importance of the area of overlap as a determinant of conductive efficiency is supported by the correlation in vesselless angiosperm plants between the height of a tree and the length of tracheid (Zimmerman, 1983).

The strength of conifer wood increases from the inside to the outside of the trunk. The outer wood is more recent and produced by a larger plant than the inner. Hence it is likely to be more greatly stressed. The greater strength relates to the greater length of the outer tracheids, to the nearly vertical orientation of the cellulose microfibrils and to the greater proportion of cellulose (Carlquist, 1988).

Vessels evolved by the breakdown of the pit membrane. Remnants of the primary cell wall as strands and bars have been observed in pores of primitive angiosperms such as *Illicium* (Carlquist, 1988). Subsequent evolution has led to the development of a single, simple, round pore. Oblique scalariform perforation plates have been retained by species from tropical montane habitats where transpiration rates may be low and the loss of conductive efficiency is counterbalanced by the increased safety of this structure. A scalariform perforation plate provides struts across the vessel and may usefully isolate a gas bubble within a single vessel element.

Vessels have evolved separately many times over. Primitive angiosperms from unstressed mesic habitats may have unspecialized vessel elements with scalariform perforation plates along with multipurpose tracheids. Seasonally dry areas have species with vessel elements with simple perforation plates for conduction and non-conducting libriform fibres for strength. The possession of xylem vessel elements confers an extra risk. Large diameter vessels are more liable to suffer catastrophic breakage of the water column. The water column is under great tensile stress. It can snap as a water-vapour bubble forms in a process called cavitation. This is likely to occur in drought or as a result of the water in the vessels freezing. Perhaps this is why the conifers, which have only tracheids, are important trees in cold dry climates.

The transition from tracheids through fibre-tracheids to libriform fibres is marked by a reduction in the number, size and elaboration of pits. Libriform fibres have small slit like pits oriented parallel to microfibril orientation, maintaining strength. Although the slit-like pore is too narrow to allow the effective conduction of water, it is enough to allow the maintenance of a protoplasm. Many fibres appear to remain alive for several years. In some cases septa are formed across the fibre after the secondary wall has formed. These septate fibres often contain starch and may have a storage

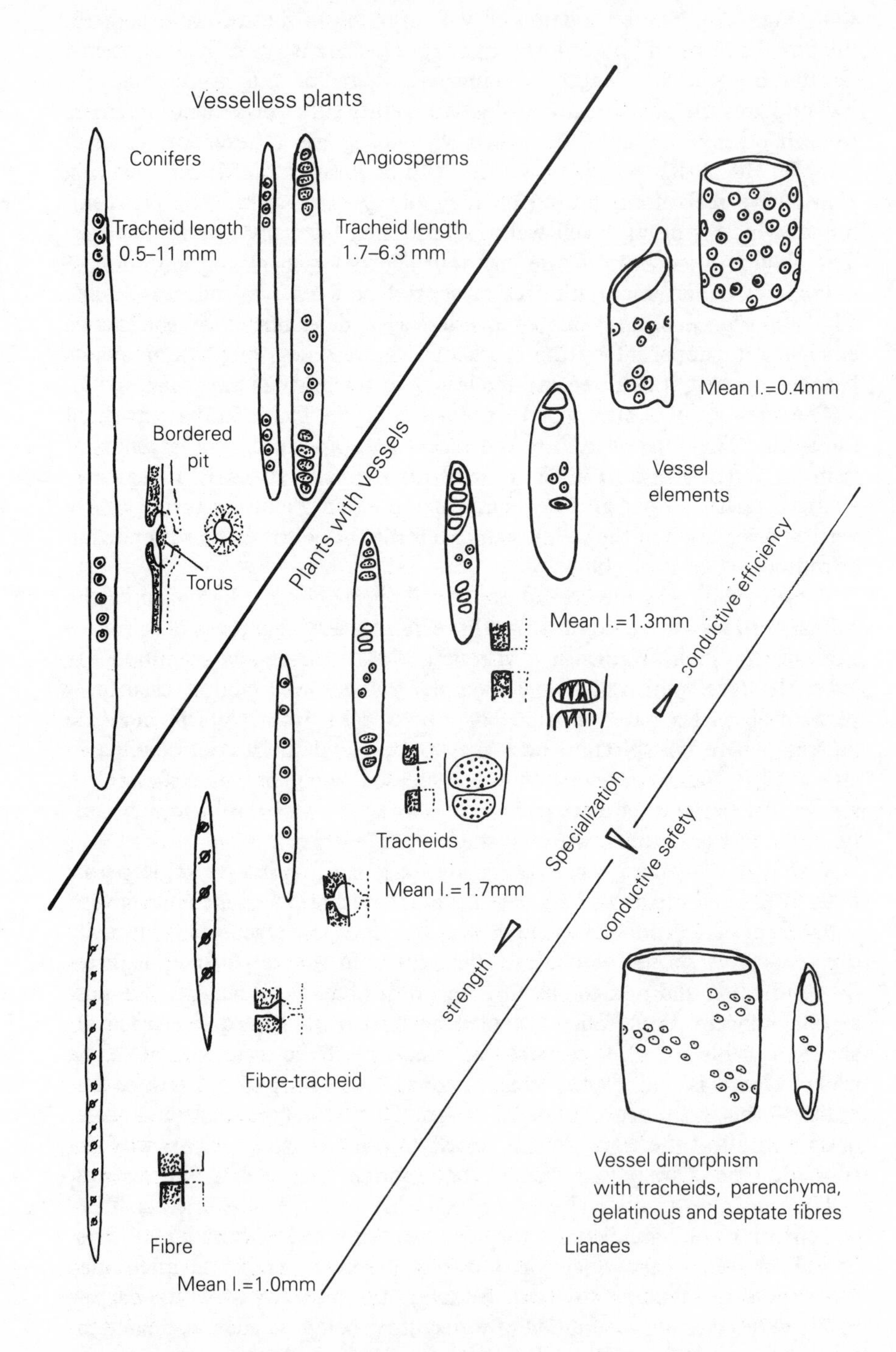

Fig. 6.2. The specialization of tracheary cells (after Bailey and Tupper (1918) and Carlquist (1975, 1988)) showing kinds of pitting.

function. The starch may be important as a source of sugars translocated into vessels to increase osmotic pressure helping to start water flow in spring or after a vessel has cavitated.

Vessel elements have a wider diameter for improved efficiency of water conduction. In tracheids, and primitive vessels, water conduction occurs through pits in the large area of overlap between the tapered cells. In the vessel element conduction is through a simple perforation plate. The vessel elements are blunt ended cylinders with a direct conduction path through them like the segments of a pipe. Vessel elements tend to be shorter than tracheids. A decrease in conducting tracheary cell length has been accompanied by a decrease in cambial-initial length and by the more regular arrangement of the cambial initials in stories. A storied cambium appears to be an advanced feature of wood but it is not always apparent in the mature wood it gives rise to, because libriform fibres derived from storied cambial initials may be intrusive, elongating out of position, inserting themselves in the stories above and below. In doing so they increase the diameter of the trunk.

The efficiency of conduction is determined both by the diameter of the vessel and the total length of the vessel itself, but not by the length of the individual vessel elements (Zimmerman, 1983). The total length of a vessel appears to vary from only a few centimetres to several metres.

There may be a relationship between vessel diameter and plant height, with taller trees having broader diameter vessels, but the pattern is complicated by the degree of environmental stress the tree is likely to suffer. Trees from temperate and xeric regions generally have vessels with a narrow diameter but this is counterbalanced by a tendency to have a greater frequency of vessels. Embolisms are not only less likely to occur in narrow diameter vessels but will also be less destructive of the total conduction of sap. Vessel element diameters ranges from less than 20 μm to over 360 μm (Metcalfe and Chalk, 1950). Species with large vessel diameters are capable of great peak water velocities (Zimmerman and Brown, 1971). Relative conductance increases to the power 4 as diameter increases.

Although the vessel diameter and vessel density are negatively correlated, the relationship is not close. Lianes with vessels as least as wide as tropical trees have a density of vessels several times higher. Vessel density varies from 3 to 20mm^{-2} in the woods of some tropical trees to 2673mm^{-2} in a boreal alpine shrub (Carlquist, 1988). Of course there are no vessels in the wood of conifers including tropical species. Conductive area (vessel density x mean vessel diameter) is low in succulents, desert shrubs and rosette shrubs.

Vessel dimorphism is present in most woody lianes. In cross section the trunks of vines have a lobed or flattened appearance. The vessel elements are dimorphic. There is relatively little xylem for the size of plant. There are a few vessels but with very wide diameters and many much narrower and slightly longer, fibriform, vessel elements. Dimorphism seems to combine the necessary conductive efficiency with safety if cavitation occurs in the large diameter elements.

In other plants different sizes of vessel element are often observed in relation to growth rings. A few very wide vessel elements are produced in spring, decreasing either gradually or abruptly in size as the season progresses. This gives rise to the well known distinction between ring porous timber as in oak, *Quercus*, and diffuse porous timber as in beech, *Fagus*. However even in so-called diffuse porous species, vessel diameter decreases as growth slows towards the end of the season. In fact Carlquist (1988) figures 25 different growth ring patterns in 15 major types depending on the presence/absence and distribution of different cell types, including ray and axial parenchyma. The early wood may be likened to the wood of a tropical rain forest tree whereas the late wood may resemble that of a desert shrub with large numbers of narrow vessels for greater resistance to embolism.

Plants of xeric habitats fall into three classes: those with safe but inefficient systems, having tracheids and narrow vessels; those with vulnerable but efficient wide vessels plus, for safety, narrow vessels and/or tracheids intermixed or adjacent to them; and those with only large diameter vessels. The last class apparently have access to ground water.

The presence of tracheids adjacent to vessel elements, called vasicentric tracheids, increases conductive safety in specialized wood. Such relatively unspecialized tracheids may have been 'reinvented' in the advanced groups where they are found, from the specialized fibre-tracheid or libriform fibre, rather than having been retained as a primitive feature. Vasicentric tracheids are found especially in families associated with arid areas, such as the Casuarinaceae, or associated with physiological drought from freezing, such as the Empetraceae, as well as in woody vines and lianes.

In most cases there is no clear relationship between the height of a plant and the length of individual fibres and xylem vessel elements. Long fibres are more obviously associated with particular taxonomic groups, especially the Scrophulariales, the Urticales and the Malvales, rather than any particular environmental factor or growth form. However, a relationship has been demonstrated within the cacti between tracheary element length and growth form (Gibson, 1973) and there is a very broad relationship between ecological category and mean vessel element length (Fig. 6.3).

6.1.3 Xylem parenchyma: rays

Most attention has been paid to the xylem tracheary elements but many wood tissues have large amounts of parenchyma, either in rays or forming axial strands. Much of this parenchyma has a storage function. Ray parenchyma also provides a radial conducting system.

Rays are files of cells which provide a radial conductive system, are a site of high metabolic activity and sites of carbohydrate and mineral storage. They arise from cells of the interfascicular region or from the procambium (primary rays) or from groups of ray initials in the cambium (secondary rays). Ray initials arise in groups by the horizontal division of a single fusiform initial or a group of initials. Rays may be homocellular (homogeneous) with only one cell type or heterocellular (heterogeneous) (Fig. 6.4).

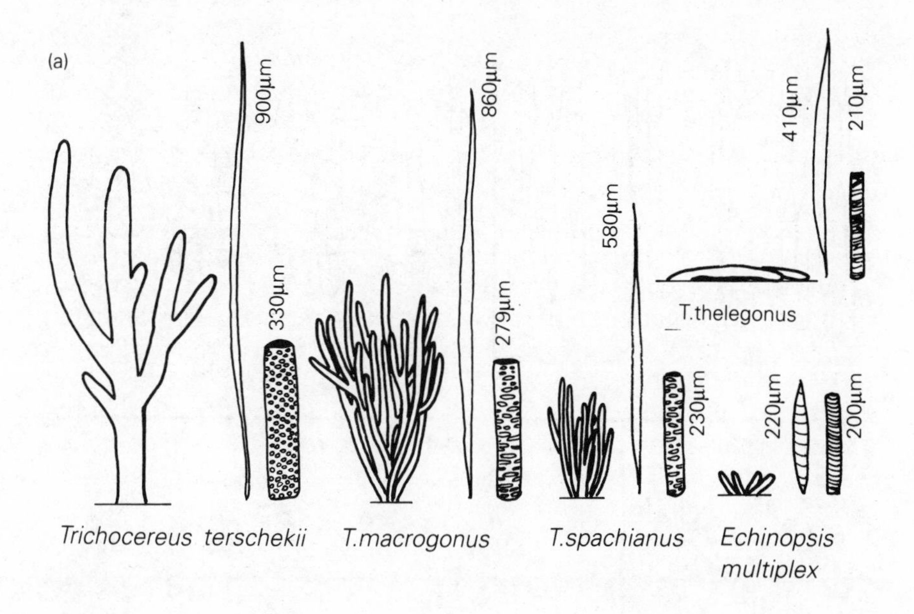

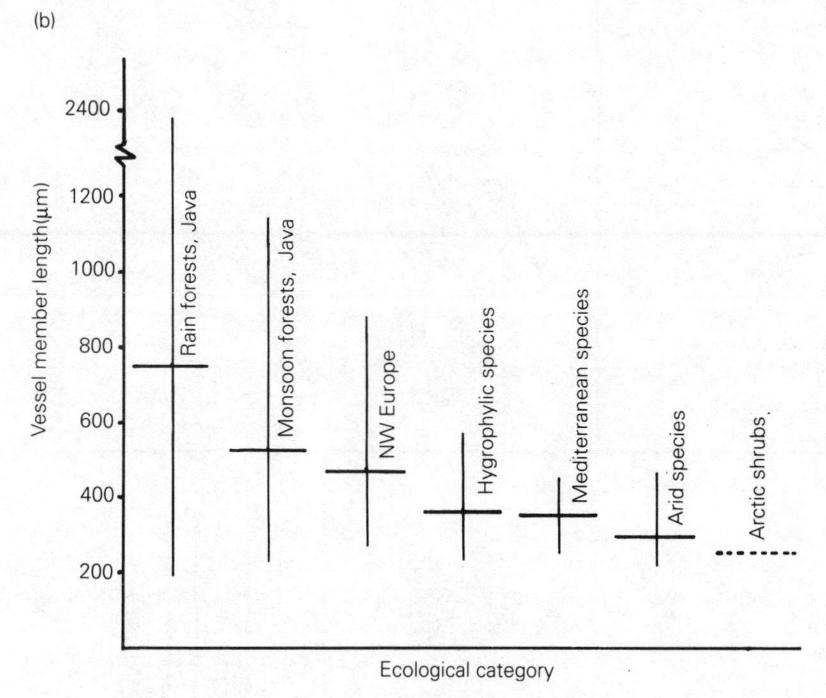

Fig. 6.3. Correlations of xylem vessel member and fibre length and diameter: (a) with height of plant and growth form in the Cactaceae (from Gibson (1973)); (b) with ecological category/vegetation type (from Baas *et al.*, (1983)) mean values and range.

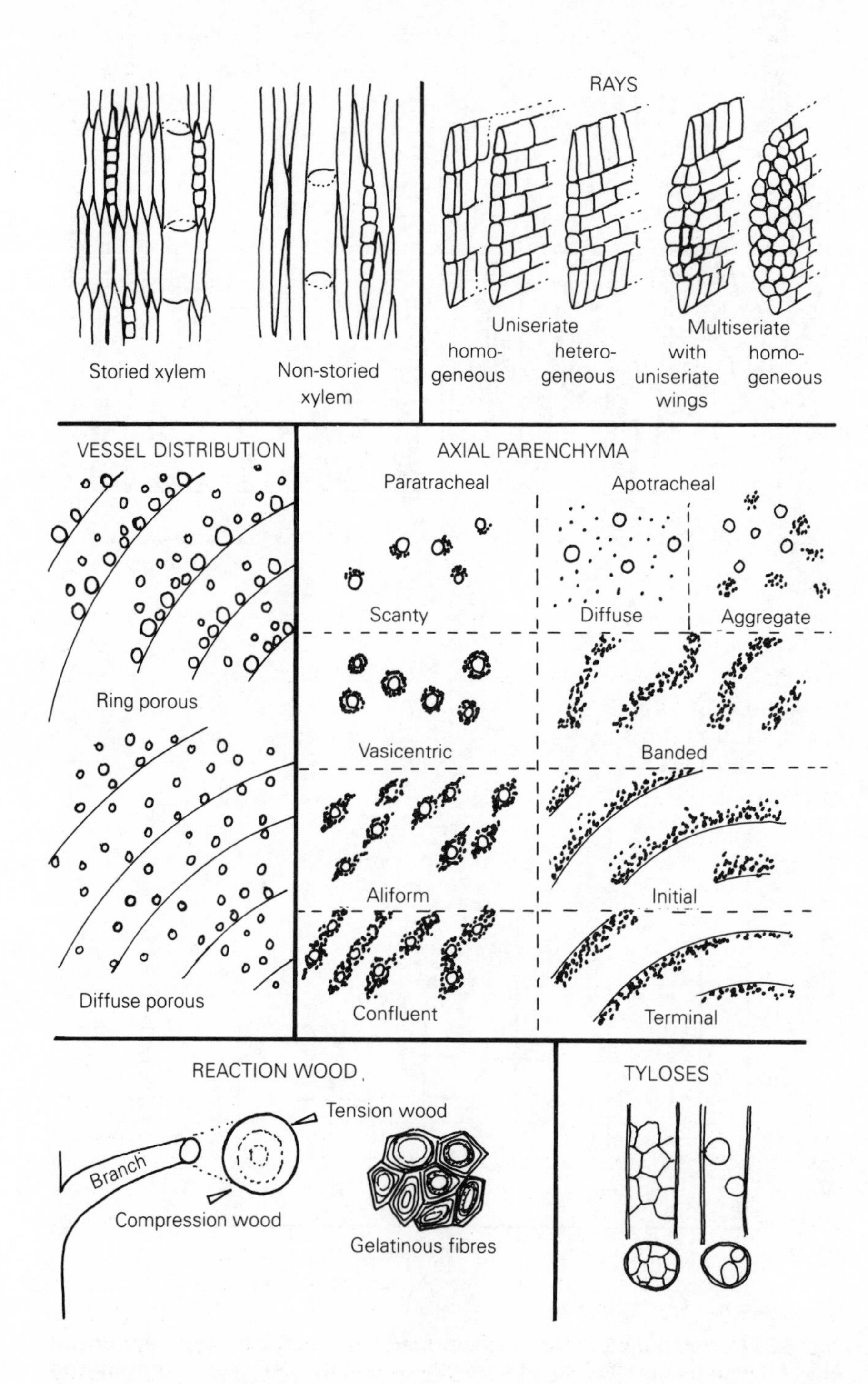

Fig. 6.4. Xylem characteristics (see text for explanation).

Rays may be made up only of parenchyma, in which case the orientation of the parenchyma may be upright or procumbent. In conifers, horizontal ray tracheids may be present along with parenchyma cells. The latter may become lignified. In gymnosperms the rays are always uniseriate, a single file of cells Both uniseriate and multiseriate rays, the latter up to 50 cells wide, are found in angiosperms.

The pattern of pitting connecting the rays to the storage parenchyma of the wood and the axial conductive system is characteristic of different groups. In gymnosperms the pitting is called crossfield pitting. It has a different appearance from axial pitting which occurs between adjacent longitudinal elements. There is a very wide pit chamber and a narrow border (a fenestriform pit). In angiosperms with heterogeneous rays it is the upright cells which form cross-field pitting with the axial system.

Axial parenchyma has been recorded as occupying up to 23% of the volume of some woods (Forsaith, 1926). It is relatively weak since its primary function is as a storage tissue. If it is produced associated with vessels it is paratracheal (Fig. 6.4). Vasicentric parenchyma surrounds each vessel element. Other kinds are scanty, aliform or confluent, each showing different degrees of development. Apotracheal parenchyma is not associated with vessels but like confluent parenchyma may be formed in bands (banded) or terminal, if produced at the end of a season. Alternatively it may be diffuse. It is the relative proportions, and pattern of distribution, of the weak parenchyma and the stronger tracheary elements or fibres which determines much of the strength characteristics of a wood. Axial parenchyma is particularly abundant in the stems of succulents where it forms a water storage tissue and also allows the action of expandable vessel element and tracheids.

Gymnosperm wood is called by the timber trade softwood but this does not mean it is weak. Although it lacks the strengthening fibres of angiosperm wood (hardwood) it also lacks their weakening axial parenchyma. Softwood here really refers to the homogeneity of gymnosperm wood, and not its strength. It is largely composed of a single cell type, the tracheid. This allows it to be worked rather easily in carpentry.

6.2 TREE ARCHITECTURE

The aerial part of a tree must do three things: it must gain height and maintain stability; it must produce a canopy, occupying space to capture light, and so competing with other trees; and it must produce reproductive structures. Forests have several or many different kinds of tree present, with different architectures. Each architecture represents an alternative evolutionary history, and a different solution to evolutionary pressures.

Many different forms, 'architectural models', have evolved. At least 23 have been described (Hallé *et al.*, 1978). The majority can be distinguished on the basis of a few basic characters; having monopodial or sympodial growth, producing terminal or lateral reproductive structures, and having radially symmetrical or flattened axes (section 2.3.1). In some models the main axis and branches have the same pattern of growth, only differing in size, while in others the main shoots and branches grow differently from each other. Models can also be distinguished by whether growth is rhythmic or continuous, whether basal (subterranean) branches are produced, whether each axis keeps the same form throughout its growth and other characters. A few contrasting models are illustrated in Fig. 6.5. Each architectural model is completed by the production of reproductive structures. It may go through phases of different forms to achieve that.

Despite the diversity of tree models which have been categorized, so far it has proved impossible to describe them as adaptations. Similar models may coexist side by side. All of the different models are found in tropical forests for example. Many are found in plants of all sizes from tall trees to herbs. Nevertheless the smaller number of models found in temperate deciduous woods does suggest that some models are selected against in some regions. The boreal forests, dominated by conifers, have even fewer models. A notable feature of the evolution of land plants has been an increase in the number of architectural models present. The earliest Devonian plants conformed to a single model, a simple dichotomously branching axis with terminal sporangia as in *Cooksonia* and *Rhynia*. By the time of the Carboniferous there were nine models with a high proportion of monopodial types or those in which the main shoots do not differ from the branches (Porter, 1989). The diversification of conifers and cycads did not greatly increase the diversity of models. Living conifers exhibit five models, and living cycads only two. It was not until the evolution of the angiosperms that many new architectural models evolved. If they represent evolutionary specializations they have not necessarily increased the adaptability of angiosperm trees.

The adaptive nature of form can be looked at in a different way. One of the most important abilities of trees is to respond and adjust to its environment. As a tree grows it passes through various environments at different levels in the canopy. It must be able to respond if it is damaged by either insects or larger herbivores, attacked by fungi or coated by epiphytes. Alternatively if it is growing in the open it must respond to the abnormal availability of light by producing a broader canopy. The way a tree responds to these various challenges is by repeating part of its normal pattern of growth, repeating part of its architectural model. This has been called reiteration. It is an expression of the modular construction of plants. It makes the detection of the architectural model difficult in some cases.

Reiteration is most easily seen where a tree loses an apical bud by accident. The trunk may fork so that effectively two young trees on one old trunk grow up. In trees which are coppiced or pollarded a whole set of young trees, like individual seedling trees, grows up from the stump. Alternatively suckers may grow up from the trunk of an old tree or from the roots, each an expression of the young architectural model. In the open, the tree may complete or fulfil its architectural model quickly by producing reproductive structures while still a small tree. Subsequently it will continue to grow repeating the architectural model in each of its branch systems. A tree growing in the open illustrates numerous repeats of the tree's basic architectural model, like a set of houses built to the same plan in a housing estate.

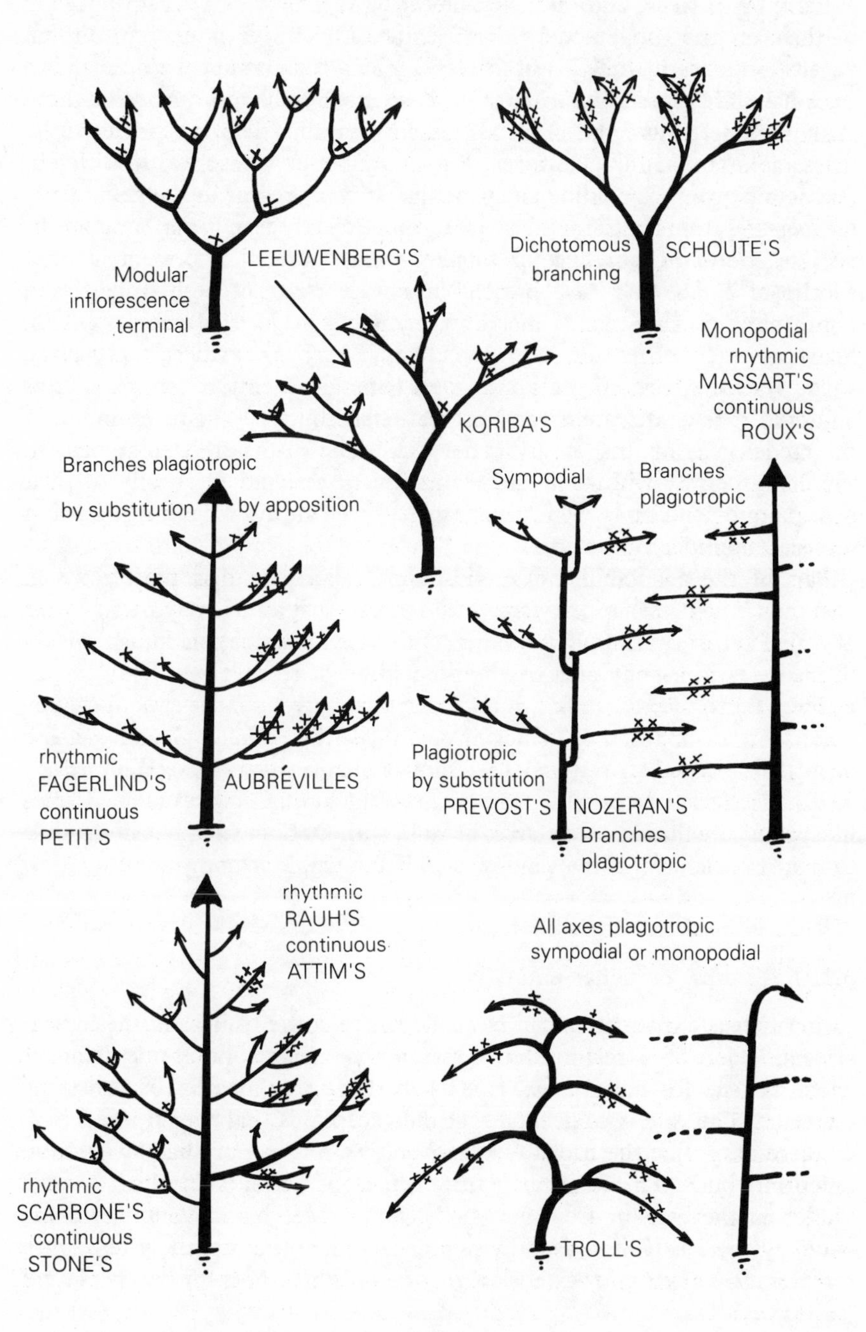

Fig. 6.5. Tree architectural models, x = position of inflorescence/reproductive axis either terminal or lateral, main axis sympodial or monopodial (after Halle *et al.* (1978)).

Early fossil trees, and most non-flowering plants, conform rather rigidly to their architectural model. The architectural model is easy to discern because it is not modified. Conifer forests look monotonous because all the trees have the same form. Almost all have monopodial growth. Many have Rauh's model where the branches are in rhythmic tiers and are basically the same as the trunk. However, Rauh's model is widespread in different taxonomic groups including angiosperms. It was present in the aerial parts of *Calamites* from the Carboniferous. It can be clearly seen in various araucarias like *Araucaria araucana* the monkey-puzzle tree. It is present in most species of *Pinus*. It is also present in a wide range of both tropical and temperate flowering plants, including many oaks (*Quercus*), maples (*Acer*), mahoganies (*Swietenia* and *Khaya*), and figs (*Ficus*). The rather unspecialized shoot system where all the apical meristems are identical confers a great ability to regenerate rapidly giving great adaptability. In some examples of the model it seems this ability is not used. The plants are stereotyped. In the angiosperms, with their greater ability to respond plastically to their own unique conditions, which makes every tree an individual, it is hard to discern the underlying model.

Part of the adaptability of angiosperms relates to their flexible xylem anatomy. The presence of xylem vessels means that a relatively narrow zone of active xylem can supply the canopy. The angiosperms can adjust quickly to increases in canopy area by the production of relatively little xylem. In conifers the active xylem has to be wider to supply a similar area of canopy because it is composed of homogeneous tracheids. A much greater increase in girth is needed to respond to a similar change in canopy (Hallé *et al.*, 1978). Pachycaul trees with massive axes, including single-stemmed trees like palms, are the least adaptable, because without branching it is impossible to increase the area of the canopy even if the trunk anatomy were to allow it.

6.2.1 Trunks, branches and twigs

In most trees a distinction can be made between the trunk and the branch system. There is a relationship between tree height and trunk diameter which is true for most trees. The height is approximately 100 times the diameter. The rule is so general that the exceptions (**pachycaul** trees) look extraordinary, like the baobab, *Adansonia*. Trunks can produce twigs from **epicormic** buds. These may arise from persistent lateral buds which become hidden in the bark or from new buds. Some trees have a long and **short shoot** system, as if there is a division of labour in the branch system, with some branches being for extension growth and the others for bearing leaves (section 2.3.1).

The diameter of twigs varies a good deal. *Betula*, birch, twigs are thin while those of oak are thick. One form may be adapted to growing fast and reproducing early so maximizing reproduction. It will have long internodes, few or slender branches and possibly produce a thin canopy. A tree like this, *Betula* for example, will be a tree of the early succession, a pioneering species, which takes quick advantage of openings in established canopy or colonizes new areas. Alternatively a tree, *Fagus* for example, will grow more slowly, producing more wood, have delayed reproduction, and produce a dense competitive canopy. This is the distinction between different kinds of life history; the result of selection for rapid reproduction in low competitive situations or committing more resources to vegetative growth in competitive situations.

6.2.2 The stability of trees

Tall tropical trees with buttresses or fluted trunks grow on thin soils. The buttresses give the trunk strength and stability while minimizing the amount of wood that has to be produced. The buttresses arise from the uneven activity of the cambium. Trees with huge crowns like the banyan, *Ficus benghalensis*, support it by the production of aerial roots which branch when they reach the soil and thicken into props.

The roots of trees are little studied. We know most about the roots of seedlings or the surface or aerial rooting systems of plants which grow in swampy areas like *Taxodium* and the mangroves (section 7.4.2). Mangroves have a massive root growth giving the tree stability in the wave exposed tidal areas and on the thixotropic muds they grow on. Foresters interested in understanding the ability of forestry trees to withstand windthrow have provided the most information. *Pinus sylvestris* has a long primary root from which many secondary and tertiary roots arise. There is very plastic growth of the smaller roots exploiting the heterogeneous soil environment. *Picea abies* has a mature root system of shallow laterals. It is liable to windthrow. The primary root stops elongating after only five years. The root system of *Abies alba* is little-branched but is very deep rooting with a dominant tap root. This restricts its ability to grow on shallow soils.

6.2.3 Alternative tree forms

Tree ferns, cycads, palms and other arborescent monocots and several fossil trees all share a similar habit, a thick trunk crowned by a rosette of large leaves. They are **monocauls**. They have been described as a number of different architectural models on the basis of a few characteristics. They include a large proportion of the nine models described from the Carboniferous.

- unbranched
  - inflorescence terminal — Holttum's model
  - inflorescence lateral — Corner's model
- one branch per module but looking unbranched (sympodial) — Chamberlain's model
- aerial branches
  - dichotomous — Schoute's model
  - sympodial with more than one branch per module — Leeuwenberg's model
- subterranean branches
  - aerial part determinate — McClure's model
  - aerial part indeterminate — Tomlinson's model

Trees with these kinds of architecture do not have strong secondary growth or may lack it entirely: they lack a well organized vascular cambium. In some a massive crown meristem grows while the plant is short. The meristem expands laterally as the crown of the trunk expands. In this establishment growth each successive node is wider so that the initial axis is obconical, like an inverted cone, as if the tree was balanced on a point. When the full trunk diameter is obtained the trunk elongates. In some palms the cotyledonary organ elongates to bury the shoot apex mitigating the potential instability of an initially obconical axis. In others the stem grows obliquely down at first and then turns up again (Hallé *et al.*, 1978). In some palms and many screw pines, *Pandanus*, the initial axis is above ground but it is supported by progressively larger adventitious aerial stilt roots (Fig. 6.6). Eventually the lower stem axis can rot away so that the mature tree is supported on a platform of adventitious roots, which in some cases even arise from branches. Adventitious roots are produced prolifically by the stem of tree ferns forming a thick mantle supporting and obscuring the trunk. Hidden in the root mantle there may be downward growing stolons which grow horizontally when they reach the soil and spread vegetatively to propagate the tree.

Some trees with these kinds of trunks do have a vascular cambium but as in the cycads it is relatively inactive. The yuccas (*Yucca*, *Dracaena*, *Cordyline*) and the grass trees (family Xanthorrhoeaceae) have a secondary wood tissue which is like primary tissue in having scattered vascular bundles. The cambium is not localized but there is a broad region which includes developing vascular bundles and parenchyma called the primary thickening meristem. It may be that these trees are limited as much by their lack of secondary roots as by the their peculiar pattern of secondary growth in the stem (Hallé *et al.*, 1978).

Anatomical similarities between unrelated pachycauls are suggested by the ability to extract sago from the trunks of palms (*Metroxylon*, *Arenga*, *Caryota*) and cycads (*Cycas*, *Zamia*). The sago is obtained from a broad starchy pith.

The Cactaceae and Euphorbiaceae are unusual in having trees without leaves, but with massive succulent stems. Arborescent succulents may be columnar or have a branched candelabra form. They may be shrubby like the platyopuntias. They may have a rosette of thick strap-like sclerified leaves. There are remarkable convergences between the external appearance of the succulents, in their growth form and pattern of ribbing. The wood tissue of the succulents has abundant axial parenchyma, septate fibres, and lignified parenchyma (sclereids). Most groups with succulents include a larger number of leafy members with a more normal anatomy, and with a vascular cambium giving rise to a cylindrical woody tissue. Even the cacti, which are almost entirely succulent, have *Pereskia*, a primitive genus of leafy cactus, with 'normal' stems.

In columnar cacti the primary vascular bundles form a helical latticework or a set of nearly discrete rods. These become lignified and a fascicular cambium develops. However an interfascicular cambium does not develop fully and some of the primary rays persist for long periods. This allows the

Fig. 6.6. Alternative tree forms; most are monocauls with a terminal rosette of very large leaves; strangling *Ficus* is unusual, starting as an epiphyte and later engulfing its support; bamboos (*Bambusa*) are woody grasses.

trunk to expand in diameter. The pith enlarges in the centre of the trunk (Gibson and Nobel, 1986). In *Carnegiea* from Arizona the rods of lignified tissue become widely spaced. If the wood develops from a latticework the primary rays become arranged in horizontal rows, as in *Trichocereus* from Argentina. In the platyopuntias the tree form arises in a quite different manner. The cladodes do not have large primary rays. There is a continuous vascular cylinder but lignified tissues are only produced from the fascicular cambium. The interfascicular cambium only produces unlignified ray parenchyma. The tracheary cells of cacti have wide banded helical thickenings conveying strength but also allowing for shrinkage and expansion. The euphorbs have pseudoscalariform pitting with possibly the same function. It is formed by the paedomorphic expansion of pitted cells. The pits are stretched sideways as the diameter of the cell expands. Similar pseudoscalariform pitting is found in the succulents of other groups such as the senecios.

The **caudiciforms** have a **caudex** (the stem and root axis together) which is swollen. Some like *Adenia* and *Dioscorea* have a swollen basal caudex from which relatively normal thin branches are produced. The pachycauls have stems and branches which are swollen at the base but tapering gradually apically. They include bottle stem trees like *Nilsonia*, *Adansonia digitata* (the baobab) and some species of *Erythrina* and *Cecropia*, which are a conspicuous feature of some vegetations. The pachycaul and caudiciform habits are widespread in different families and conform to several different architectures. Some species are poorly branched or unbranched. In some respects they are semisucculents. Some even have rather mesophyllous leaves. However, like the true succulents the trunk thickness is the result of having very wide axial parenchyma bands, as in *Adansonia*.

Some bananas (family Musaceae) are the largest living herbs. The largest *Musa ingens* from Papua New Guinea has been recorded at 15 m high (Argent, 1976). They are constructed on Tomlinson's model, but the trunk is a massive fleshy corm at soil level and the aerial part of the plant does not have a true stem. The pseudostem is made up of spirally arranged leaf bases tightly wrapped round each other. Some gingers (family *Zingiberaceae*) produce huge compound leaves from a massive underground rhizome e.g. *Alpinia boia* which may grow up to 10 m high.

Bamboos are 'woody' grasses (family Poaceae). The subfamily Bambusoideae includes herbs, shrubs, tall trees and climbers. It probably evolved as a group of savanna grasses which were displaced by the evolution of more advanced grasses. The subfamily survived, and preserved its primitive floral condition, by diversifying into forest and aquatic environments. One of the specializations which evolved was the woody habit in the tribe Bambuseae, the bamboos (Clayton and Renvoize, 1986). They have evolved woodiness without a vascular cambium. Small vascular bundles are crowded around the periphery of the axis with larger bundles towards the hollow centre (Esau, 1965). Strength is conferred by the presence of a large amount of extraxylary fibres. Living septate fibres fulfilling the dual role of conveying strength and storage are common.

An important feature of bamboo trees is that the aerial portions (culms) are vegetatively determinate: they develop as a very short stem which then abruptly elongates (Hallé *et al.*, 1978). They have McClure's model. They arise from an underground axis which branches to produce a kind of underground trunk. The branching may be sympodial (pachymorph) producing a thick congested underground shoot system. The rhizomes turn up at the tip to produce an aerial culm and the rhizome is continued by a lateral bud. The aerial culms are crowded together like the culms of a tufted grass. Another kind of branching is monopodial (leptomorph). Here the rhizome is elongated, like a rhizomatous grass producing culms from lateral branches. Tillers at the base of the culm may produce a subsidiary clump.

The culms have a clearly differentiated trunk and plagiotropic branches. The culms and branches have a conspicuous segmented appearance with leaves and branches arising at each node. The trunk is tapering, a result of the extension growth which may be likened to the opening of the concentric segments of a telescope. Rather like some palms, the trunks achieve the necessary aerial diameter as rhizomes before they turn upwards to produce the aerial culm. In *Bambusa arundinacea*, a commonly cultivated bamboo, the culms rise to 25 m in height with a trunk diameter of 30 cm. *Dendrocalamus giganteus* is the tallest grass reaching up to 35 m. The preformation and extension growth allows very remarkable rates of increase in height of up to 1 m per day. Only the distal part of the trunk bears proper leaves. Below, the leaves consist of a broad leaf sheath with a rudimentary blade. The branches are determinate and analogous to complex pinnate leaves. The lower may be spinous. Secondary branches may also be present.

The bamboo pattern, of a subterranean axis giving rise to determinate aerial axes, is widespread in plants but not common in trees. It is found in the monocot families Marantaceae and Costaceae. The arborescent horsetails, the calamites of the late Carboniferous period, have a somewhat similar growth form to the bamboos, though the aerial part followed a different architectural model.

6.3 LEAVES

It is evident that all the many differences in leaf shape cannot be described as being adaptive. There is so much variation within species that seems to make no difference to the performance of the tree that much variation must be of relatively little importance. Nevertheless it is possible to regard some of the variations as adaptations because they correlate with particular environments. For example the leaves of trees and shrubs from an undisturbed rain forest understorey, which probably represent a whole range of species, are usually very similar; typically they are lanceolate with a simple outline. Those in clearings, from early in the succession, are very variable with many lobed and compound leaves (Janzen, 1975). Cordate leaves are commoner on lianes. Narrow leaves tend to occur on stream-side plants (Richards, 1952). Leaves with drip tips are common in tropical forests.

Lobed and compound leaves may be the result of selection for increased surface area and rapid growth. Extended growth of lateral meristems around each main vein produced a palmate lobed leaf and ultimately a palmate compound leaf. Alternatively, pinnately compound leaves are like throw-away branches and are more common in deciduous trees. The main branches of the tree move on and are not burdened by many old and relatively unproductive twigs. Compound leaves maximize leaf area while limiting either the mechanical disadvantages of large simple leaves or the production of woody tissue. Ash, *Fraxinus excelsior* is a pioneer species of secondary habitats with compound leaves. Rapidly growing species like the poplars, *Populus*, which have simple leaves, do actually shed twigs very readily. These twigs are thin and not very woody.

The leaf margin is one area which is very variable. Toothed margins are very common in humid environments where mean annual temperature is less than 13°C (Wolfe, 1979). The toothed margin probably creates turbulence breaking up the air flow over the leaf so that the boundary layer of high humidity air around the stomata is destroyed and transpiration can take place effectively. This hypothesis is supported by the distribution of toothing in some individuals. The understorey leaves which grow in very sheltered conditions have a toothed margin, whereas the higher canopy leaves are entire (Longman and Jenik, 1987).

The adaptiveness of leaf size is complicated. There is a complex relationship with temperature and humidity. Leaves are large in humid, shaded forests. Smaller leaves are found in lower humidities and higher light levels. Leaf size is somewhat correlated to leaf texture but not only the largest leaves but also the smallest tend to be thick evergreen leaves. The small sort are the typical sclerophyll of mediterranean type environments with large amounts of strengthening to prevent wilting. They have a multiple palisade layer and only small intercellular spaces.

The close relationship between average leaf shape and environment has allowed the reconstruction of palaeoenvironments from the spectrum of leaf shapes present in a fossil flora (Upchurch and Wolfe, 1987). Characteristics used are leaf size, leaf margin, leaf texture, presence of drip-tips, and leaf shape.

6.3.1 Leaf abscission

Dead leaves burden the branch system and are foci for infection and decay. The majority of angiosperm trees from areas with winter frosts are deciduous. Deciduousness is also a strategy in areas with summer drought. Only a few gymnosperms, *Larix*, *Metasequoia*, *Taxodium* and *Ginkgo*, are deciduous. In tropical dicot trees, growth is often periodic with new leaves flushing at a particular time though not synchronously with other trees. New leaves may be flushed as old leaves are abscised or old leaves may only abscise after the new leaves have become photosynthetic (Longman and Jenik, 1987). In conifers and monocaul trees like the palms and tree ferns and a few other species like the mangrove, *Rhizophora mangle*, leaves are produced more or less continuously. Interestingly large monocot trees do not have the ability to abscise their leaves. Their leaves are generally much larger and sometimes slower growing and it is not energetically advantageous to be deciduous. They also have massive stem/branch systems which can bear the weight of dead leaves. The dead leaves may confer some protection to the plant. They clothe the trunk in cycads and tree ferns.

The abscission zone has a separation and a protective layer. The separation layer has at least two layers of parenchymatous cells. In abscission, either the middle lamella between them dissolves or there is a break in part of the primary cell wall as well. Sometimes there is dissolution of whole cells. In or below the protective layer a scar tissue forms.

6.3.2 Life in the shade, the leaves of the underflora

The growth limiting factor for many plants growing below the tree canopy is the amount of photosynthetically active radiation (PAR) they receive. Many have dark green, chlorophyll-packed leaves to maximize reception of light. Many species of the underflora show remarkable leaf plasticity depending on whether leaves are in the light or shade. Shade leaves are larger, thinner and deeper green. There are large air spaces in the mesophyll and palisade cells are funnel-shaped rather than columnar (Packham and Willis, 1977).

There are several unusual shade plant adaptations in rainforest herbs (Lee, 1986). Many have a red colouration in the mesophyll layer directly under the palisade layer which acts to back scatter light increasing the efficiency of photoreception. Another common adaptation is the obvious satiny or velvety sheen of many shade plants due to the presence of epidermal cells shaped liked lenses which focus light on the chloroplasts of the palisade below. *Fittonia*, has upper epidermal cell-like lenses called ocelli, which seem to be designed to focus light on the palisade cells below. In the Bignoniaceae, Melastomataceae and Cyperaceae the plants often have a blue sheen due to interference, possibly from the epidermal wall. This acts like the antireflection coating of camera lenses to increase the absorbance of the wavelengths greater than 700 nm which are more abundant in the understorey.

6.3.3 Leaf mosaics

It is not sensible to talk about the shape of leaves without describing how they are held or arranged on the branch system. For example if the leaf is held from below, a peltate or deeply cordate leaf form makes more mechanical sense (Givnish, 1986).

Leaves are said to form a mosaic which maximizes the reception of light. This is not to say that leaves fit together like the pieces of a puzzle to form a continuous surface. The leaves overlap each other, shading each other. It is possible to imagine a growth form which minimizes this shading but the canopy is not like the tiled roof of a house which is static. At any one moment we see the canopy in the middle of a continuous process of growth.

Leaves are different sizes, of different ages, stages of development and positions on the branches. A perfect leaf surface would be difficult to achieve. Nevertheless in some trees the leaf mosaic is very complete. Dominant climax trees may have a monolayer of leaves which covers a total area not much greater than the shaded area of ground beneath the tree. In *Fagus grandifolia* the leaves are simple and the branching system and leaf arrangement is strongly plagiotropic. A well organized monolayer can intercept light and outshade competitors very effectively (Horn, 1975).

However, many tree leaves can achieve near maximal photosynthetic rates at 25% of full sunlight. It may, therefore, be more advantageous to have a deep canopy, a multilayer in which each layer captures only a portion of the light. Productivity is maximized by this strategy, which is important in pioneer or colonizing species. Lobed leaves and a more even distribution of leaf-bearing twigs is an effective way of distributing light through a canopy. This also helps to explain some of the difference observed in leaf shape within a single tree. Shaded leaves from lower down in the canopy are less lobed. They catch as much as possible of the light that is left by the canopy above. A fast growing tree like the poplar, *Populus tremula*, achieves a more even distribution of light through its canopy by having flat flexible petioles. The leaves vibrate up and down in the slightest breeze, letting light through. The pioneer tree *Betula* also has very mobile leaves on thin petioles.

A very important advantage of multilayers is that it is easier for a multilayer to maintain even temperatures. The thermal energy of light is dissipated over many more leaves. Lobed leaves can maintain their temperature closer to air temperature easier than simple leaves with the same area. Lobing brings more of the lamina close to the leaf margin, narrowing the boundary layer for more effective cooling transpiration. Small simple leaves are also effective heat dissipators. Trees and shrubs of hot arid environments, like the olive, have small sclerophyllous leaves evenly distributed through the branch system not just confined to its distal part. Some eucalypts hold their leaves vertically increasing the penetration of the canopy by light. These adaptations are important in areas where it is not possible to maintain low temperature by transpiration.

6.4 BARK AND PERIDERM

The bark consists of the **periderm** and the other tissue layers, including the phloem, exterior to the cambium. A periderm is produced not just over the root and shoot systems of woody plants but may be produced elsewhere, as in the winter bud-scales of some plants, where a protective layer is needed. The periderm arises from a lateral meristem, the **phellogen** (the cork cambium) which produces the **phellem** towards the outside. In some species the phellogen also produces a narrow band (1–4 layers) of tissue called the **phelloderm** towards the inside.

The phellogen arises initially in the primary tissues of the cortex. Subsequently, as the initial bark is pushed outwards by the production of xylem and phloem, and it is stretched, new phellogens arise in the phloem. A balance between the production of phloem and the rate of phellogens arising maintains the bark at a constant thickness. In parts of a plant which are wounded a phellogen may arise in areas which were once deep inside the plant, including the secondary xylem, to protect and cover the wound. First, the outer wound layer becomes suberized and lignified forming a closing layer. Then a phellogen becomes active in the layer immediately below.

The phellem consists of tightly packed layers of cork cells and may be very thick as in the 'cork' of *Quercus suber* which is harvested for wine bottle corks. The cork cells are suberized so that the periderm is impermeable to water, CO_2 and O_2. The suberin occurs in layers alternating with wax on the inside of the primary cell wall. The cork cells are close-fitting bricks forming a tight impermeable wall. Two types of phellem cells are commonly found; hollow air-filled thin-walled cells and thick-walled anticlinally flattened cells filled with dark resins or tannins. These two types can occur on the same plant. In *Betula* they occur in alternating layers so that the bark peels in sheets like paper. In places cork cells, called complementary cells, are produced loosely to form patches called **lenticels** which act as pores for gaseous and water exchange. In young roots lenticels are produced in pairs either side of laterals. Elsewhere they may be arranged regularly in rows or irregularly.

The periderm of different species is very varied. There are many different colours and textures. Some species shed their bark frequently. Others maintain the same phellogen for their whole lifetime of many years. The regular shedding of bark provides a mechanism of shedding epiphytic lichens, bryophytes and higher plants. It is a striking observation in tropical forests how some trees are covered with epiphytes of all sorts and others are clear. A heavy load of epiphytes potentially places a tree under severe mechanical stress. As a tree increases its girth the bark is liable to split, providing many irregularities for the lodgement of the propagules of epiphytes. Regular shedding of bark maintains a smooth continuous surface.

The chemical nature of barks is also important in determining the epiphytic load of a tree. In *Citrus*, orange trees are good hosts for epiphytes but lemon trees are not. In the Mexican cloud forest, in a range of oak species with the same bark texture some have no epiphytes and others many (Frei, 1973). *Quercus magnoliaefolia* whose bark contains gallic and elagic acids only has a few lichens. The bark of *Q.scytophylla* has a detrimental effect on orchid seedlings though some epiphytes will grow. *Q.castanea* which lacks chemical deterrence has orchid, bromeliad, moss and lichen epiphytes.

The dead part of the bark, the outer bark or rhytidome is very variable. Those species which retain their phellem build up a thick insulating layer protecting the twig or trunk from extremes of temperature, including frost and fire. The thick layer is strengthened and prevented from falling off by the inclusion of many phloem fibres. Hard barks have many fibres and hard phellem cells. In trees like oak the phloem has many long fibres which give the bark strength allowing a thick bark to build up which becomes deeply fissured as the girth of the tree increases. Soft barks consist of phloem parenchyma and soft phellem cells. In some trees which produce a powdery bark the phellogen forms in small patches and there are few fibres. In trees with scaly barks like eucalypts the phellogens form in large patches so that the periderm consists of overlapping scales. Flakes break off where there are thin walled parenchyma or phellem cells. In species with ring barks the periderms are complete cylinders.

The different colours and smells of barks of different species may relate to the proposed function of bark as a repository of waste compounds. Shedding or partitioning waste compounds in the bark is an alternative to shedding them in leaves or storing them in the heartwood. Many bark compounds may have a role in deterring wood boring insects.

The monocots are primarily a herbaceous group and yet many including the palms have evolved a kind of periderm like that of the dicots and gymnosperms. In others the periderm is less organized. Many long-lived monocots and ferns do not produce a periderm but the epidermis becomes sclerified. The stem and branches of many tree species, especially in the monocots and ferns are not protected by a periderm but by leaf bases or a dense covering of dead leaves or a dense mat of hairs. The tree ferns Cyathaceae have a trunk covered with a dense mantle of adventitious roots only the lower of which reach the ground.

Some barks are so thin and translucent that chlorophyllous cells in the cortex can carry out photosynthesis. This is the case in stem succulents. Some species lack a bark but have a double epidermis. Below the epidermis there is a collenchymatous hypodermis. In some columnar cacti the **hypodermis** may be 10 cells thick. It gives a tough flexible outer layer to the cactus. It is traversed by air passages leading to the stomata. In old plants the skin of epidermis and hypodermis is replaced by a periderm.

6.5 EPIPHYTES

Epiphytes are plants which grow on other plants. Epiphytism is very widespread. About 10% of all vascular plants, approaching 30 000 species, are epiphytes (Madison 1977). Epiphytism is found in 65–9 families of angiosperms (Madison 1977, Kress 1989). Epiphytism is especially important in monocots of which 30% of species, mostly orchids, are epiphytic (Fig. 6.7). Other important families are the Bromeliaceae, Gesneriaceae, Bignoniaceae, and Araceae, Loranthaceae, and Piperaceae. A total of 30% of all pteridophytes, especially ferns, are epiphytes. Bryophytes are particularly important epiphytes in temperate or subtropical climates.

Epiphytism has evolved many times over even in the same family. However, having arisen, epiphytic taxa have speciated markedly so that there are many large primarily epiphytic genera. Apart from the parasitic kinds, epiphytes are not usually confined to a particular host. There are exceptions. For example in Madagascar the orchid *Cymbidiella pardalina* is restricted to the fern *Platycerium madagascarense* which is itself an epiphyte (Dressler, 1981).

The major advantage of epiphytism is the ability to grow in an area of enhanced light without the necessity of growing a large woody trunk. Many

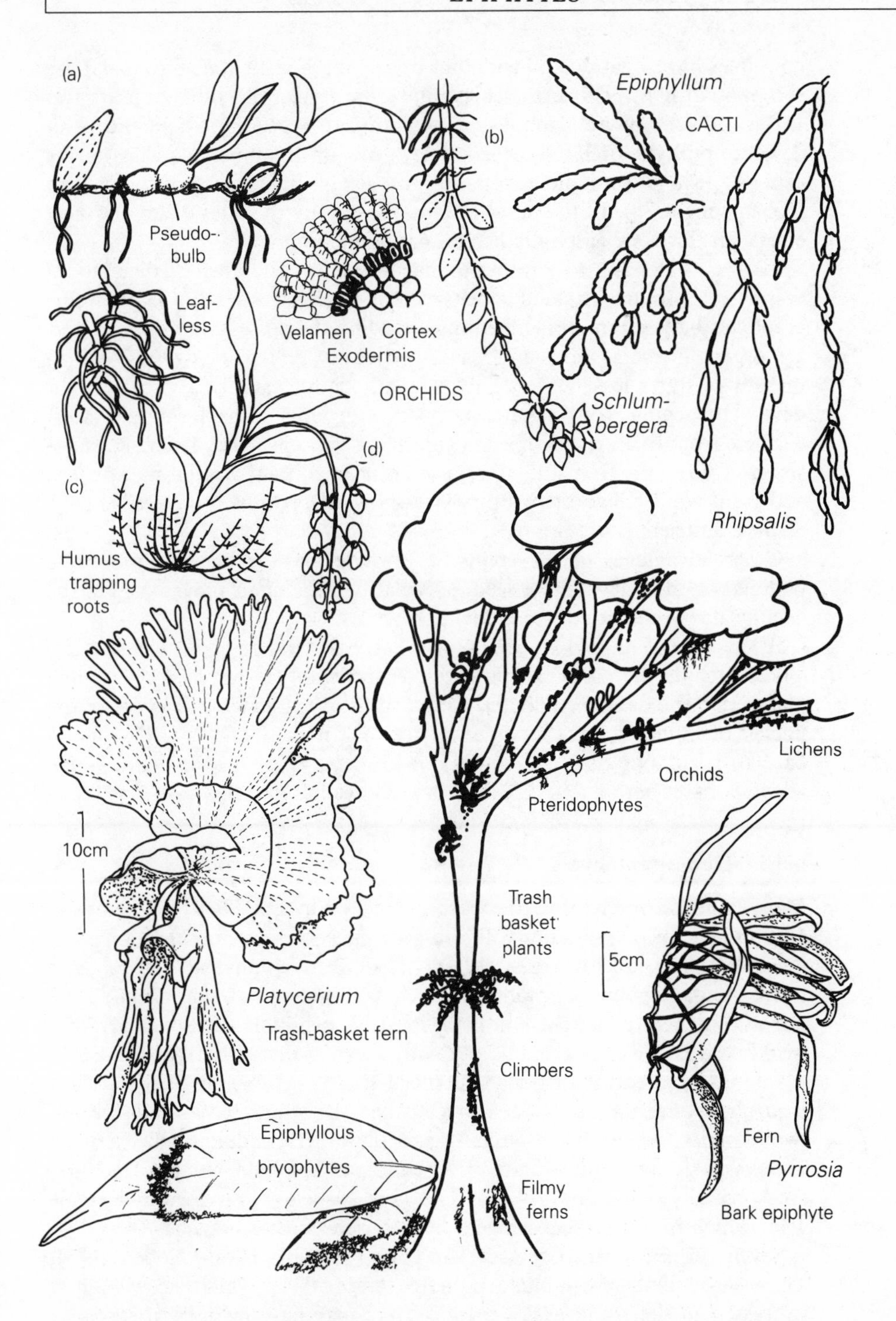

Fig. 6.7. Epiphytes. Different forms have a characteristic distribution in the canopy. Orchids, ferns and cacti are important epiphytes. The orchids (a–d) illustrate a range of adaptations.

epiphytes have a flattened, pendulous or arching branch system to maximize reception of light. The ferns achieve the same result with their large fronds. In fact most epiphytes cannot tolerate complete exposure to sunlight. Over 75% of epiphytic orchids, for example, grow on the inner branches of the canopy, 48% in a middle zone, and only 4% on outer parts of the canopy (Johansson, 1975). In the outer canopy the epiphytes suffer water and heat stress. In this zone epiphytic lichens are abundant.

Lack of water is a major environmental factor. The greatest diversity of epiphytes is found in cloud forests where water stress is the least. Many epiphytes have xeromorphic features including the possession of fleshy storage organs in roots, stems and leaves and have CAM photosynthesis which minimizes water loss. Others have adaptations for capturing rain water or dew. The regime is also nutrient poor. Nutrients come from rainwater, leachate from bark, leaves and nitogen-fixing cyanobacteria, from decomposition of the host, from the excreta of animals and ant and termite debris, and from wind blown dust. Epiphytes show adaptations to take advantage of these nutrient sources.

Other advantages of the epiphytic lifestyle may be better exposure to pollinators, more favourable seed dispersal because of air movements in the canopy and avoidance of terrestrial herbivores.

All degrees of epiphytism are known. Some species are obligate epiphytes. These include the parasitic epiphytes. Many climbers begin as soil rooted plants and end as holoepiphytes. Opportunistic epiphytes will grow in any pocket of humus whether it is on a tree or a rock. Plants which grow on bare rock (lithophytes) will often also grow as epiphytes on bark. Some adaptations fit both habitats. Many bryophytes fall into this category.

6.5.1 Humus epiphytes

Many humus epiphytes are opportunistic epiphytes. They may grow at ground level or on cliffs but will also grow in pockets of humus trapped in the boles and trunks of trees. Others are obligate epiphytes. They show a range of adaptations for epiphytism such as a measure of xeromorphy, like leaf succulence, a flattened growth form, a pendulous habit and rather weakly developed root systems, but with a strong mycorrhizal association.

Interesting, because of their superficial similarity to some of the earliest land plants, are the humus epiphytes called the whisk ferns. *Psilotum* has two variants, one erect and arching and the other pendulous. The related *Tmesipteris* is normally a flattened, pendulous epiphyte but also includes erect, radially symmetrical plants. The stem has leaves traversed by a single vein, which may represent winged, lateral, determinate branches.

Seven per cent of the Cactaceae are epiphytes (Gibson and Nobel, 1986). There are climbers, humus epiphytes and others which grow tightly adpressed to the trunk (*Selenicereus*). They may have cylindrical stems or be flattened with 2–4 ribs. Two ribs is the commonest pattern. In the genus *Rhypsalis* there is a broad range of stem form. The buds (areoles) are confined to the edge of the ribs: they lack leaves, and most lack even spines. The flattened stem of these forms cannot really be regarded as being succulent. It seems to be an adaptation to maximize light reception under the canopy. The flattened species grow in fairly wet conditions. In some such as *Strophocactus* the parenchyma may be lignified (Gibson and Nobel, 1986). Epiphytism has evolved separately in North America and South America. The North American flattened species like *Eccremocactus* have indeterminate stems but the South American species, the Christmas cactus *Schlumbergera* and *Rhypsalis*, have a series of determinate cladodes. In the mainly North American genera *Cryptocereus*, *Disocactus*, *Eccremocactus*, *Epiphyllum* and *Nopalxochia* the flattened stem has evolved several times in parallel. Like many angiosperm epiphytes, including orchids, these cacti have have large tubular bird-pollinated flowers.

Other important families of humus-rooted epiphytes are the Ericaceae which even has shrubby epiphytes, the Gesneriaceae, the Melastomataceae, the Piperaceae and the Rubiaceae. Strangling figs in the Moraceae start life as humus epiphytes in the canopy but send roots down to the soil. The roots encircle the trunk of the host tree anastomosing with each other as they grow in size. The canopy grows larger, shading out the host. Eventually the host is overcome and dies so that the fig stands self-supporting. Stranglers are mainly figs but also occur in other families and include *Schefflera* (the umbrella plant, family Araliaceae) commonly grown as a house plant.

6.5.2 Trash basket, tank epiphytes and ant plants

Many epiphytes trap leaf litter, dust and water accidentally in the spaces between the leaves and stems and especially at the base of the leaves. Some are specially adapted to do this. The leaves or fronds are placed close together and their bases overlap to create a trash basket, a trap for leaf litter. The rosette of trash basket ferns like *Asplenium* may completely encircle the trunk. Roots grow up into the basket to take advantage of its water and nutrients. Some ferns, especially *Platycerium* species, have shield fronds modified to grow back to clasp the supporting branch or trunk and create a trap. Similarly some orchids have negatively geotropic basal aerial roots which form humus collecting baskets.

Bromeliads create a tank in the rosette of leaf bases. Instead of roots specialized trichomes are found on the leaves for absorbing water. The tanks of *Tillandsia* may contain an aquatic epiphyte, *Utricularia nelumbifolia*, the bladderwort. *Utricularia* also includes true epiphytes, with runners and tuberous branches, living in humus and the mat of epiphytic mosses.

Other taxa create traps for water and humus in different ways. Many epiphytes in a range of families have a close association with ants. In *Hydnophyton* in the Rubiaceae, there is a swollen grooved tuber with extensive spiral chambers inside. On the upper surface of this structure there are large pores which catch rain water. In lowland habitats the chambers are occupied by ants which may provide some nutrients for the plant. Related to *Hydnophyton* is *Myrmecodia* (Fig. 8.4). It has spiny tubers with a complex pattern of chambers. There are narrow chambers with warty walls in which debris accumulates. Larger smooth-walled chambers house ant broods. Both these kinds of chambers are connected to the exterior by a honeycomb series of ante-chambers with suberized walls. The relationship with the ant is more constant than in *Hydnophyton* and the tubers develop from the swollen hypocotyl in the absence of ants. The ant species is normally a single species, *Iridomyrmex cordatus*. There are also two fungal associates found regularly in the chambers.

Another kind of ant plant is *Dischidia* in the Asclepiadaceae. Some species have ordinary succulent leaves. In others the succulent leaves are larger and have a chamber connected to the outside by a pore. Adventitious roots grow into the chamber. Both *Dischidia* and *Myrmecodia* are found in areas which are normally inimical to epiphytes, on trees like *Casuarina* which grow in nutrient poor, dry areas. Hollow ant chambers are found in some epiphytic ferns. There are hollow rhizomes in some Polypodiaceae. The fern *Solanopteris* has hollow tubers filled with water. The orchids, *Schomburgkia* and *Laelia*, have hollow pseudobulbs. However in both ferns and orchids it is more common for ant gardens to be produced. The ants make a humus-based nest which is reinforced by the roots of the epiphytes. In some relationships two different genera of ants occupy the football-sized ant nest produced, one in the centre and one nearer the surface.

6.5.3 Bark epiphytes

Bark epiphytes have to tolerate low levels of water availability and low nutrients. Bark epiphytes include a high proportion of obligate epiphytes. Many show some form of xeromorphy like leaf succulence. Some produce resting tubers. Some of the most highly adapted bark epiphytes have CAM (crassulacean acid metabolism) photosynthesis.

Some of the most higly adapted bark epiphytes are orchids. They are either monopodial or sympodial (Goh and Kluge, 1989). Monopodial ones with lateral inflorescences are more commonly humus epiphytes. One extreme monopodial form, as in *Chiloschista* has a vestigial stem system, only scale leaves, and profuse photosynthetic roots. Sympodial orchids are more commonly bark epiphytes. They have a compact habit with inflorescences terminal. Growth is continued by axillary buds producing a clustering of flowering stems. Aerial roots are not as frequent as in monopodial forms but pseudobulbs, formed from a single or several internodes of the thickened stem are very common. The pseudobulb has a ribbed outline allowing it to contract and expand as it loses and gains water like the ribs of barrel cacti.

The roots of orchids (family Orchidaceae) and arums (family Araceae) are convergent in the possession of a velamen. This is a multilayered epidermis composed of long hyaline cells exterior to an exodermis where the cells have conspicuously thick walls. The hyaline cells, which are dead at maturity, have a spirally or reticulately thickened wall which prevents them from collapsing when they dry out. When it rains they fill with water except in areas called pneumatodes (pneumatothode) where there are cells with especially dense thickenings and water repelling oil droplets. The pneumatodes are located above thin walled aeration cells which interrupt the exodermis. They allow the aeration of the root. The evidence for the function of the velamen for taking up water and nutrients is conflicting (Goh and Kluge, 1989). Water and nutrients are said to be absorbed by the cortex through thin walled 'passage' or transfer cells in the exodermis (Benzing, 1989). The relationship between transfer cells and air cells is obscure. In some cases the uptake of water from the velamen by the cortex is slow. A dry velamen is also functional as a protective layer helping water conservation.

Tillandsia are bromeliad aerophytes adapted for absorbing water from the atmosphere. They have CAM photosynthesis. They are covered with peltate trichomes which allow water absorption over the whole surface (Fig. 6.8). The trichome absorbs water especially from dew at night. The outer surface is cutinized but as water becomes available the centre of the shield-like trichome lifts into a dome sucking in water underneath where it is absorbed into the uncutinized surface of the stalk of the trichome (Benzing, 1976). The most extreme species of *Tillandsia* and the most widespread is *T.usneoides*, Spanish moss. It is a true aerophyte which lacks roots altogether. It looks like and takes its name from the lichen *Usnea*.

6.5.4 Epiphytic bryophytes: leafy liverworts

The small size of bryophytes adapts them as epiphytes par excellence. They find nooks and crannies in the surface of trees in which to grow; microhabitats not available to other plants. Their system of **rhizoids** can penetrate into tiny cracks. Most importantly some bryophytes have a remarkable ability to survive temporary desiccation; they are poikilohydric. Reproduction by the production of small air-dispersed spores favours the colonization of their hosts.

In temperate regions bryophytes are the predominant epiphytes. Some bryophytes can only be considered as weak epiphytes because they are confined to the base of the tree, on the exposed roots or buttresses. In lowland tropical rain forests, which do not have a diverse or rich bryoflora, this is the habitat with the greatest diversity of bryophytes (Schofield, 1985) many of which will also occupy the forest floor.

Leafy liverworts (order Jungermanniales) are important epiphytes. They are delicate plants with leaves only one cell thick (unistratose). They do not survive long periods of dehydration but in humid and wet areas like cloud forests they are important epiphytes. Leafy liverworts are flattened and can grow over the surface of trunks and branches rather like **pleurocarpous** mosses (section 7.2.2). The leaves are in threes but the plant is flattened with two ranks of leaves large and overlapping and the third situated as a rank of underleaves called **amphigastria**. The underleaves are often divided and numerous rhizoids arise from their base, as in *Lophocolea* (Fig. 6.9). In a few species, shoots of different sorts are produced. In *Bazzania* there are downward growing 'flagella'. The leaves are often modified in different ways to trap water. In *Trichocolea* they are finely divided. In *Diplophyllum* the

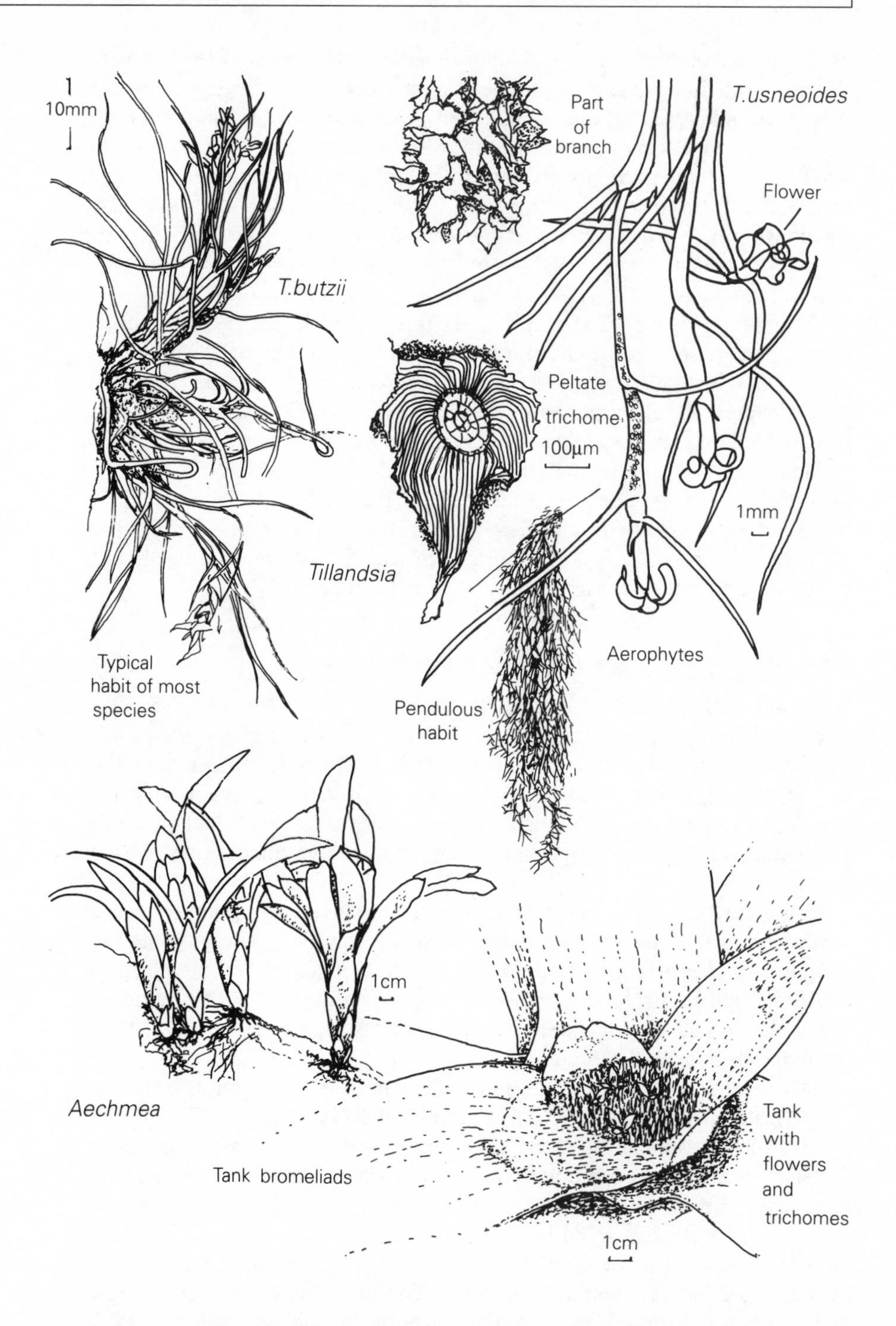

Fig. 6.8. Bromeliads: *Tillandsia* are aerophytes absorbing water through peltate trichomes which cover the surface. Most bromeliads have a tank formed by overlapping leaf bases.

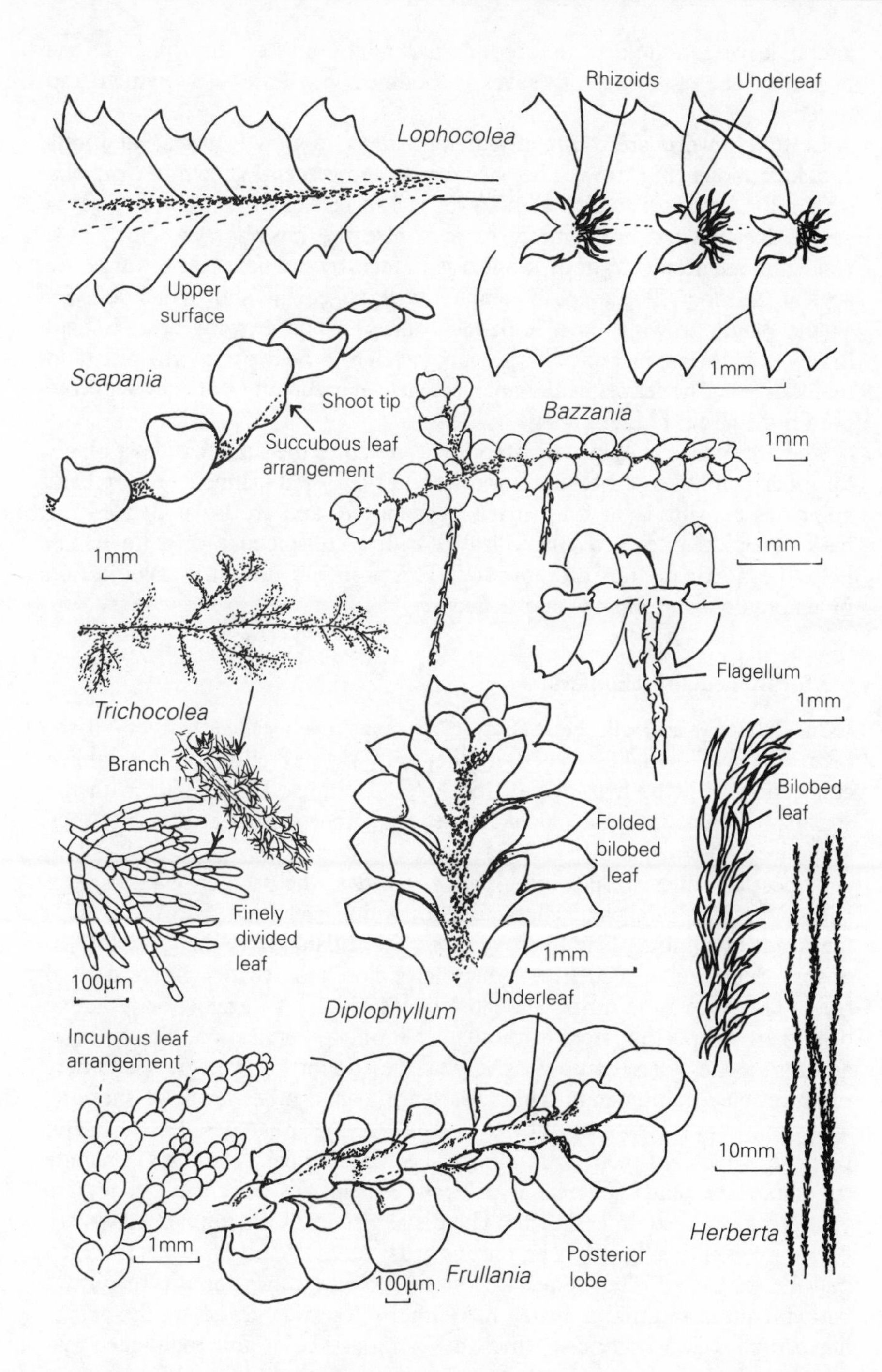

Fig. 6.9. Leafy liverworts: ectohydric plants that show many adaptations for the quick absorption of water. Many grow epiphytically. *Herberta* is peculiar in having the upright habit of an acrocarpous moss.

lateral leaves are folded to produce a kind of pocket. In *Frullania* the posterior lobe of the lateral leaves is modified into little flasks which trap water.

Leafy liverworts are commonly divided into two growth forms, **incubous** and **succubous** (Fig. 6.9). The succubous arrangement, shown in *Scapania* is found in liverworts which climb a vertical surface like a tree trunk. Leaves are held somewhat horizontally to maximize the reception of light. As a result the anterior margin of the leaf is hidden by the next leaf towards the apex of the shoot. The alternative incubous arrangement, shown in *Frullania*, adapts plants growing on a horizontal surface. The growing apex is held lower than the main body of the shoot which is held up on rhizoids and underleaves. The leaves at the apex are held horizontally but now with the posterior margin hidden.

The prevalence of epiphytic bryophytes modifies the surface of host plants for other epiphytes, especially important in their establishment. Many bark epiphytes actually grow on a mat of bryophytes and are isolated from the bark. Tropical trees often have leaves with a conspicuous drip tip which helps the lamina to drain off rainwater. This probably minimizes colonization by epiphyllous bryophytes and lichens.

6.5.5 Lianes and climbers

Lianes (lianas) and climbers (Fig. 6.10) have been called hemiepiphytes because they begin life rooted in the soil. As they climb they establish connection with the host or with pockets of humus. The contact with the soil may become insignificant or be lost altogether so that at maturity they are holoepiphytes.

Almost all lianes are flowering plants. *Gnetum*, the peculiar gymnosperm with some flowering plant characteristics is one exception, and there are a few kinds of climbing fern like *Lygodium*. Some lianes have the growth form of very slender elongated trees. Many have close relatives which are normal trees. The simplest climbers are those lianes which lean against or scramble over their supporting trees without any intimate connection. Some lianes are more or less carried up passively as the supporting tree grows. Others produce long arching stem which reach up to find other levels of support. They may have hooks or thorns derived from leaves, petioles or lateral branches which aid their scrambling. The climbing palms (rattans), include the important genus *Calamus* with about 370 species worldwide. They are very common in south east Asia. The distal pinnae of the pinnate leaves are backward pointing spines. The rattans grow very fast and have stems which may be well over 100m long. The stems provide canes for the furniture, basket making and mat industries. Another interesting group are the climbing bamboos like *Dinochloa* which has a zig-zag culm and roughened leaf sheaths to aid climbing.

A closer connection to the host is achieved by the vines with tendrils modified from leaf or stem or which twine around the supporting tree. Unlike lianes they may not conform to any simple architectural model. They

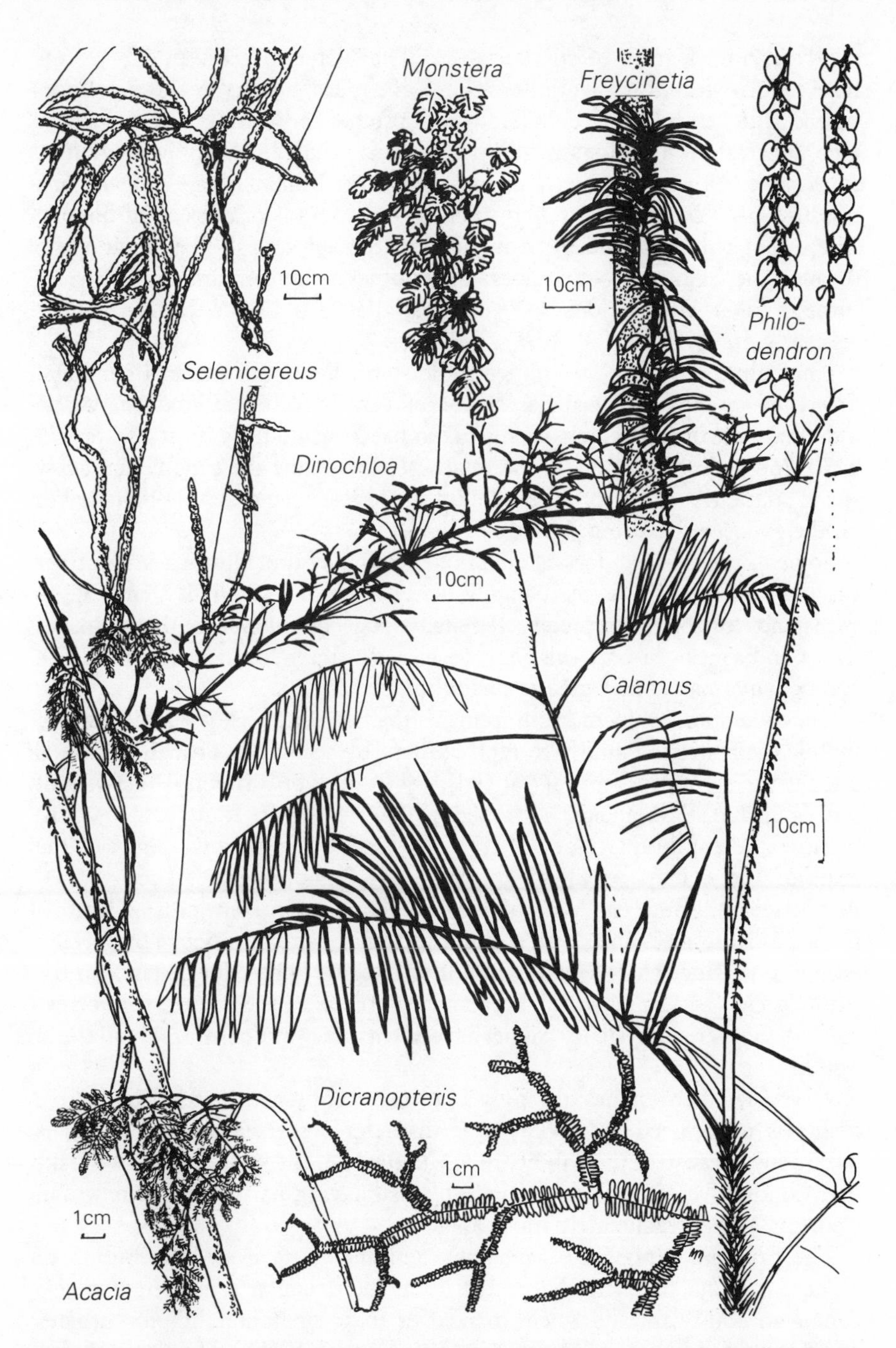

Fig. 6.10. Lianes, climbers and scramblers: *Selenicereus* (cactus), *Dinochloa* (bamboo), *Calamus* (palm) and *Freycinetia* (screw pine) illustrate how many different groups have produced lianes or climbers. In *Acacia* lianes are not common and closely related to normal *Acacia* trees, emphasizing how lianes are essentially trees with very narrow trunks. *Dicranopteris* is a fern which scrambles near the ground in tropical areas.

show varying degrees of specialization. The Nepenthaceae, pitcher plants have tendrils extending from the midrib of the leaf. An even closer connection is achieved by those climbers which produce adventitious roots. These may penetrate the bark as in the ivy, *Hedera*, and the climbing pandanus, *Freycinetia*. Others like the familiar houseplants *Monstera* and *Philodendron* produce large corky aerial roots which take advantage of pockets of humus. These root climbers tend to be highly adapted epiphytes. They migrate up through the canopy. Root climbers may obtain significant amounts of water through their adventitious roots so that they can become holoepiphytes relatively easily.

Some climbers grow flat, hugging the stem. Ferns are unusual climbers. *Stenochlaena* the vine fern has slender green rhizomes. *Lygodium* has an indeterminate frond which produces pinnae continuously as it grows forward. Most climbers grow in the relative shade below the canopy and show shade adaptations. Ivy produces lobed shade leaves and simple light leaves on the exposed flowering shoots.

Some of the most important adaptations of lianes and climbers are in their vascular anatomy. The free hanging lianes must have a pliant stem able to withstand torsion movements. Parenchyma is abundant in the stems of lianes and vines. In part this may be because fibres are not required. The parenchyma may confer greater flexibility.

The xylem and phloem has to remain functional at a great age because of the plant's restricted ability to replace them by secondary growth. There is a great variety of anatomical patterns, the results of differential activity of the cambium. Many have a ribbed xylem (lobed in transverse section) because the cambium ceases activity in places. The furrows between the arms of the xylem are filled with phloem (Bignoniaceae, Apocyanaceae, Acanthaceae). Some have only two lobes giving a flat stem pressed against the supporting tree. In others an interfascicular cambium does not develop except to produce extra separate bundles. In some lianes, for instance in the Sapindaceae, and in *Gnetum*, successive cambia are produced in the cortex, so that they are polystelic. Some have **intraxylary phloem** or **bicollateral** bundles.

A very narrow stem has to supply a profuse canopy with water. Conductive ability is maximized by having large diameter vessel elements but this is hazardous because of the liability of cavitation, i.e. the water columns breaking. Conductive safety is maintained by also having narrow diameter vessel elements and/or **vasicentric** tracheids.

The preponderance of parenchyma and the more even distribution of phloem through the stele that results from the irregular cambium may also confer an ability for the xylem to recover from cavitation. Photosynthates are distributed throughout the stem (Carlquist, 1988). The parenchyma provides sites for starch storage, since lianes have no other area where it might be stored, but in addition this source of soluble sugars may be important in the recovery of cavitated vessel elements. Sugars transferred into the vessels will increase osmotic pressure thereby encouraging the flow of water back into them. The parenchyma also provides relatively unspecialized cells which may allow regeneration of the vascular tissue through the formation of successive cambia or regeneration after wounding (Dobbins and Fisher, 1986).

6.5.6 Stem parasites

Stem parasites have some characteristics of root parasites, in the development and form of the haustoria which links them to the host and in the reduced photosynthetic capacity of some them. Like root parasites their host provides them with water and nutrients. They have numerous stomata so that a high transpiration pull draws water and nutrients from the host. However it is possible that many stem parasites have evolved from and share many features with ordinary epiphytes. In the field it may be difficult to distinguish them. There are two main kinds of stem parasite: the large bushy aerial **hemiparasites** of trees, especially the mistletoes in the order Santales (families Loranthaceae, Viscaceae and Eremolepidaceae); and the dodders *Cassytha* and *Cuscuta*.

In the mistletoes, the hypocotyl elongates from the seed bearing a tiny radicle on its tip. The radicle tip is covered with papillae which secrete a glue when they touch the surface of the host. The radicle then enlarges to form a cup like sucker from which it penetrates the host tissue. Below the epidermis a branched green callus is formed. In many tropical mistletoes (Loranthaceae and Eremolepidaceae) such as *Plicosepalus* epicortical roots are produced which scramble along the branch, dodder-like, producing secondary haustoria wherever they touch.

One very unusual feature of tropical mistletoes (family Loranthaceae) is the way some mimic the leaf shape of their host. *Amyema linophyllum* is a parasite of *Casuarina* and has leaves shaped like *Casuarina* branches. Other species mimic the phyllodes of *Acacia* (Barlow and Wiens, 1977). Over 75% of Australian mistletoes may be mimics. *Dendrophatae shirleyi* mimics three different kinds of hosts with either flat linear–lanceolate leaves, thick rounded leaves or linear compressed leaves. The mimicry is a kind of camouflage which may protect the palatable mistletoe from predation by possums or butterflies. Alternatively it may encourage the search behaviour of the mistletoe bird which is the major dispersal agent of the seed. It eats the berry-like fruit but the seeds pass through its gut safely. The bird has a larger target since it can search for the host tree.

The evolution of mimicry suggests a close specificity between mistletoe and host. This is certainly true for some host-parasite pairs but many mistletoes, although they have their preferences, may be found on a wide range of hosts. An alternative hypothesis suggests that the mimicry is the result of plastic development, a result of the physiological influence of the host and is not necessarily adaptive (Atsatt, 1983). At its simplest this may explain the differences in leaf succulence of mistletoes growing on different species of mangrove. They are least succulent in the salt-excluder *Rhizophora* and most succulent in the salt accumulator *Lumnitzera*. The physiological influence may be a more subtle hormonal one.

The remarkable similarity between the two genera of dodders *Cuscuta* and *Cassytha* is only superficial, a result of the reduction of the plant to a yellow twining stem with tiny scale-like leaves. Numerous haustoria are produced from the stem. The stems may be produced so profusely that the host is covered. Both *Cassytha* and *Cuscuta* are **holoparasites** since the stem normally lacks chlorophyll. However, the seed germinates in the soil and there is a short green twining phase before parasitic contact is made. *Cassytha* is a rather peculiar member of its primitive family of trees and shrubs, the Lauraceae. Like the tropical mistletoes and many tropical epiphytes it produces a red fleshy berry which is attractive to birds. *Cassytha* is a perennial. If the host enters a dormant phase the stems become green or it follows the host into a period of dormancy. *Cuscuta* is mainly an annual, in its own family the Cuscutaceae, but is closely related to the Convolvulaceae, which includes many twining and climbing herbs. It mostly parasitizes herbs but *Cuscuta nitida* can perennate within the host's tissue, sending out new shoots in Spring (Visser, 1981). There is little host-specificity in either genus.

Adaptive growth forms: the limiting physical environment

7

7.1 WATER RELATIONS
7.2 THE BRYOPHYTES; NON-VASCULAR PLANTS
7.3 GAS RELATIONS
7.4 AQUATIC PLANTS
7.5 NUTRIENT RELATIONS
7.6 SURVIVING ENVIRONMENTAL EXTREMES
7.7 LIFE FORMS

SUMMARY

This chapter is about the adaptations of plants to the physical environment where it appears to be the principal factor limiting growth. If most plants are characterized as mesophytes, there are other plants which show peculiar adaptations to the stresses of particular environmental conditions. Light is one aspect of the environment which may be limiting. Some shade adaptations have been described in Chapter 6. Plants adapted to harsh environments are sometimes stressed when placed in what might seem more equable conditions.

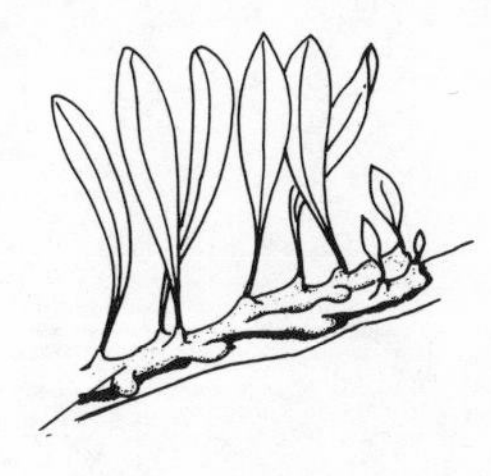

7.1 WATER RELATIONS

Plants can be divided into two sorts on the basis of their water relations with the atmosphere. There are **poikilohydric** plants which are in equilibrium with the atmosphere, dependent on the degree of atmospheric humidity for the degree of hydration of their cells. Alternatively there are **homoiohydric** plants which are independent of air humidity. They usually have roots through which they obtain water.

Poikilohydric plants are not necessarily confined to humid areas. They can survive dry seasons in a highly desiccated state. They include many bryophytes. Few vascular plants are poikilohydric. Notably only a very few pteridophytes occupy droughted areas. However, they are probably restricted as much by their reproductive requirements as much as by the growing conditions. Most, like the Sonoran desert fern, *Notholaena parryi*, occupy shaded habitats and coastal deserts where there is substantial dew fall (Nobel, 1987). *Selaginella lepidophylla*, the resurrection plant has a similar ability to recover from drying like that of many mosses.

Homoiohydric plants may be **xerophytes** adapted to dry areas. They restrict water loss by having a thick cuticle, sunken stomata and in other ways. Many have CAM photosynthesis (section 7.1.1). There is a wide range of xerophytic life forms: **sclerophylls** with small hard leaves; sclerophylls with large thick leathery leaves; plants without leaves; succulents; shrubs and trees which escape drought by being deciduous; phreatophytes which are very deeply rooted into the water table; and ephemerals, which avoid the periods of drought as seeds. Different strategies have evolved in related species. Most African *Acacia* trees lose most of their leaves at the beginning of the dry season but the phreatophyte *Acacia albida* flushes its leaves at this time. Succulent xerophytes stand somewhat outside the classification of plants into poikilohydric and homoiohydric plants. They have their own store of water.

Seasonally dry areas like the mediterranean region have a high proportion of **geophytes**. The plant dies back to a subterranean storage organ. A bulb is formed from tightly packed swollen leaf bases. They may be difficult to distinguish from corms which are mainly the swollen basal region of the stem. Tubers are the swollen apices of stems and roots. Rhizomes may be swollen. There may be a well-developed taproot. The term caudiciform has been applied to plants which survive periodic drought as a massive swollen perennial water storage organ formed from the stem and root (Rowley, 1987). Unlike stem and leaf succulents they put out new photosynthesizing organs each growing season. Even bryophytes with their own particular kind of axis can produce a kind of resting tuber as in the thalloid bryophytes, including *Anthoceros*, *Fossombronia* and *Geothallus*.

Probably the most important class of drought avoiders are the ephemeral plants. They can complete their entire life cycle following the rains, in a very short period, in four weeks in one species.

7.1.1 Succulents

Succulence has evolved in quite unrelated families like the Cactaceae, Euphorbiaceae, Agavaceae, Asclepiadaceae, Aizoaceae, Chenopodiaceae, Crassulaceae, Asteraceae and Apocynaceae (Fig. 7.1). Arborescent succulents have been described in section 6.2.3. Smaller succulents may be barrel shaped or globose, lacking leaves but having spines, with the stem ribbed and having the ability to expand and contract; sometimes called cereiform after the genus of cactus called *Cereus*. Alternatively, there are leaf succulents with normal stems but fleshy leaves. One remarkable form is the pebble plant, *Lithops*, with only two leaves buried in the soil surface. There has been remarkable convergence so that not just a kind but the spectrum of kinds of succulents is repeated in different families. Succulence is important in epiphytes and terrestrial plants where it provides a means of water storage. In halophytes it is a consequence of living in soils with a high salt concentration, and the compartmentalization of excess salt away from the cytoplasm (section 7.4.1).

The epidermis is thick and clear. The outer wall is impregnated with a thick waxy cuticle. Below the epidermis there is a thick and flexible hypodermis. Below this there is a chlorenchyma connected to stomata by

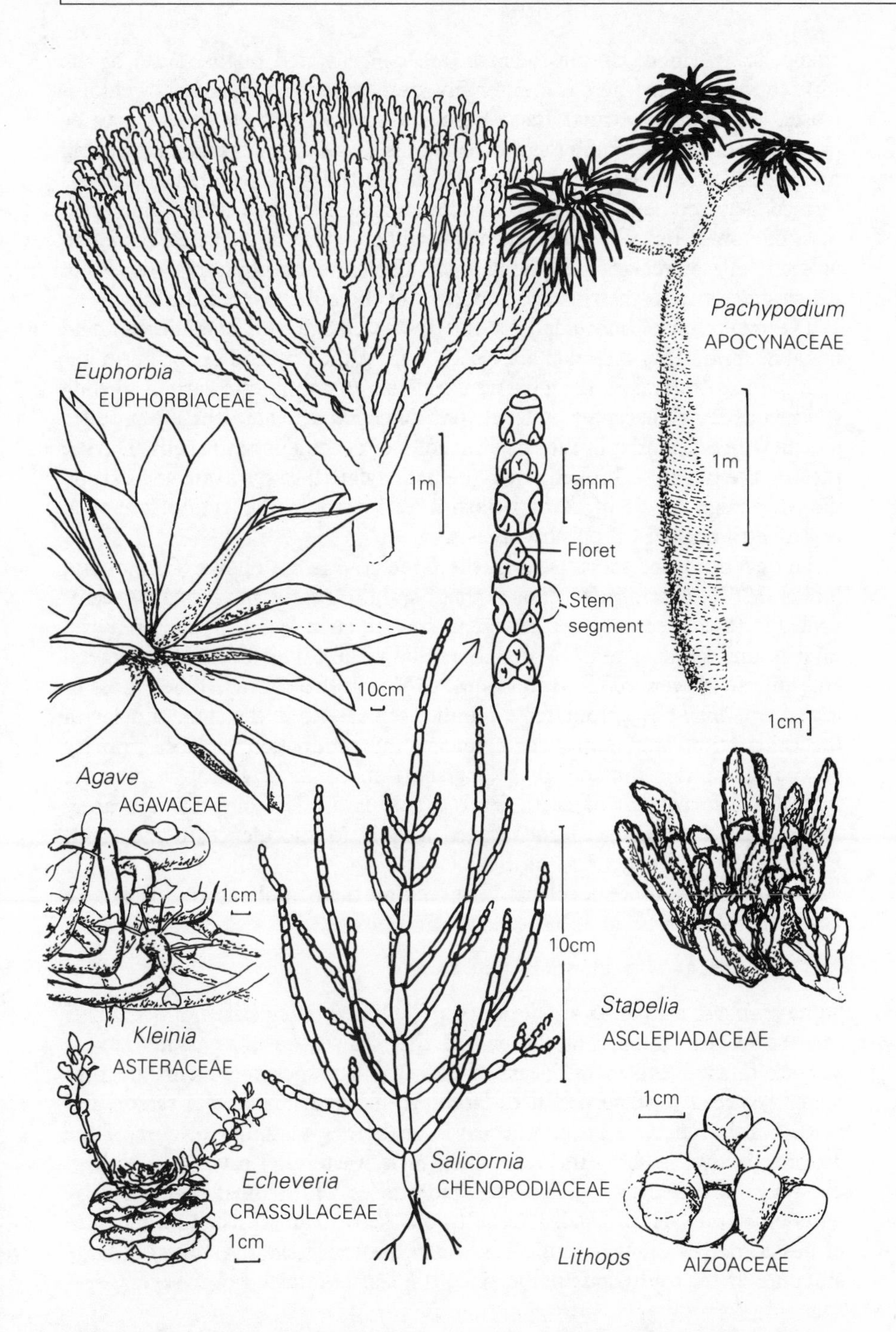

Fig. 7.1. Succulents, xerophytes and halophytes: succulence has arisen in many different families, and there are many remarkable examples of convergence. *Stapelia* in the Asclepiadaceae has the same cereiform habit as some cacti and euphorbs.

channels. In ribbed species stomata are concentrated on the sides of the ribs. In the interior there is a water storage parenchyma which lacks chloroplasts. As well as vascular traces reaching across the cortex there may be mucilage cells. In *Sterculia* the mucilage cells break down at maturity leaving mucilage filled canals. In the cactus *Nopalea* mucilage is secreted into canals formed between cells where the middle lamella is broken down. The great mass of storage tissue develops from a peripheral zone of the apical meristem or from an intercalary peripheral meristem. A subprotodermal meristem which gives rise to the ribs and tubercles of the outer part of the stem.

The evolution of succulence is closely associated with crassulacean acid metabolism (CAM). Stomata are closed during the day and opened at night to minimize water loss through transpiration. Atmospheric carbon dioxide is taken up by an acceptor molecule phosphoenolpyruvate (PEP) to produce malate which is stored in the large vacuoles of succulent tissue cells. During the day the malate is broken down (decarboxylated) to pyruvate and carbon dioxide. High levels of carbon dioxide dissolved in the cytosol are then passed into normal C3 photosynthesis.

The evolution of succulence in the Cactaceae is as follows (Gibson and Nobel, 1986). A primitive genus in that family, *Pereskia*, has leaves. It seems probable that in some ancestral cactus lineage such leaves became reduced in size, and became more fleshy. Eventually the leaves became ephemeral and minute or were reduced to spines. Glochidia are barbed spines. Loss of leaf auxins may have promoted a number of changes in the stem: a delay in the completion of the vascular cylinder; the production of large primary rays; and the development of a long/short shoot arrangement to give the typical areole pattern of cacti. Each areole is a short outgrowth bearing spines. All these changes encouraged a shift to succulence in the stem. Finally CAM photosynthesis arose.

The distribution of succulents is limited by their intolerance of frost.

7.1.2 The leaves of plants in arid areas

Some plants of arid areas are deciduous. Perennial grasses are not deciduous but the leaves die back at the end of the wet season. The plants have a tussock form. Deep in the heart of the 'dead' tussock meristems are protected. *Encelia farinosa* produces large green leaves in the wet season and small highly reflective pubescent leaves in the dry season. However, if the growing season is short the loss of leaves is wasteful. Leaves are retained but they have thick waxy lignified epidermises with stomata protected by hairs or in pits. Hard small sclerophyllous leaves can withstand a measure of desiccation without wilting. It is worth the energetic investment of large amounts of sclerophyllous tissue since the leaf is retained over several seasons.

One of the most important problems that desert plants face is the toleration of very high temperatures in the sun without the ability to carry out transpirational cooling. Large horizontal leaves maximize the reception of thermal energy. Leaves which are small and held vertically are more able to remain near air temperature without transpiration. In the vertical leaves of *Eucalyptus* the palisade layer is found on both sides of the leaf. Some plants have succulent leaves. The leaves of *Agave* and *Yucca* are large and fibrous with a thick skin.

7.1.3 The roots of plants in arid areas

Succulent plants tend to have a shallow root system. In cacti, in the dry season the shallow root system dies back leaving a few weakly branched roots which are well protected by a thick periderm. Following rains, rain roots are produced rapidly from cortical buds, elongating rapidly to take advantage of soil water. Tussock grasses have a remarkable ability to extract water from very dry soils. The uptake of nutrients from dry soils is difficult but experiments on rye-grass, have shown that it can take up water in a moist subsoil and divert it to secrete it in a dry but nutrient rich topsoil (Nambiar, 1977).

In some species of arid environments like *Eucalyptus* and grasstrees (*Kingia*) there are few roots in the uppermost 0.2 m and more profuse development between 0.2 and 0.7 m (Lamont, 1983). Where the roots reach the water table there is a greater concentration of lateral ones. Phreatophytes like *Prosopis* and *Tamarix* are plants which are very deep rooting. They are often found growing along the seasonal water courses, called khors in Africa, and in some ways defy the seasonal drought. Phreatophytes usually have small xeromorphic leaves which can rehydrate rapidly. Some have green stems like *Cercidium microphyllum* from southwestern USA.

In arid areas tree species are widely spaced to exploit large volumes of soil. There is evidence of allelopathic interaction between some species, which maintains the spacing and reduces competition for water.

7.2 THE BRYOPHYTES; NON-VASCULAR PLANTS

One of the most important adaptations of bryophytes is their small size. Even where they grow in the same place as other plants, their own micro-environment is very different. Linking together all bryophyte gametophytes is the fact that they are all to a greater or lesser extent poikilohydric. They have only a very limited ability to control their uptake and loss of water. Another feature which links them is the absence of stomata, which are only present in some of their sporophytes. They have several life-forms which intergrade with each other.

1. Thalloid (liverworts and hornworts)
2. Leafy (liverworts and mosses)
   (a) Procumbent or prostrate, branched
      (i) Mats – adpressed creeping shoots
      (ii) Wefts – loosely interwoven shoots often richly and regularly branched
   (b) Pendant and trailing

(c) Upright
  (i) Turf – not crowded shoots, unbranched or only weakly branched
  (ii) Cushions – more or less hemispherical, shoots radiating from a central point
  (iii) Dendroids – 'tree-like' often arising from a creeping horizontal stem
  (iv) Bog moss (*Sphagnum*)

Particular forms characterize particular habitats. Epiphytes tend to be mat, weft or pendant forms. Turf forms tend to be found in open habitats. Some species are plastic. One major distinction, between upright and adpressed mosses correlates with the position of the reproductive structures: acrocarpous mosses are upright with, usually single terminal sporophytes; pleurocarpous mosses are mostly prostrate with several 'lateral' (i.e. terminal on short side branches) sporophytes.

Bryophytes have traditionally been called non-vascular plants. However some, the endohydric mosses, do have a vascular system which is directly comparable to that of the other land plants, the tracheophytes. Other bryophytes are called ectohydric because water is not conducted internally by a specialized tissue. In terms of their water relations bryophytes can be divided into three sorts: the thalloid hornworts and liverworts; the leafy ectohydric liverworts and mosses; and leafy endohydric mosses. These sorts overlap, so that the Metzgeriales are liverworts with diverse forms from clearly thalloid to clearly leafy. Intermediate forms between ecto- and endohydric, the mixohydric mosses are also defined by some bryologists.

7.2.1 Thalloid liverworts

Thalloid liverworts can be placed in four different orders.

The Metzgeriales have a primitive and simple thallus. They are represented by the fossil *Hepaticites (Pallaviciniites)* from the Upper Devonian. *Pellia* is a living representative (Fig. 7.2). The thallus is thin, one cell thick at the margin but has a kind of thickened midrib, several layers thick. There are numerous rhizoids arising underneath, especially from the midrib region. The rhizoids are usually unicellular and smooth, though the tip of the rhizoid may be swollen or branched. There are no intercellular spaces and all cells of the thallus have chloroplasts. The thallus grows from an apex which branches dichotomously occasionally. Lobed leafy forms within the order illustrate the way leafy liverworts of the order Jungermanniales may have evolved (section 6.5.4). The thalloid representatives have many similarities to some of the gametophytes of pteridophytes.

Another liverwort order is the Monocleales from New Zealand and South America. They have a large thallus up to 20 cm long and 5 cm wide. The thallus is rather like *Pellia*, though the uppermost cells have many chloroplasts while lower layers have few but many starch grains. Brown oil bodies are present.

Another order the Sphaerocarpales, the bottle liverworts, have peculiar

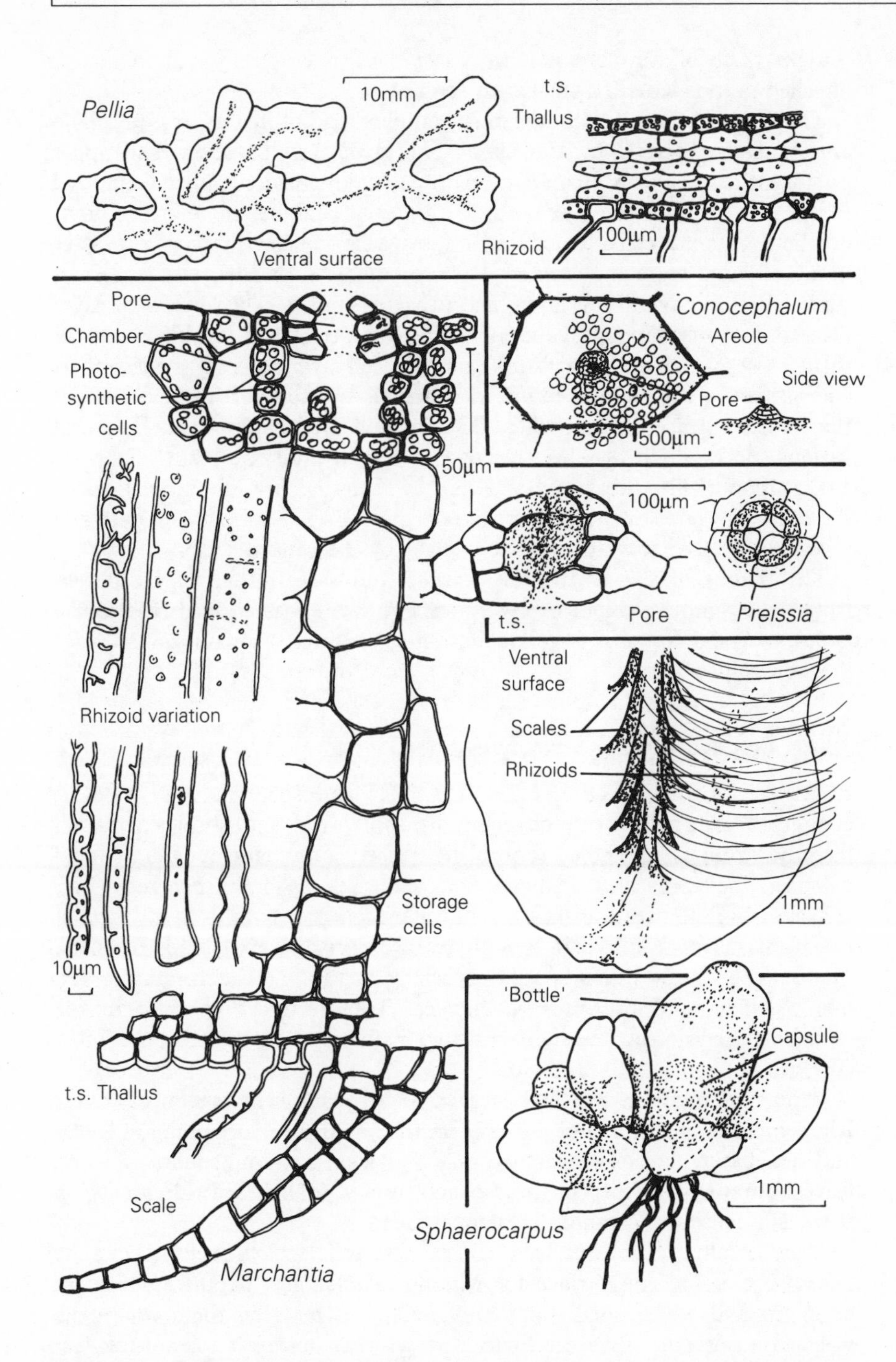

Fig. 7.2. Thalloid liverworts: *Pellia* has a simple thallus; *Marchantia, Conocephalum* and *Preissia* have complex chambered thalli and rhizoids of several sorts; *Sphaerocarpus* is peculiar in producing bottle-shaped structures inside which the capsules arise.

upright thalli of various sorts. In *Riella* there is a spirally wound thallus attached on one side to a thickened stem.

The Marchantiales exhibit the most complex kind of thallus. It is multilayered and chambered. In *Marchantia* two layers can be seen. The upper photosynthetic layer is chambered with each polygonal chamber connected to the exterior by a complex pore. Within each chamber there are columns of photosynthetic cells. Below the photosynthetic layer there is a massive tissue of large storage cells and a few oil bodies. In *Preissia* the pore is surrounded by several cell layers and is barrel shaped. The lowest cell layer projects into the pore and gapes open when the thallus is turgid. When the thallus loses water the pore shrinks and this cell layer seals off the chamber. The surface of the plant is cutinized. Rhizoids and rows of scales arise from the ventral surface of the thallus. The rhizoids of the Marchantiales are of various sorts, which may be present together in the same plant. They are either smooth, like those in *Pellia*, or have internal peg-like projections of the cell wall. The smooth ones penetrate the soil. The tuberculate ones run together like a wick which is held in place by the ventral scales.

The thallus of the hornworts, Anthocerotopsida, is not very different from some simple thalloid liverworts except that it has a lobed rosette-like appearance and contains large mucilage-filled cavities, probably a device for the storage of water.

7.2.2 Ectohydric leafy bryophytes

In leafy liverworts (section 6.5.4; Fig. 6.9) and some mosses, water is conducted externally over the plant, by capillarity, and absorbed or lost directly through the whole of the plant surface. The plant body is fine and delicate. The leaves are usually only one cell thick, as in *Plagiothecium* (Fig. 7.3) allowing the easy transmission of water from the exterior to all parts of the plant. Many ectohydric bryophytes grow in dense cushions or mats. Many are prostrate. This is seen especially in the **pleurocarpous** mosses, like *Hypnum*, and the Jungermanniales. The liverwort *Herberta* in the Jungermanniales grows upright in a turf as if it was an **acrocarpous** moss (Fig. 6.9).

Various adaptations can be observed to help the transmission of water. These include the folding of leaves (*Fissidens*), and the overlapping of leaves and leaf bases. Various structures act as wicks, including hairs, divided leaves (**paraphyllia**), tufts of rhizoids and tufts of branches. In *Aulocomnium* there is a dense felt coating the stem surface.

Many ectohydric mosses have conical leaf cells which project from the lamina increasing the surface for wetting. *Tortula* has papillae which are cutinized and which shed water into the spaces between them where the water is absorbed. Many ectohydric mosses are amazingly tolerant of desiccation. *Tortula muralis* can survive for 10 months without water and then revive within a few hours. Tolerance of desiccation is helped if drying is slow. Growing in tight clumps and cushions helps slow the process of drying. The presence of long, hyaline leaf tip hairs (**aristae**) as in *Rhacomitrium* also

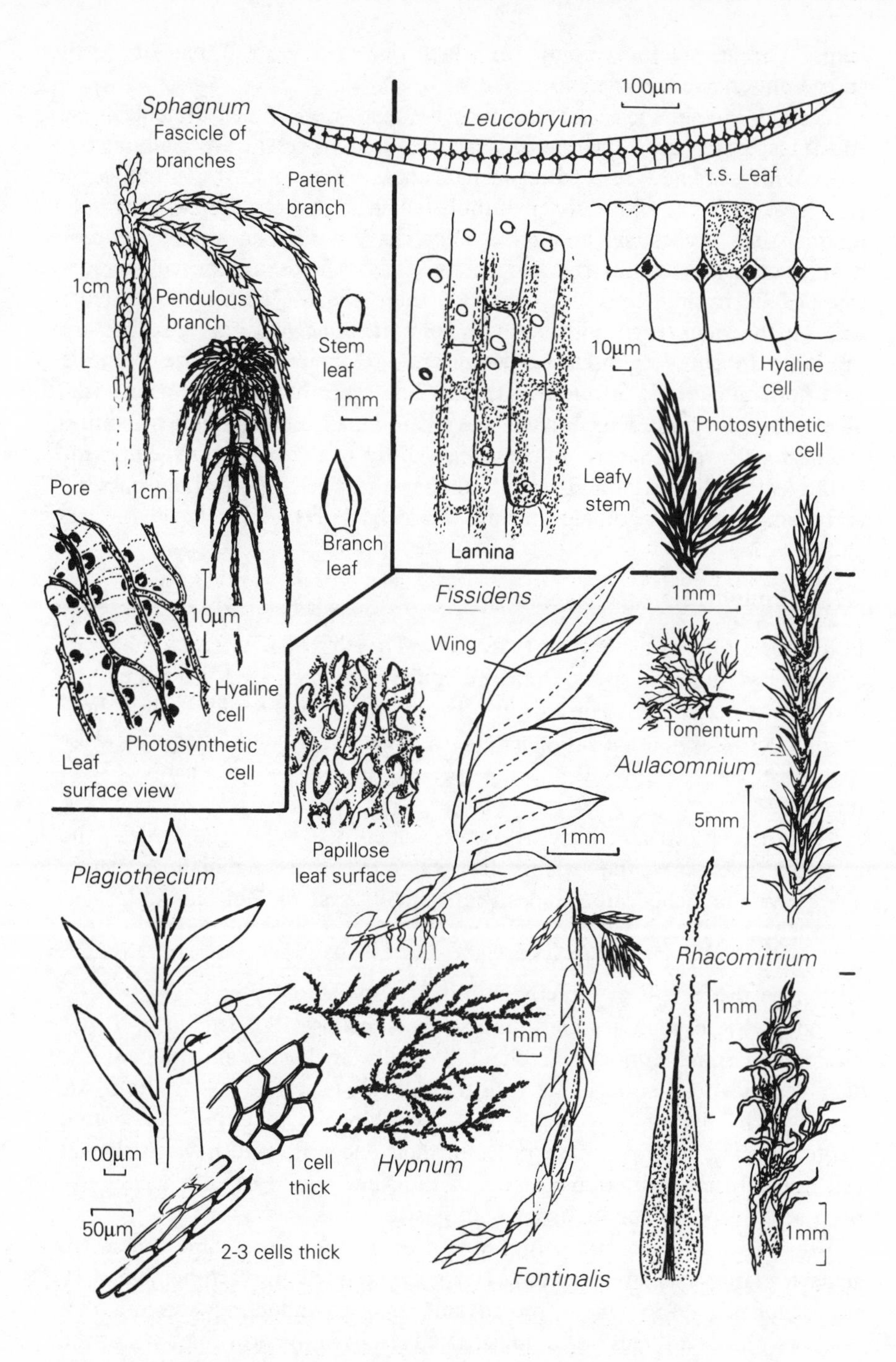

Fig. 7.3. Ectohydric mosses: *Leucobryum* and *Sphagnum* have leaves with large hyaline water storage tissues; many mosses have a papillose surface or a tomentum which helps them absorb water.

helps. The aristae form points on which dew can form. When dry they spread out and reflect the sun.

In many mosses the lower part may be dead. Cells which are colourless and transparent called hyaline cells are frequently present especially at the bases of leaves. They are dead cells which act as a water reservoir. Frequently they have a pore to allow any air bubble which forms to escape as they take up moisture. *Sphagnum* and *Leucobryum* have a regular arrangement of hyaline cells throughout the leaf so that each photosynthetic cell is surrounded by hyaline cells. In *Sphagnum* there are also large hyaline retort cells on the stem, each with a single pore. In *Sphagnum* the branches are produced in bundles, fascicles. There are two kinds of branches. One is held horizontally and is photosynthetic. The other hangs down as a wick. The stem and pendulous branch leaves are small and clasping providing capillary pathways. Species of *Sphagnum* differ in the relative development of the leaf hyaline cells and in the number and size of pores they possess, correlating with how aquatic they are (Daniels, 1989).

7.2.3 Endohydric bryophytes

Endohydric mosses are robust (Fig. 7.4). They have an extensive network of capillary rhizoids arising from the epidermis which feed water into the cortex of the stem. The rhizoids are covered with small papillae which help the uptake of water. In the centre of the stem is a conducting strand or stele. Leaf traces are found in some species but do not always connect with the stele.

There is one order of leafy liverworts which is possibly endohydric, the moss-like liverworts, the Calobryales, which even have a radial symmetry. They have a branched subterranean rhizomatous system. Rhizoids are absent but there is a fungal associate which grows outside the rhizome as well as within the cortex, where there are abundant mucilage cells. These cells may encourage the fungus or directly help in the uptake of water.

Endohydric bryophytes are found in the subclasses Polytrichidae, Tetraphidae and some Bryidae. Tracheid-like cells, hydroids, and phloem-like cells, leptoids, can be recognized in the central stele. The xylem like tissue is called the hadrom (hydrome) and the phloem like tissue the leptom (leptome). The hydroids have pores in their end walls and have thickened cell walls. In the subterranean part of the stem a kind of endodermis has been recognized. A conducting strand is also found in the seta.

The presence of a conducting system evidently allows some mosses to achieve a large size. Free standing *Dawsonia superba* can reach heights of 60 cm (Schofield, 1985). One of the advantages of a conducting system is that it allows photosynthates to be translocated easily to the growing apex of the plant. It also allows the body of the plant to be thicker since diffusion alone from the surface is not the only source of water and nutrients. However, as well as absorbing water through their rhizoids, endohydric mosses also absorb water through the surface of the leaves and are therefore more

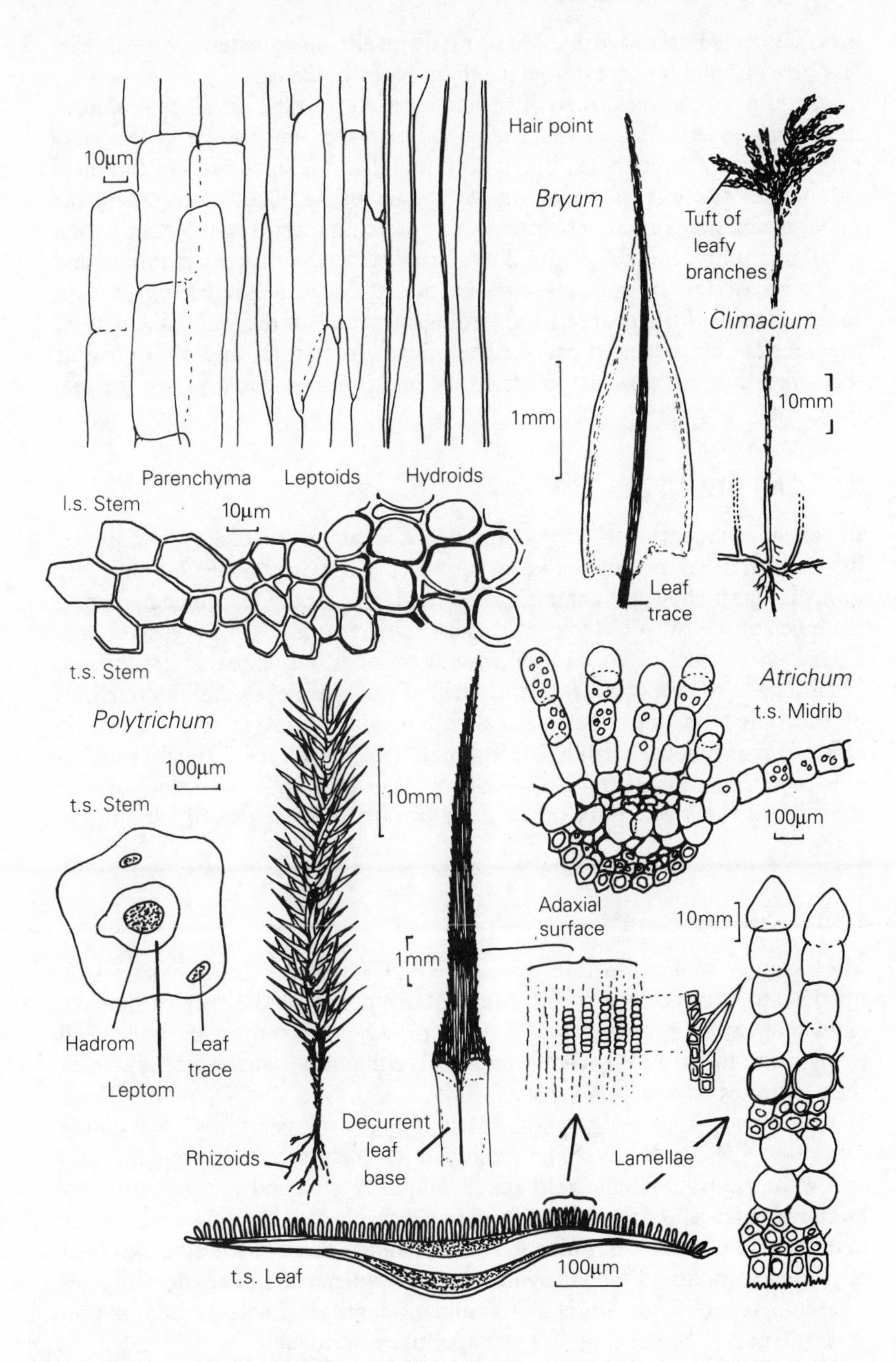

Fig. 7.4. Endohydric mosses: *Polytrichum* and *Atrichum* have chlorophyllous lamellae which run down the stem; the upright shoots of *Climacium dendroides* arise from a creeping horizontal stem.

properly called mixohydric. Many of them also have extensive decurrent leaf bases providing an external capillary path for water.

The kind of complex internal structure of intercellular air spaces connecting to the exterior via stomata of other plants is not known in the leafy gametophyte of bryophytes but can be found in the apophysis of the moss capsule. In the leaf of *Polytrichum* a system of lamellae, cutinized in the upper region may function to maintain a humid atmosphere analogous to that in the mesophyll of other plants. The abaxial surface is strongly cutinized and as the leaves dry the margins curl over and the apex of the lamellae touch, thus shutting off the exposed non-cutinous surfaces of the leaf. In *Atrichum*, the lamellae are concentrated on the midrib, the costa, and it is possible that here they function in the transmission of water down or up the leaf (Fig. 7.4).

7.3 GAS RELATIONS

In hot environments the lower solubility of CO_2 and the necessity of minimizing exposure to atmosphere to prevent water loss may make CO_2 limiting to growth. Plants show leaf anatomies and stomatal adaptations which maximize the efficient use of CO_2. One of the most important is CAM photosynthesis.

Oxygen is most often limiting in aquatic or waterlogged environments. Waterlogged soils may be black and smell of sulphide because of the activities of anaerobic reducing bacteria. An environmental limit to growth is the toxic effect of manganous, ferrous and sulphide ions which are formed in anaerobic conditions. Plants from aquatic habitats or growing in waterlogged soils show adaptations which maximize the diffusion of oxygen to all parts of the plant.

7.3.1 Kranz anatomies

Most plants from temperate regions have C3 photosynthesis where CO_2 uptake and transport is by diffusion. Many plants in the tropics have an extra 'loop' in the carbon-fixing cycle where CO_2 is taken up in the mesophyll by pyruvate to form a 4-carbon compound, either aspartate or malate, which then migrates to the bundle sheath. This is a kind of CO_2 pump which increases the efficiency of photosynthesis at higher temperatures, overcoming the lower CO_2 solubility. The ability of C4 plants to scavenge for CO_2 at low concentrations allows them to minimize stomatal aperture thereby minimizing transpirational loss of water. C4 plants also have a greater ability to utilize high levels of illumination. It also allows them to counter the effect of photorespiration. C4 photosynthesis is advantageous in hot, dry climates where there are high levels of illumination but in cooler shady humid environments it loses some of its competitive advantage.

C4 photosynthesis is associated with a particular kind of leaf anatomy which has been called Kranz anatomy because of the prominent bundle sheath (Fig. 8.2). *Kranz* is the German for wreath. Non-kranz plants have irregular chlorenchyma cells with more than four cells between the sheaths of adjacent vascular bundles. Kranz plants have only 2–4 cells between adjacent sheaths. There are several kinds of C4 photosynthesis and anatomies. In the grasses, for example, there are two main kinds of anatomy: parenchyma sheath (PS) and and mestome sheath (MS) associated with aspartate and malate as the 4-carbon compound respectively. There are a number of variant anatomies of each (Clayton and Renvoize, 1986). PS grasses have a chlorenchyma which radiates strongly around a double bundle sheath. The outer sheath arises from the mesophyll and the inner one, the mestome sheath, is a kind of endodermis around the vascular bundle. Two subvariants are those with chloroplasts to the inner or the outer margin of the sheath cells. They have Nicotinamide Adenine Dinucleotide Malate Enzyme (NAD-ME) and Phosphoenolpyruvate carboxykinase (PEP-CK) as the decarboxylating enzyme respectively. The MS grasses have rather irregular chlorenchyma and lack the outer parenchyma sheath.

7.3.2 Stomata

Stomata (singular = stoma) are present in all land plant groups except liverworts. Thalloid liverworts with a complex chambered thallus have analogous structures. There is a remarkable similarity between the stomata of very different groups. The pore has beneath it a substomatal chamber which connects to intercellular air spaces. The pore is surrounded by two guard cells which arise from the division of a single cell. They are usually elliptic to sausage shaped. The pore is formed in the central portion of the wall between the guard cells where the middle lamella breaks down. The cell wall around the pore is thickened and has a cuticularized ledge. The pore opens when the guards cells inflate, bulging away from each other. The stomata of grasses and sedges are dumbbell shaped and thickened on both sides of the thin part connecting the ends. These kind of stomata open by the ends inflating.

In most cases the stomata are surrounded by special epidermal cells called subsidiary cells which are part of the stomatal complex. These may look like other epidermal cells (anomocytic) or be variously distinguished or arranged. In some cases the subsidiary cells arise from the same meristematic cell as the guard cells (mesogenous).

Stomata may be found on all parts of the aerial part of the plant. In leaves they are often confined or in a much higher frequency on one surface of the leaf, usually the abaxial surface but the adaxial surface in grasses and floating plants. They are not found in roots. The stomata of the mosses and hornworts are found in the sporophyte only.

The guard cells of bryophytes are apparently largely non-functional (Martin *et al.*, 1983), because the aperture of the stoma does not change very much. Those of pteridophytes are sluggish. However, in the gymnosperms and angiosperms the guard cells are highly responsive to environmental changes and endogenous signals. They will react to changes in light opening slowly at dawn and closing at dusk. Cloud cover may induce partial closure. They will respond rapidly to changing internal CO_2 concentration, changes in temperature, changes in leaf turgor and in air humidity. Stomata also respond to endogenous rhythms. In the tropics it is not unusual for stomata to close briefly at midday. CAM plants show a reversed daily response, opening at night.

Surrounding the surface of the plant there is a shell of humid air called the boundary layer. The rate of transpirational water loss from the stomata when they are open depends in part on the steepness of the humidity gradient of the boundary layer. Stomata, especially in xerophytes, may be protected in chambers, pits or grooves in the surface of the plant or surrounded by epidermal hairs making the boundary layer thicker. In the marram grass, *Ammophila arenaria*, the stomata are in grooves in the adaxial surface. As the leaf dries it rolls up to protect the stomata further. In some species of *Carex* the stomata are abaxial and leaf rolling exposes them further.

Stomata can provide the pores through which water is exuded in hydathodes, and nectar in nectaries.

7.4 AQUATIC PLANTS

Hydrophytes are aquatic plants. They may be rooted or free floating but they all rely on water for mechanical support.

Rooted aquatics often have a rhizome with many adventitious roots. Leaves with long petioles may arise from the rhizome. The long petioles allow the leaf laminas to float on the surface. The rhizome and root system is particularly well developed in the sea-grasses which are subject to tides and wave action.

In still or gently flowing water, the floating leaves are commonly rounded in outline with the petiole inserted centrally. Stomata are present only on the upper surface. Examples are the aquatic ferns *Marsilea* and *Regnellidium* and the water-lilies. The huge leaves of *Victoria amazonica* are strengthened by a strong ribbing along the veins (Fig. 7.5). In faster flowing water or tidal systems surface leaves are absent.

Submerged leaves are often finely divided and linear as in the popular aquarium plant *Cabomba*. In the fern *Pilularia* the filiform fronds lack a lamina altogether. The cuticle is absent or weakly developed and there are no stomata. This increases the surface area available for gaseous exchange, and the surface area volume ratio. Mechanical stresses in flowing or running water also encourage a linear growth form. The sea-grasses like *Zostera* and *Posidonia* have relatively tough, flexible ribbon-shaped leaves. In the Mediterranean the tough fibrous leaves get broken off and rolled up into fibrous balls called 'nuns' farts' in some parts of Spain where they wash up on the beach. The leaves are relatively thick with large amounts of aerenchyma. Chloroplasts are confined to the epidermis (Dawes, 1981). Some aquatic plants like the aquatic buttercups, *Ranunculus*, have plastic development and produce both submerged and floating leaves.

Many free-floating aquatics are small so that surface tension and a film of air is enough to support them (Fig. 7.6). *Lemna*, duckweed, has a floating lobed thallus or frond with downward hanging roots. The related *Wolffia*,

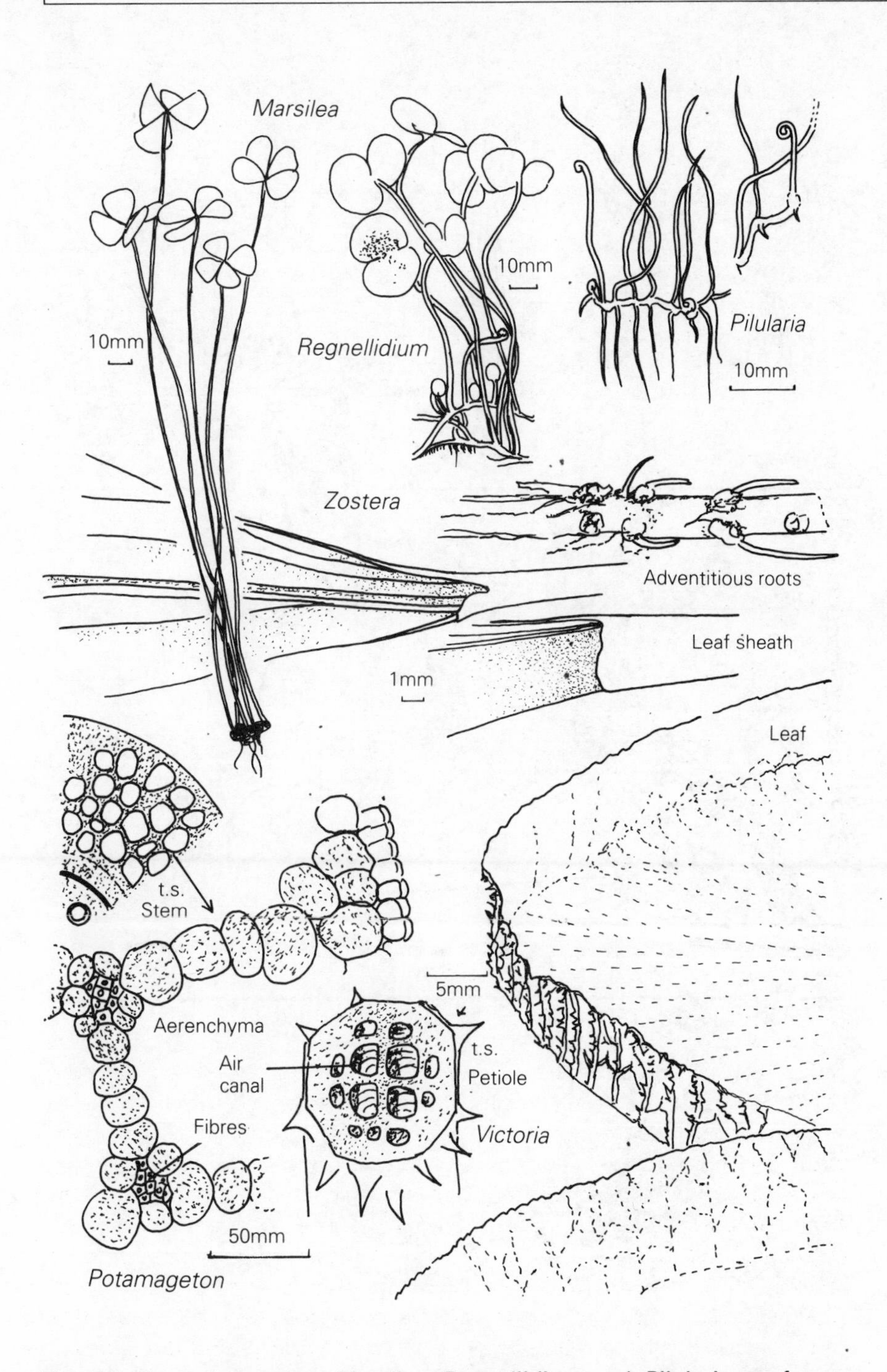

Fig. 7.5. Rooted aquatics: *Marsilea*, *Regnellidium* and *Pilularia* are ferns; *Zostera* is a marine aquatic; *Potamageton* and *Victoria* have aerenchyma in their stems or petioles; the huge leaves of *Victoria*, which may have a diameter of 2 m, are strengthened by a pattern of ribs.

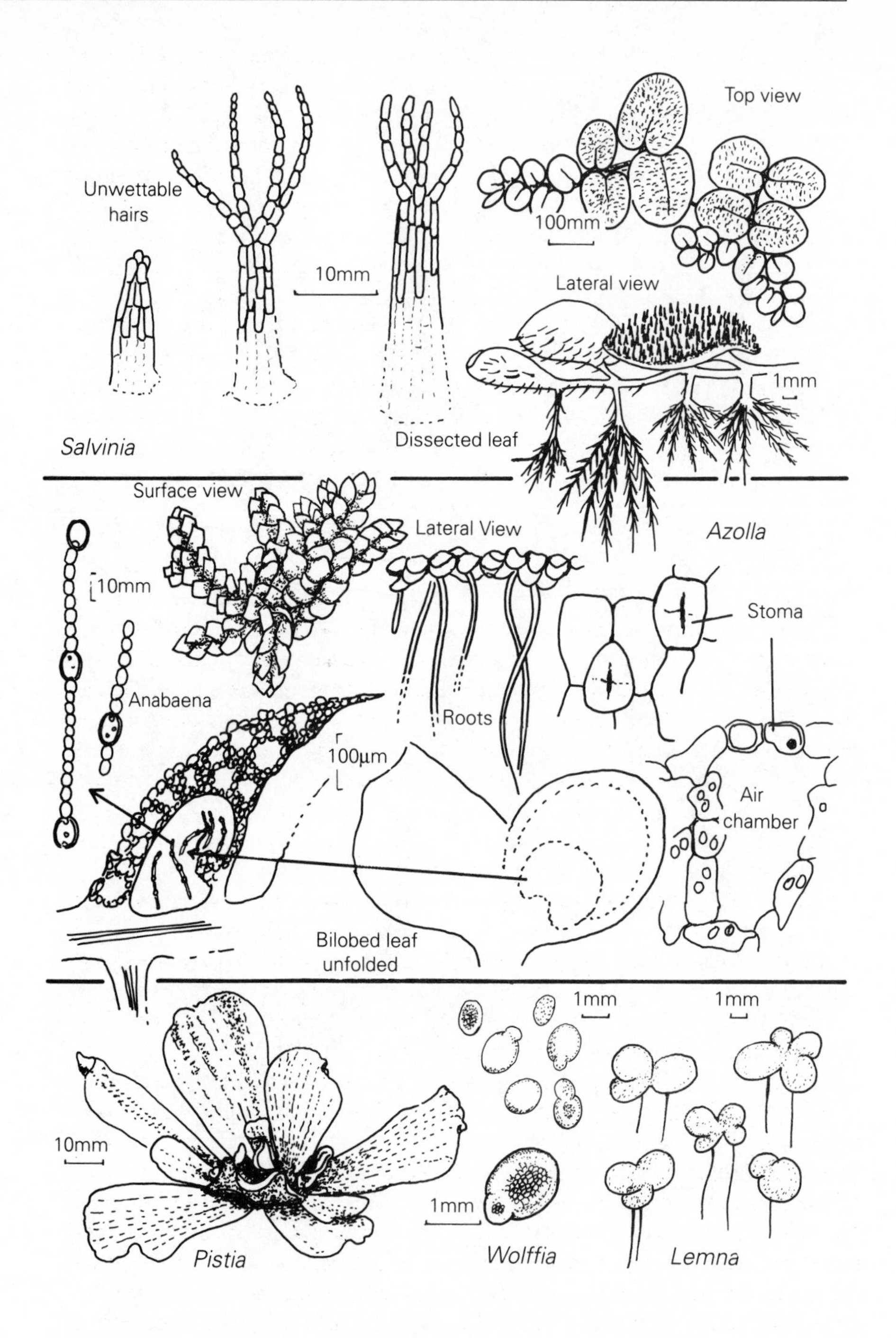

Fig. 7.6. Floating aquatics: *Salvinia* and *Azolla* are ferns; *Azolla* has the blue-green alga *Anabaena* as a symbiont; *Lemna* and *Wolffia* are tiny aquatics in the same order as *Pistia*, the Arales.

is the smallest angiosperm. It is subspherical without roots. Other important genera are the aquatic ferns *Salvinia* and *Azolla*. *Salvinia* lacks roots. It has leaves in threes one of which is submerged and finely divided, looking like a root but acting more like a sea anchor than an absorptive root. The surface of the floating leaves have branched unwettable hairs which hold a film of air so that *Salvinia* is unsinkable. *Azolla* has a folded bilobed leaf and roots. Air is trapped between the lobes and in aerenchyma. The stomata look rudimentary.

Other free floating aquatics are larger. In *Eichornia*, the water hyacinth, the petiole is inflated as a float. *Pistia* relies on broad flat unwettable leaves. They have short, determinate roots which hang down and act as a sea anchor to stabilize the plant. A common feature of free-floating aquatics is that they have stems which disarticulate to allow free floating parts to disperse. This allows them to reproduce vegetatively very effectively. *Eichornia* has become a troublesome weed in many waterways where it has been introduced.

7.4.1 Halophytes

Halophytes grow in saline conditions. Most halophytes are saltmarsh or mangrove species. Mangroves are trees which grow in the intertidal zone. They come from a wide range of families and most have fully inland relatives. Important taxa are the Rhizophoraceae and Sonneratiaceae and the genus *Avicennia* in the Verbenaceae. There is one important species of mangrove fern called *Acrostichum*. Saltmarsh species come from a wide range of families but the Chenopodiaceae, which includes *Salicornia*, are particularly important. *Salicornia* is unusual in including many annual species. Other saltmarsh plants like *Limonium* or *Spartina* are perennial and produce extensive rhizomatous growth which anchors the plant.

Coastal soils have a high salt content but halophytes are adapted in various ways to their saline environment. Salt may be excluded; the aerial surface of the plant has a thick tough epidermis and the root cells actively exclude salt. The mangrove species *Rhizophora* and *Avicennia* have an ultrafilter in the roots where sodium is actively pumped out. The saltmarsh grass *Puccinellia* has a double endodermis in its roots (Crawford, 1989).

Salt glands which actively excrete salt are found in the leaves of a range of halophytes in the families Plumbaginaceae, Tamaricaceae and Chenopodiaceae, and the mangroves *Avicennia* and *Sonneratia*. The gland may be hair-like as in *Atriplex*. This gland has a temporary life releasing a concentrated salt solution by rupturing. A two-celled gland is found in the saltmarsh grass *Spartina*. In *Avicennia* it consists of 2–4 collecting cells, a stalk cell and eight secretory cells. In the Plumbaginaceae it is a complex 16-celled structure buried in the surface of the leaf. The gland cells lack a vacuole. Casparian-strip-like regions isolate the cell wall, requiring all passage of materials to be through the **symplasm**.

Many halophytes are succulent or have other xerophytic characteristics. *Salicornia* has tiny scale leaves pressed close to a branched succulent axis which is a combination of the stem and petiole. The succulent tissue has cells with large vacuoles in which salt solution is stored. The saltmarsh plant *Suaeda vera* has succulent leaves with large water-storing mesophyll cells. The degree of succulence may increase over the growing season as salt accumulates (Crawford, 1989). Some mangroves accumulate salt in the tissues, especially the bark and leaves. Others like *Lumnitzera* accumulate salt in a large-celled hypodermis (Saenger, 1982). Some species have isobilateral leaves with a large undifferentiated water storing mesophyll. Large storage tracheids may be present. They are irregular with spiral or reticulate thickening and flange-like connection to the hypodermis.

Succulence is only a temporary refuge from salinity. It works in annual species, and in perennial species through leaf fall. Many saltmarsh perennials die back to a rhizome or rootstock in winter. In mangroves growth and leaf loss is a continuous process.

7.4.2 Aerenchyma

There is a strong species zonation in saltmarshes and mangrove swamps. Marsh, saltmarsh and mangrove muds are fine-grained, semi-fluid and ill-consolidated. Though nutrient levels are high, with abundant humus and calcareous material, the soils have a low oxygen content or are anaerobic. The most important environmental factor for mangroves is the low oxygen content of mangrove soils. Hydrophytes, marsh plants and mangrove species contain a large proportion of aerenchyma in their stems and roots which provides a path for atmospheric oxygen to the oxygen-starved areas (Fig. 7.5). Herbaceous bog or marsh plants like sedges and rushes also have aerenchyma in the root cortex, sometimes forming a complex network of air channels. Roots of woody plants growing in waterlogged soils often have large air cavities in the stele.

Aerenchyma arises in different ways in different groups. In *Caltha palustris* the aerenchyma forms by cells splitting apart, schizogeny, from the enlargement of normal intercellular air spaces. The cells may become star shaped or lobed as in *Juncus*. In others like *Ranunculus flammula* the aerenchyma forms by the breakdown of cells, lysogeny, leaving lines of broken cells as the walls of the air channels (Smirnoff and Crawford, 1983).

The mangrove environment is unstable because of wave action and so species on the margin of the swamp have massive root and lower stem platforms supporting a relatively small canopy (Fig. 7.7). The trunks may have plank-like buttresses or stilt roots which are covered with lenticels. The stilt roots reach down to about 30 cm below the surface where they branch to produce thick anchoring roots and very fine absorptive roots. In the aerial portion of the roots chlorophyllous cells are present, the cells have thick cell walls and **trichosclereids** are present. Here only 5% of the root is gas space. Below ground the roots have a spongy parenchyma and 50% of the root is gas space. Aerial roots hanging down from the branches also increase the surface area for lenticels. Where the aerial roots enter the mud they produce a tuft of lateral rootlets.

Surface roots which are exposed in part of the tidal cycle are present in

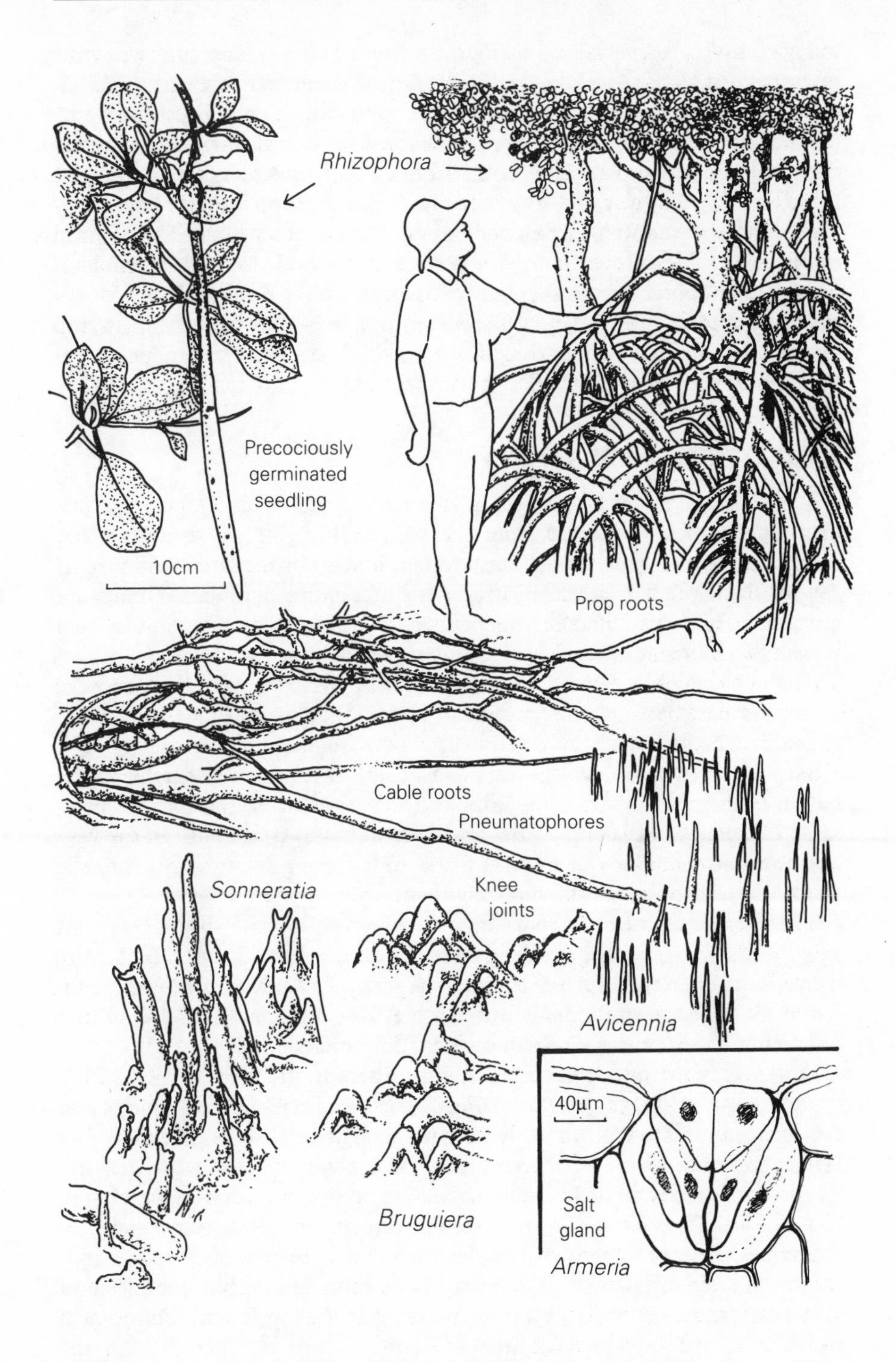

Fig. 7.7. Adaptations of coastal plants: root systems of mangroves; vivipary in *Rhizophora*; and one kind of salt gland (*Armeria*).

many relatively unspecialized mangroves. Some like *Aegiliatis* prefer to grow on a rocky surface. *Avicennia*, *Rhizophora* and *Lumnitzera* are more specialized. They have a system of main cable roots which may occasionally be exposed and small downward growing anchoring lateral roots. A horizontal densely branched superficial network of fine absorptive roots arises from the cable roots. From the cable-roots negatively geotropic **pneumatophores** may arise. These are short stiff roots of various forms, either finger-like, conical or plate like. They project above the surface of the mud. As mud accumulates the pneumatophores grow keeping the apex above the mud surface. Conical pneumatophores are found in *Taxodium distichum*, the freshwater swamp cypress. In some plants like *Bruguiera* the cable roots bend to produce knee-like joints, with many lenticels, above the surface of the mud.

7.5 NUTRIENT RELATIONS

Typically dicotyledon herbaceous plants have a single large taproot or primary root which is derived from the radicle (Fig. 7.8). The taproot has numerous lateral roots which bear other higher order laterals. In some species the taproot is weak or atrophies but a number of sinker roots are produced from the laterals. Monocotyledons usually have a fibrous root system of abundant adventitious roots produced from rhizomes.

The soil is exploited in a very organized way by some species which have a very regular 'herringbone' pattern of root branching and long root hairs. However, the soil is a very heterogeneous environment. Most species have a root system which develops very plastically to exploit the patchy distribution of nutrients so that it is difficult to observe an overall root architecture. The density of laterals varies at different depths in the soil. In the very low nutrient conditions of tropical forest soils roots grow upwards into the trunks of dead but still standing neighbours scavenging nutrients.

The average diameter of roots varies between those with magnolioid root system and those with graminoid systems (see section 2.3.8). Root hairs increase the epidermal surface of the roots many times over; 5.8x in *Leucadendron laureolum* from the fynbos of South Africa (Lamont, 1983). Root hair production is suppressed in high nutrient conditions.

Different temperate tree species adopt different strategies for exploiting the soil in a deciduous woodland. *Fagus sylvatica*, exploits a volume of soil very concentratedly with short densely branched lateral roots (Packham and Harding, 1982). However it is sensitive to drought. *Fraxinus excelsior* has longer, moderately branched laterals and exploits a greater volume of soil. It is a pioneer species and forms secondary woodlands (Rackham, 1980). In most trees the root system is very shallow. This is easy to observe in wind-thrown species of all sorts. The bulk of the roots are within one metre of the surface and most of the absorbing roots are in the top 15 cm (Zimmerman and Brown, 1971). However, the roots may reach out many times the diameter of the crown. Root competition in the crowded conditions typical of a wood may greatly curtail this lateral expansion.

Relationships between the roots of land plants and other soil organisms

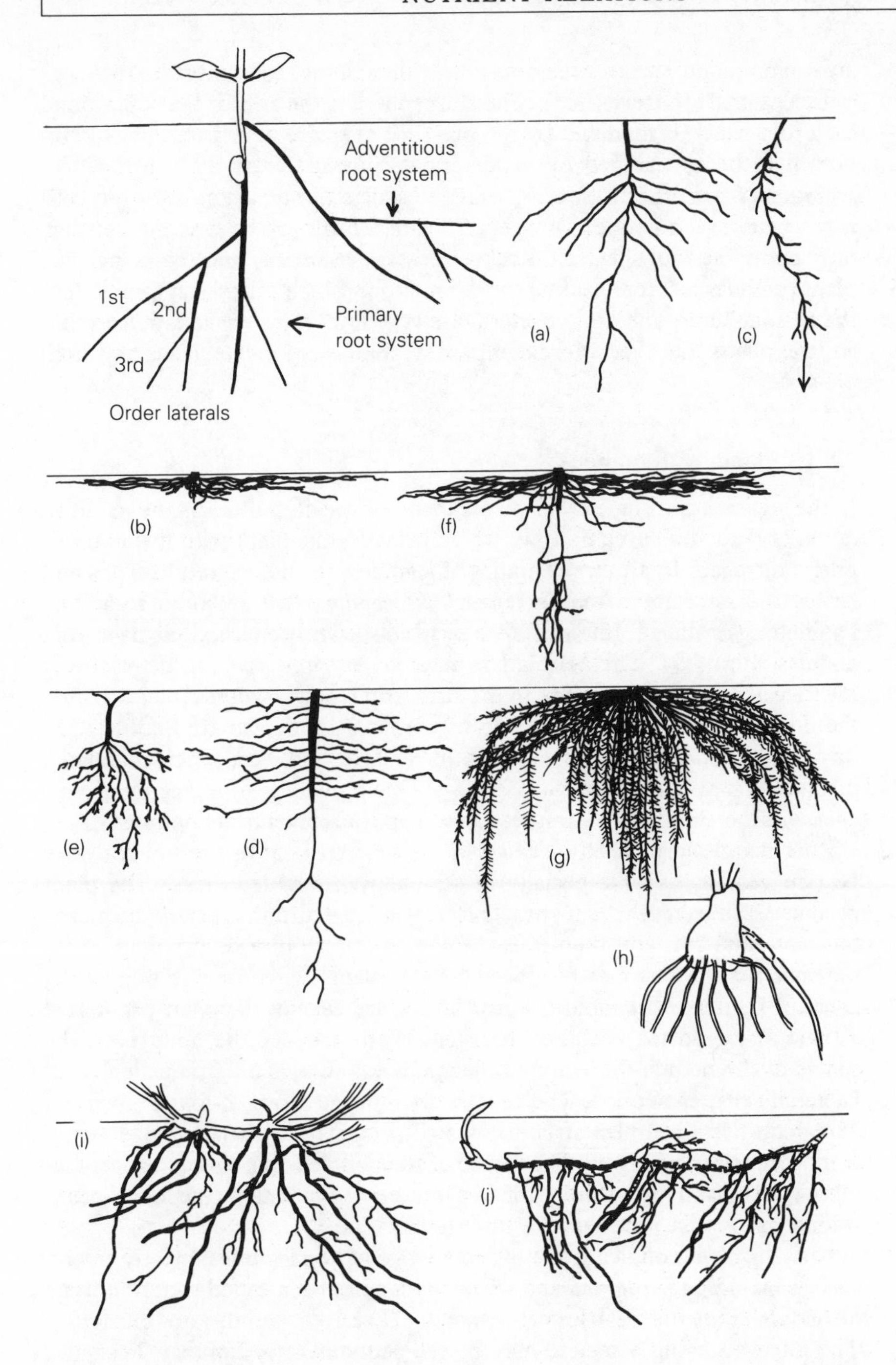

Fig. 7.8. Root systems (from Cannon (1949)): (a)–(f) mainly or only primary root systems; (g)–(j) mainly adventitious root systems. (a) diffuse, (b) superficial roots only, (c) deep rooting with weak laterals, (d) like 1 but with a strong tap root, (e) rooting from a subterranean caudex, (f) superficial roots and some strong deeper roots, (g) dense herringbone pattern characteristic of crop grasses, (h) arising from a storage organ, (i) arising from a stolon, (j) arising from a rhizome.

are so common that the situations where they do not exist must be regarded as exceptional (Jeffrey, 1986). The rhizosphere is the area of the soil around each root which is modified by the presence of the root. A particular microflora may be encouraged by exudates like mucigel from the root apex or secretions from other root cells. Simple leakage of nutrients from root cells may encourage microbes. A special relationship may be present between nitrogen-fixing root surface microbes such as *Azotobacter* and the plant. The plant provides a carbon source for the microbe which releases ammonia into the **rhizosphere**. The cyanobacteria are very important for aquatic or semi-aquatic plants like rice. More complex symbioses are mycorrhizae and nodule-systems.

7.5.1 Plants with nodules

In the legumes and some other plants the presence of **nodules** is the manifestation of a microbial relationship which provides the plant with fixed atmospheric nitrogen. In all three families of legumes, including trees, shrubs and herbs, the associate is the bacterium, *Rhizobium*. In the Mimosaceae and Papilionaceae almost 100% of species which have been checked have root nodules (Fig. 7.9). The association may be very important in determining the success of the Mimosaceae in the semi-arid tropics, where *Acacia* is often the dominant tree. A total of 30% of the Caesalpiniaceae are nodulate. In this family nodulation is uncommon in rainforest species. Often nodulated plants also have a mycorrhizal associate. One non-legume, the genus *Parasponia*, in the elm family Ulmaceae, also has *Rhizobium* in its nodules.

Root nodules are usually located in the upper 10 cm of the soil and may be renewed each year especially in seasonally droughted areas. The plant produces a substance which attracts *Rhizobium*. In turn the *Rhizobium* makes the root hairs curl and then infects them. Infection threads proliferate into the root tissue. Flavones produced by the plant switch on the nodulating genes of the *Rhizobium* and in turn the cortical cells of the plant proliferate as part of the cortex becomes meristematic to produce the nodule. In the centre of the nodule there are swollen cells which contain strangely shaped bacterial cells, bacteroids. The cells are rich in the nitrogen fixing enzymes. The nodule is a complex structure which protects the nitrogenase enzymes from atmospheric oxygen. The centre of the nodule is pink with leghaemoglobin which has a high affinity for free oxygen and maintains the low concentrations of oxygen necessary for nitrogen fixation.

Non-rhizobial nodules are found in a range of non legumes, mostly woody species including *Casuarina* and *Myrica*. This has been called the *Alnus* type of nodule from the best known example. The root nodules are clustered. The nitrogen fixing associate may be the actinomycete, *Frankia* (Postgate, 1978). This kind of association has evidently evolved many times over or has been lost many times. It is not even constant within a single species. The nitrogen fixing blue–green algae *Anabaena* and *Nostoc* are common associates of land plants. *Nostoc* is found in the massive coralloid nodules of the surface roots of cycads. *Gunnera*, the familiar flowering plant with huge

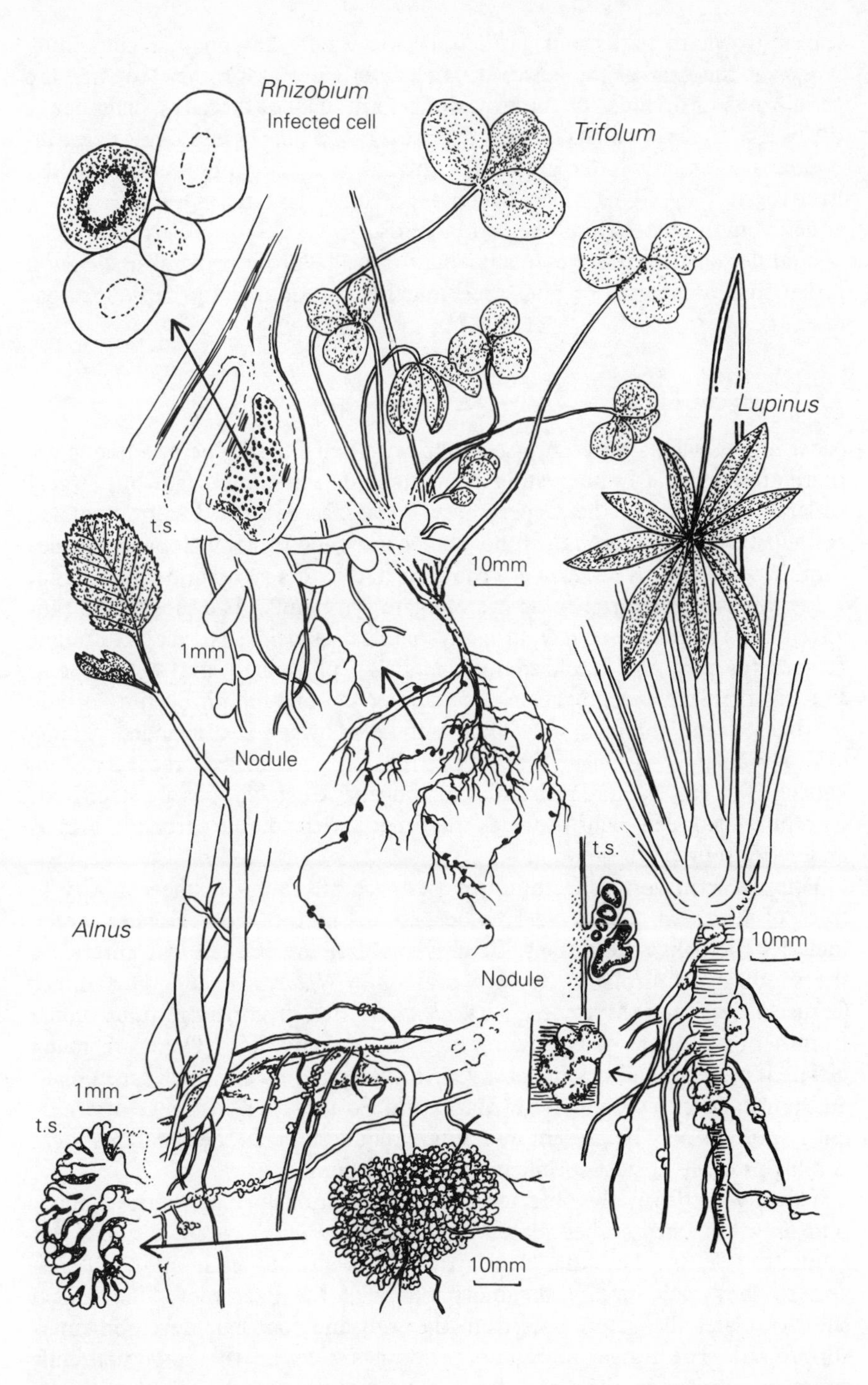

Fig. 7.9. Root nodules associated with nitrogen fixing symbioses: *Trifolium* and *Lupinus* are legumes.

leaves grown in parks and gardens beside water, has nodules containing *Nostoc* at the base of the leaves. Cyanobacteria are also associated with the aerial roots of orchids. In the tiny water fern, *Azolla*, there is a chamber at the base of the leaf which contains filaments of the blue–green alga *Anabaena*. *Nostoc* is commonly found among the rhizoids and scales underneath thalloid liverworts. Free living Cyanobacteria are important fixers of nitrogen in aquatic environments like rice paddy fields.

Nodules on the roots of plants may indicate a mycorrhizal association rather than a bacterial or blue–green algal association, as in the Podocarpaceae.

7.5.2 Mycorrhizae

Most higher plants have an association with soil fungi. The exceptions are therefore particularly interesting. They include some important but largely ruderal families like the Cyperaceae, Polygonaceae and Brassicaceae (see below). However, there is still no information about many tropical families (Smith, 1980). A mycorrhiza is a root infected with a symbiotic fungus (Fig. 7.10). Most mycorrhizal fungi are obligate symbionts. The most important advantage for plants of having a mycorrhizal associate is that the fungus forages for rare and localized nutrients like phosphate and stores them. Ectomycorrhizae are important in obtaining soil organic nitrogen especially in stressed and infertile soils. The uptake of water is also aided. Plants with vesicular–arbuscular (VA) mycorrhizae have enhanced recovery from wilting. The root is more resistant to pathogens. Unstable soils are stabilized. In return the green plant provides the fungus with soluble carbohydrates as an energy source.

Ectomycorrhizae or sheathing mycorrhizae are found in the majority of trees of northern temperate areas especially in the order Fagales which includes the oaks and birches. Ectomycorrhizae are present in conifers. In the southern hemisphere they are present in *Nothofagus* and the families Casuarinaceae and Myrtaceae (*Eucalyptus*). Ectomycorrhizal fungi come from a broad range of basidiomycetes and ascomycetes. There are many different patterns and colours of mycorrhiza depending on which green plant and fungal species are involved (Marks and Foster, 1973). A range of different symbionts may be present on a single root system. In the eucalypts there is a high degree of host specificity in the symbiont.

The mycorrhizae, the infected roots, are variously black club shaped, pinnately branched, tuberculate, Y-shaped (in *Pinus*), coralloid and other kinds (Fig. 7.10). The fungi are capable of producing auxin and cytokinin and so they may modify the morphology of the host root. The fungal sheath isolates the young root from the soil, and root hair development is suppressed. The fungal mycelium penetrates between the epidermal cells and then ramifies in the intercellular spaces of the cortex. The network of hyphae in the cortex is called the Hartig net. The hyphae do not penetrate past the endodermis.

Endomycorrhizae involve a more intimate relationship between fungus

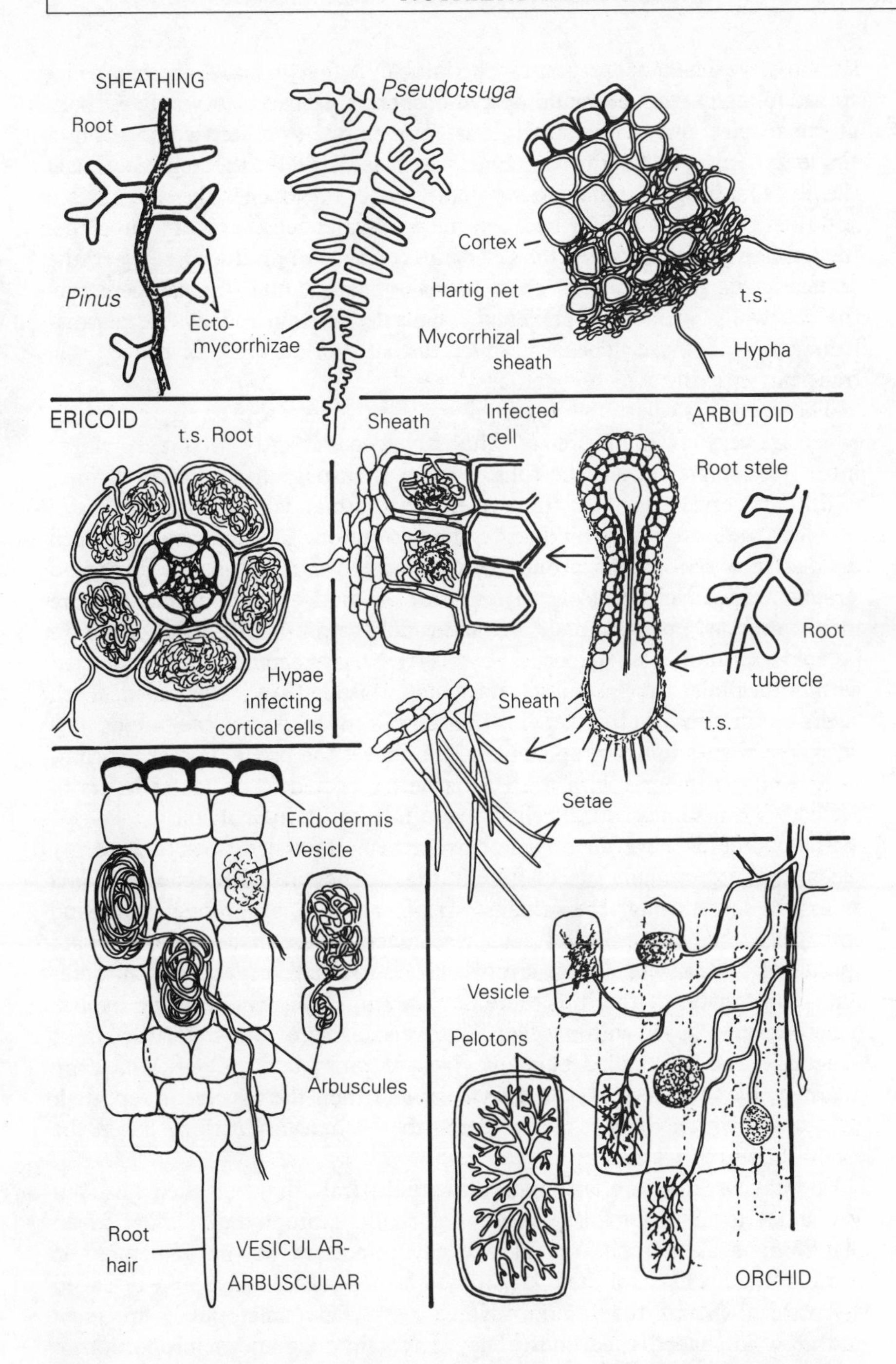

Fig. 7.10. Mycorrhizae, representing a symbiotic relationship between a fungus and plant roots: sheathing adapted from Zak (1973) and Meyer (1973); ericoid adapted from Read and Stribley (1975); Arbutoid adapted from Rivett (1924); vesicular-arbuscular adapted from Mosse (1963); orchid from Hadley and Williamson (1972) and Jackson and Mason (1984).

and host. Vesicular arbuscular mycorrhizae (VA mycorrhizae) are very widespread in many families including gymnosperms, ferns and bryophytes, both in the tropics and temperate regions. The fungal symbiont comes from a single zygomycete family, the Endogonaceae. Usually there is no fungal sheath. The fungal hypha flattens slightly where it touches the root surface and then penetrates into or between the epidermal cells. Then in the cortex the hyphae ramify between the cells and vesicles are produced between the cortical cells, pushing them apart to make space. Other hyphae penetrate the cell walls, producing branched arbuscules surrounded by the plasmalemma or tonoplast. The arbuscule, the site of nutrient exchange, is a transient structure which eventually lyses.

The order Ericales, the heathers and heaths, often grow on acid soils which are very low in phosphate. They have peculiar and varied mycorrhizae involving septate fungi. The fungal symbiont usually has a wide tolerance to different ericaceous hosts in species rich heathlands (Robinson, 1973). It provides the host with amino acids. The ericaceous mycorrhizae have been divided into ericoid and arbutoid kinds named for the genera *Erica* and *Arbutus* respectively. Ericoid mycorrhizae, a kind of endomycorrhiza, are more common. For example, *Calluna* produces thin 'hair-roots' which have no epidermis and a single cortical cell layer which becomes heavily infected with intracellular hyphal coils (Jackson and Mason, 1984). More cortical cell layers are present in other genera. In arbutoid mycorrhizae, the fungus may be shared with ectomycotrophic trees. It involves the production of a sheath, extracellular haustoria with the Hartig net restricted to the outer layers of the cortex but some cortical cells are also filled with hyphal coils.

Many orchids have an obligate mycorrhiza (Dressler, 1981). The tiny seeds cannot germinate successfully in the absence of the fungus. A hypha enters the seed through the suspensor region and penetrates the germinating embryo. As the embryo grows out new infections are made. The fungi are species of *Rhizoctonia*. There are two layers to the root. There is an outer fungal host layer. Within this there is a digestion layer where fungal hyphae penetrate and grow within cells. The hyphae form intracellular coils or irregular structures called pelotons (Jackson and Mason, 1984) which are digested, by a process like phagocytosis, and then the process is repeated. In orchids which die back to a stem-tuber a new mycorrhiza has to be established each season.

Endomycorrhizae are an important carbohydrate drain on their host but VA and ericoid mycorrhizae may be partially saprophytic. Similar green plant/fungus associations are particularly important in the gametophytic phase of the life cycle of the pteridophytes. Such relationships may be called mycorrhizal though they do not involve roots. The gametophytes are small and grow in shaded conditions. They may gain a significant proportion of their energy requirements from their associate fungus which is living as a saprophyte on the dead tissue of other plants. The mycorrhizal gametophytes of *Psilotum* and *Lycopodium* are largely or entirely saprophytic.

The root systems of grasses, especially cultivars grown in high nutrient conditions, are only weakly mycorrhizal. Mycorrhizae or nodule systems are absent from the families Brassicaceae Polygonaceae, Cyperaceae and Proteaceae. The first two taxa have many ruderal, weedy species, crucifers and knotweeds, which grow in nutrient-rich areas. Many species of the Cyperaceae produce rootlets which are carrot-shaped (dauciform) and densely covered with long root hairs when growing in low nutrient situations. In the Proteaceae the primitive genus *Persoonia* has rootlets with a dense covering of long (6 mm) root hairs (Lamont, 1983). More advanced genera like *Banksia* have densely packed root clusters concentrated in the upper few centimetres of the soil. The rootlets are arranged in longitudinal rows and have a dense covering of root hairs.

7.5.3 Saprophytic plants

Saprophytes lack chlorophyll and rely entirely on a fungal associate to provide them with a carbon source as well as other nutrients. Saprophytism in higher plants is an extreme kind of mycotrophy. There is a close taxonomic relationship between saprophytic groups and strongly mycotrophic groups. Many normal orchids live saprophytically for a long period as seedlings, relying entirely on their mycorrhizal associate. Albinistic variants of non-saprophytic *Cephalanthera* and *Epipactis* have been observed to survive until flowering. There are also specialized orchid saprophytes, *Galeola*, *Corallorhiza*, *Neottia* and others. The roots are absent or poorly developed and leaves are rudimentary. They have a much branched stem with a coral-like appearance which is hidden in dead wood. The stem only becomes apparent at flowering. The mycorrhizal associate can digest lignin.

The family Monotropaceae is sometimes included within the Ericaceae. The association that *Monotropa* has with its fungal associate, a kind of arbutoid mycorrhiza, may be better classified as a kind of indirect parasitism since the fungal associate has a mycorrhizal association with forest trees.

7.5.4 Parasites

Holoparasites do not require light since they obtain all their energy from the host. They also obtain water and mineral nutrients from the host. **Hemiparasites** are photosynthetic though their photosynthetic rates may be low and respiration rates high so that they may rely on the host for significant quantities of fixed carbon. There is traditional distinction between obligate and facultative parasites. Facultative parasites may be grown independently in experimental conditions. They have an effective, though poor root system. The more obligate parasites have weaker root systems even lacking root hairs. Parasites may be divided into stem and root parasites.

Parasitism has evolved many times over. It is present in 17 families of flowering plant and over 3000 species (Stewart *et al.*, 1989). Parasites include many herbaceous species but also some large trees. The principal adaptation is the organ of penetration and attachment, the **haustorium**. This is a complex structure formed from the intimate union of host and parasite tissue. Solute transfer is aided by the parasite having a higher transpiration rate than the host creating a suction pressure between parasite and host. Haustoria are very variable in morphology between species. Even within the single genus *Striga* the haustorium may be simple as in *S.hermontheca* or massive and swollen as in *S.gesneriodes* (Fig. 7.11). The haustorium taps into the xylem or the xylem and phloem of the host. The mistletoe, *Anothofixus*, which is an obligate epiparasite taps into the phloem of its mistletoe host, *Amyema*, while the *Amyema* only taps the xylem of its own host, *Casuarina*.

The host–parasite relationship has been studied in detail in *Striga* which is a parasite of many crops including sorghum and maize. *Striga* seed is stimulated to germinate in the presence of exudates from the host roots. The site of primary attachment is the primary haustorium. Sometimes lateral roots establish secondary haustoria. Root parasites often show a measure of specifity between the host and the parasite, and within individual *Striga* species, even between cultivars of the crop host and biotypes of the parasite.

In many groups similar evolutionary trends in parasitism can be observed. They involve reduction of the parasite, the loss of its photosynthetic ability, a reduction in leaf size, increased self pollination and greater production of smaller seeds. In the Scrophulariaceae and related families, even within single genera, there are parasites which look scarcely any different from non-parasitic plants and others which are obviously highly parasitic. So *Striga hermontheca* is a leafy outbreeding plant with showy flowers, *S.asiatica*, is an inbreeder but also leafy and *S.gesneriodes* has small scale like leaves with a reduced chlorophyll content (Fig. 7.11). The related genus *Orobanche* is devoid of chlorophyll. In the Balanophoraceae reduction has gone further: unisexual flowers are grouped on a swollen receptacle which is all that is ever seen of the plant. Plants are dioecious or diclinous and female flowers consist only of a pistil with a peltate stigma. The holoparasitic order Rafflesiales shows a great reduction. *Cytinus* produces a short stem directly from the root of the host. It consists of a few bracts and a large inflorescence only. *Rafflesia arnoldii* from Sumatra has the largest flower of any plant, over 1 m in diameter. Only the flower is ever seen above ground. The rest is part of the haustorium within the root of the liane host. Another species, *Pillostylis*, is only a few millimetres high. In the peculiar Hydnoraceae, only the apex of the flower appears above the soil.

The distinction between stem and root parasite would seem fairly arbitrary except that different taxonomic groups have specialized in either habit. This may indicate different evolutionary origins for each kind of parasitism. The possibly greater specifity and wider taxonomic distribution of root parasites is interesting. Root parasites may have evolved either from direct root fusion or by sharing a mycorrhizal association, both of which require a close physiological compatibility. Stem parasites may have evolved from epiphytes which generally do not have species specific relationships. *Cuscuta* and *Cassytha* are both taxonomically isolated: there are no linking forms between *Cuscuta* and other twiners in the Convolvulaceae and *Cassytha* is particularly isolated in the Lauraceae.

The Santalaceae, which are related to the mistletoes, are particularly

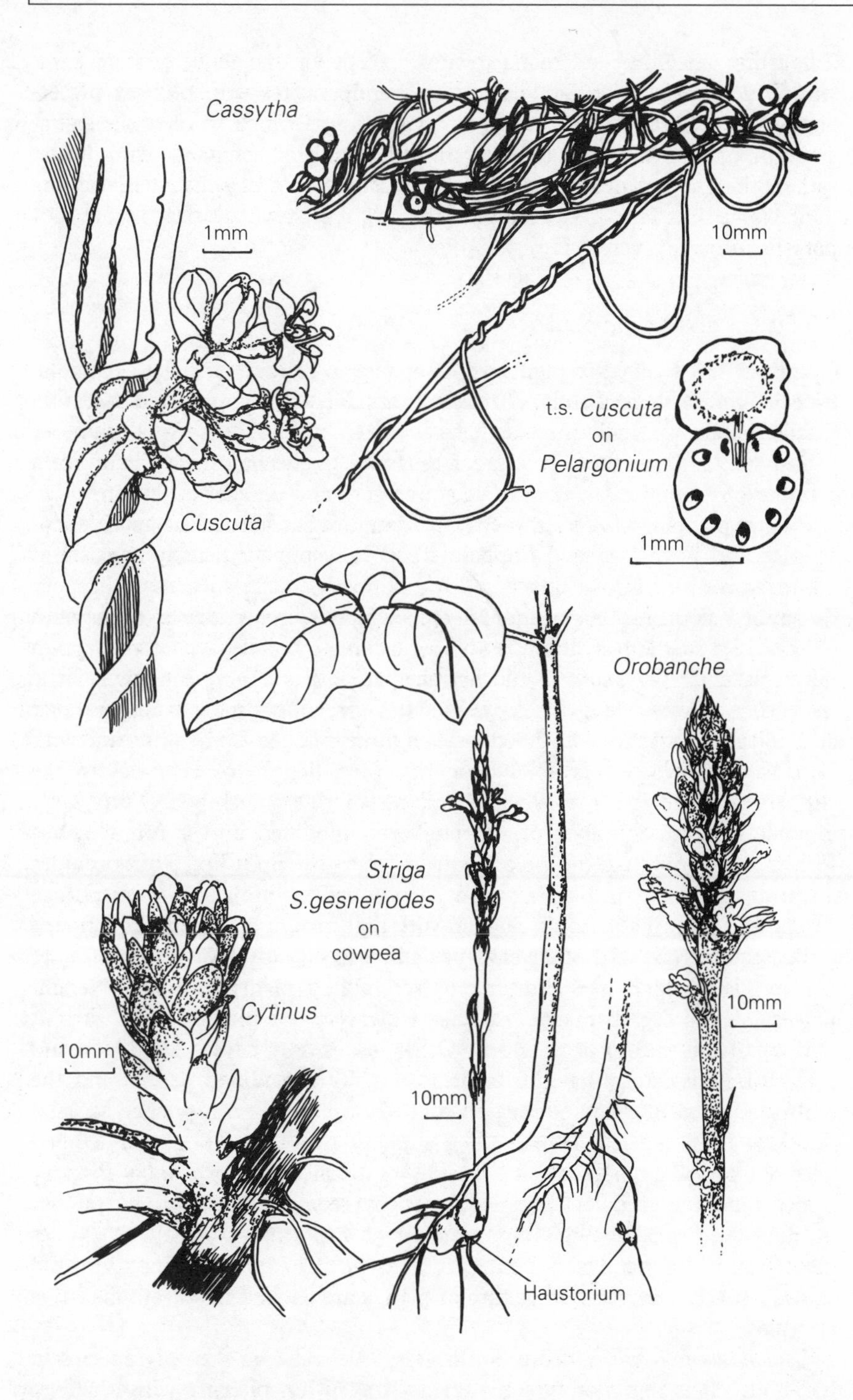

Fig. 7.11. Parasites: *Cuscuta* and *Cassytha* are stem parasites in distinct families but have converged remarkably in their habit; *Cytinus, Orobanche* and *Striga* are root parasites.

interesting since they are root parasites, except for one genus of stem parasite, *Dendrotrophe*. The Santalaceae are hemiparasites with no host preferences (Visser, 1981). *Thesium* has wandering roots which traverse the soil at a depth of a few centimetres establishing multiple contacts with hosts, rather like the epicortical roots of the mistletoe *Plicosepalus*. Perhaps the Santalaceae illustrate one example of the origin of stem parasitism from root parasitism or vice versa (Kujit, 1969).

7.5.5 Carnivorous plants

Carnivory has evolved in plants as a response to nitrogen-deficient habitats, especially water-logged soils. Nitrogen is obtained by trapping and digesting small animals especially insects. It has evolved only in a very small number of families and yet there is a great diversity of mechanisms. There are a number of examples of remarkable convergent and parallel evolution.

Pitcher plants are found in the related families Sarraceniaceae and Nepenthaceae and in the isolated Cephalotaceae. The pitcher usually consists of an attractive nectiferous hood or lid, a slippery collar, an area of slippery downward pointing hairs, and an area which secretes digestive enzymes (Fig. 7.12). At the base there is an area lacking a cuticle, where absorption takes place. *Heliamphora* in the Sarraceniaceae has a simple pitcher, little more than a folded leaf (Slack, 1988). In *Sarracenia* a more complex hood and collar is present. The hood prevents the pitcher from filling up with rain water. An insect is led by an increasing density of nectaries to the slippery collar region from whence it plunges into the chamber. *Darlingtonia* is similar except the apex of the pitcher is modified into a dome with a forked nectiferous tongue poking down from its upper lip. This provides an attractive first landing spot but downward pointing hairs discourage escape and soon the insect is clogged with digestive fluid. The fluid contains protease, invertase and other enzymes and also digestive bacteria.

Nepenthes, which is a climbing or scrambling plant, in family Nepenthaceae, is similar in many respects. However, the pitcher looks as if it develops from the tip of a tendril. Other leaves only have a leaf blade and a tendril. It is likely that the blade is actually a modified petiole and the tendril is a continuation of this, so that like *Sarracenia* the pitcher actually develops from the leaf lamina. The slippery rim of the pitcher is ribbed with downward nectiferous ribs. Inside the pitcher there is a waxy slippery upper zone and a lower digestive zone covered with glands. The pitcher produces the volatile alkaloid coniine which has a mousy smell. It entices insects and also paralyses them (Mody *et al.*, 1976). The digestive fluid shows a marked increase in acidity as glands are excited to secrete digestive enzymes.

Cephalotus follicularis, from South West Australia, is the only species in its family. It is remarkably convergent. It even has pitchers with a ribbed rim like *Nepenthes*. However the pitcher has evolved in quite a different way from the basal parts of the leaf. Like some of the other species it has ridges up the pitcher which serve to direct an insect up towards the opening.

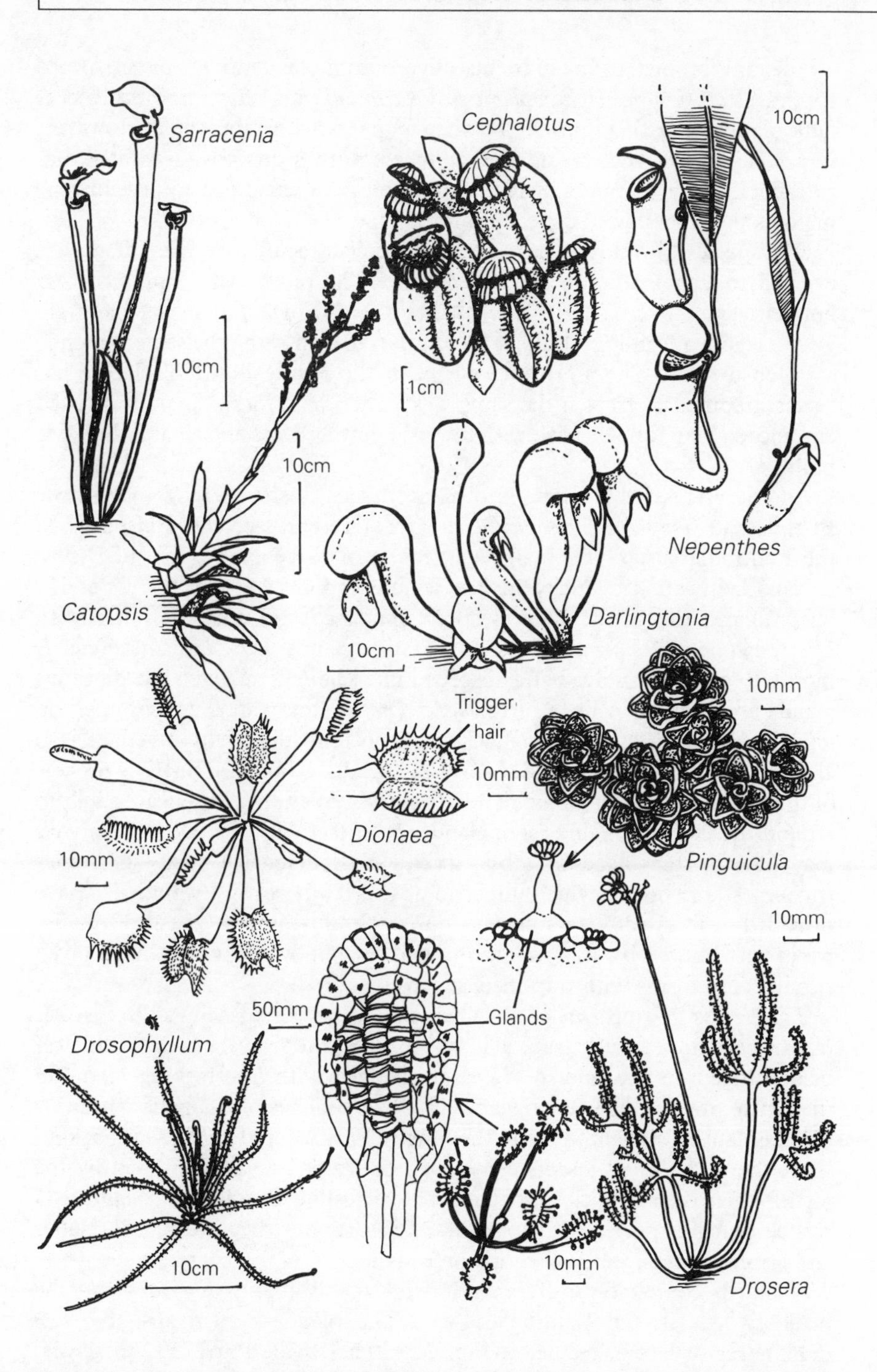

Fig. 7.12. Carnivorous plants: *Sarracenia*, *Cephalotus*, *Nepenthes*, *Darlingtonia*, and *Catopsis* are pitcher plants; *Pinguicula* and *Drosera* are active flypapers; *Drosophyllum* is a passive flypaper; and *Dionaea* is a spring trap plant. See also Fig. 2.5 for *Utricularia*.

The tank bromeliads includes insectivorous and facultatively insectivorous species like *Catopsis*. *Heliamphora* and *Sarracenia* may have first evolved as tank plants since they initially fill their pitchers with rain water. However, *Nepenthes* and *Cephalotus* fill their pitchers with secretions. *Nepenthes* has extrafloral nectaries and a strong association with ants, and so possibly the pitchers developed first as pouches for ants.

One species of *Sarracenia*, though superficially similar to the others, has evolved to trap insects in a different way. The pitchers of *S.psittacina* lie horizontally. At the base of the pitcher the downward pointing hairs are well developed forming a kind of no return trap, following the same principle as a lobster pot. The aquatic *Genlisea*, in the family Lentibulariaceae has leaves modified to trap in the same way. There is a forked spirally twisted entrance, a long tubular neck with inward pointing hairs and a basal digestive bulb.

Another remarkable example of convergence is shown by *Byblis* in the Byblidaceae, *Drosera*, *Drosophyllum* in the Droseraceae, and *Pinguicula* in the Lentibulariaceae. They all have leaves, or leaves and stems in *Byblis*, covered with stalked sticky glands which trap insects like fly paper. In its struggles the insect collects sticky mucilage and suffocates itself. In *Drosophyllum* and *Pinguicula* sessile digestive glands then start to secrete. A mosquito can be dissolved and absorbed in 24 hours. In *Byblis* the digestive glands are sunken in little furrows. In *Drosera* the glands are grouped on long stalks forming tentacles. Each gland secretes the sticky mucilage trap and the digestive enzymes and absorbs the products of digestion. The process of digestion is aided in *Drosera* and *Pinguicula* by the way the leaves roll up around the insect bringing more glands in contact with the body. Carnivory may have first evolved in those species with defensive sticky glandular trichomes. Trichomes which impeded or killed a herbivore would also have provided dead insects as a nutrient source. Gradual improvement of the fly-paper mechanism by leaf rolling or tentacular movement may have given rise to those forms with trap mechanisms.

The Venus fly-trap *Dionaea muscipula* is in the same family as *Drosera*. It traps insects in a sprung trap. The two lobes of the trap are parts of the leaf blade hinged on the midrib. They are fringed with interlocking teeth. In the centre of each lobe are trigger hairs. Two touches of the hairs within 20 seconds cause a signal to pass to the hinge areas. A rapid change in turgidity shuts the trap. Small insects can escape but large ones are trapped by the teeth. Then the lobes close together and digestion begins! *Aldovandra* is a related aquatic species with traps like *Dionaea*, but they are much smaller and arranged in successive whorls on a stalk.

A fourth completely different kind of trap is found in *Utricularia* the bladderworts, in the Lentibulariaceae. The bladders are traps of a few millimetres or less in diameter (Fig. 2.5). The smallest are 250 μm across. The unsprung bladder is concave. Water has been absorbed creating a negative pressure inside the trap. The trapdoor is shut but so delicately sprung that when it is triggered it opens inwards so that there is an inrush of water carrying the hapless prey. With the suction released the trapdoor then springs back closing the bladder. Water is withdrawn by four-armed glands and then digestive enzymes are secreted.

7.6 SURVIVING ENVIRONMENTAL EXTREMES

One of the most important adaptations which plants have for coping with environmental extremes is not a morphological or anatomical adaptation but a physiological one. Plants accumulate compounds which increase the stability of vital enzymes. The heat shock response of plants involves the synthesis of two groups of heat shock proteins of high and low molecular weight. The protein content of frost-hardy plant changes in quality and quantity as they 'harden' towards the approach of winter. Proline and glycine betaine, the polyols, mannitol and sorbitol, are implicated variously in adaptation to freezing, drought stress, anaerobisis, and high salinities.

Many adaptations considered to help a plant survive fire would also be effective against cold or against drought. In all cases a living core of the plant is protected so that resprouting can occur from resting buds. The buds are protected by being below the ground in resting organs, or in aerial parts of the plant by being beneath the bark or budscales.

Where there is a single over-riding environmental influence the plant communities may be readily characterized because they have a high proportion of plants with a particular kind of adaptation. These plant communities provide many very remarkable examples of convergent evolution with different distantly related taxa adopting a similar form.

7.6.1 Cold

In the cold polar regions of the Arctic and Antarctic seed plants are rare or absent but there may be a luxuriant vegetation of bryophytes (together with lichens and algae). In the cold-frigid region of continental Antarctica seed plants are entirely absent, except in the coastal part of the peninsula. The remarkable ability of mosses to withstand the cold is another feature of their ability to withstand desiccation (Longton, 1988). Being poikilohydrous, with little or no cuticle, mosses readily lose water from the cytoplasm as external ice is formed. Initially the cytoplasm is protected from freezing by the presence of solutes. The antifreezant solutes become more effective as they are concentrated by desiccation. The extremely low internal water content which results does not allow large damaging internal ice crystals to form. Rehydration occurs rapidly and metabolic activity is resumed.

A feature of survival in cold environments is not just the degree of cold but the number of cold–thaw cycles a plant experiences. Reviving after a period of stress depletes the energy store of the plant and cannot be repeated very often. A blanket of snow insulates plants from too many freeze–thaw cycles.

There is a remarkable convergence in the giant rosette trees of tropical high altitudes; the genera *Puya* (Bromeliaceae) and *Espeletia* (Asteraceae) in the Andes, *Senecio* (Asteraceae) and *Lobelia* (Campanulaceae) in East Africa, and *Argyroxiphium* (Asteraceae) in Hawaii (Meinzer and Goldstein, 1986). They suffer an extreme, but very regular, diurnal temperature fluctuation. Freezing temperatures last only a few hours. The presence of the thick covering of marcescent leaves insulates the plants. A thick stem pith provides a reservoir of water for the growing apex during the period when low soil temperatures limit water availability. Leaf pubescence isolates the leaves from the low air temperatures so that they are more easily warmed by incident radiation.

7.6.2 Fire

A high proportion of plants of some plant communities, like eucalypt savanna, are fire-adapted. They have a thick fire-resistant bark. Tussock grasses may be burnt back but meristems deep within the tussock are preserved. The grass trees *Xanthorrhea* and *Kingia* are protected by densely packed persistent leaves. Many species have deep, protected epicormic (epicortical) buds whose sprouting is initiated by exposure to heat. *Pinus palustris* from southern USA stays in a long-leaved dwarf form until exposed to heat. Flowering is encouraged in some species like *Anigozanthos*, the kangeroo paw, of western Australia. Seeds are released from the cones of *Banksia* after fire.

The prevalence of fire in natural communities prior to the evolution of man is difficult to ascertain. It seems likely that the periodicity of fire has increased relatively recently. However, there is considerable evidence for 'wild fire' in very early plant communities (Cope and Chaloner, 1980).

Many fire-adapted species were preadapted by mechanisms evolved for resistance to frost and drought. Nevertheless many, especially Mediterranean-type plant communities are the result of long-term adaptation to fire. As Cope and Chaloner (1980) have pointed out fire-adapted plants are not necessarily just fire-resistant they may also be highly flammmable because of a high content of gums and oils; they encourage wild fire. Periodic fire, which they survive, eliminates competing species which are not fire-resistant.

7.6.3 Budscales

At the end of the growing season, in many perennials under seasonal regimes, a protected resting bud is produced. This is a portion of stem with very short internodes and crowded specialized leaves, the bud scales or cataphylls. The budscales have short petioles or are sessile and are wrapped tightly together to protect the apical meristem. The epidermis of the budscales may be thick-walled. In many species a corky periderm is produced, and resins and waxes glue the scales together, making the bud impermeable and protecting it from frost and predation.

In species which lack specialized budscales the young unfolded leaves may be protected with a covering of hair, scales or by mucilage. A familiar example is the crozier, the tightly rolled frond, of ferns. The stipules may protect the young leaves as in *Ficus*. Alternatively the base of an already developed leaf may protect the bud. The leaf goes through a large part of its development within its protective covering and when released it unfolds into the full leaf.

7.6.4 Underground survival organs

Swollen underground storage organs and their associated buds enable plants to survive drought, fire or frost and to re-establish quickly when conditions ameliorate. They are protected by scale leaves or a corky periderm. Many species have adaptations for burying the storage organ (Fig. 7.13); they may be positively geotropic (e.g. *Polygonum viviparum*) or produced at the ends of positively geotropic shoots (eg. *Solanum tuberosum*, *Tulipa sylvestris*). Bulb and corm geophytes may have contractile roots which pull them down into the soil. Rhizomes are the least adapted storage or survival organs. They also function as agents of lateral spread (Chapter 8).

7.7 LIFE FORMS

There are several ways in which the life form of plants may be categorized. One of the most useful ways has been the life form classification of Raunkiaer (1934) which divides tracheophytes into different categories on the basis of where the buds are situated in the harsh season and how they are protected (Fig. 7.14). These categories overlap with other categories: xerophytes, sclerophylls, epiphytes, parasites and others. Some tracheophytes do not easily fit into the scheme. Some, like the Podostemaceae, a peculiar family of angiosperm aquatics, are more easily characterized by a system designed for the bryophytes. Raunkiaer's lifeforms can be summarized as follows.

1. Phanerophytes (trees and shrubs > 25 cm tall)
   (a) Evergreen trees without a bud covering
   (b) Evergreen trees with a bud covering
   (c) Deciduous trees with a bud covering
   (d) Shrubs 25 cm–2 m tall
2. Chamaephytes (woody or semiwoody perennials, with resting buds < 25cm above ground)
   (e) Shrubs or semi-shrubs which die back to the resting buds
   (f) Passively decumbent shrubs
   (g) Actively creeping or stoloniferous (procumbent) shrubs
   (h) Cushion plants
3. Hemicryptophytes (die back in harsh season, with resting buds at the soil surface)
   (i) Protohemicryptophyte
   (j) Partial rosette plant
   (k) Rosette plant
4. Cryptophytes (with buds below ground or in water)
   (l) Geophytes

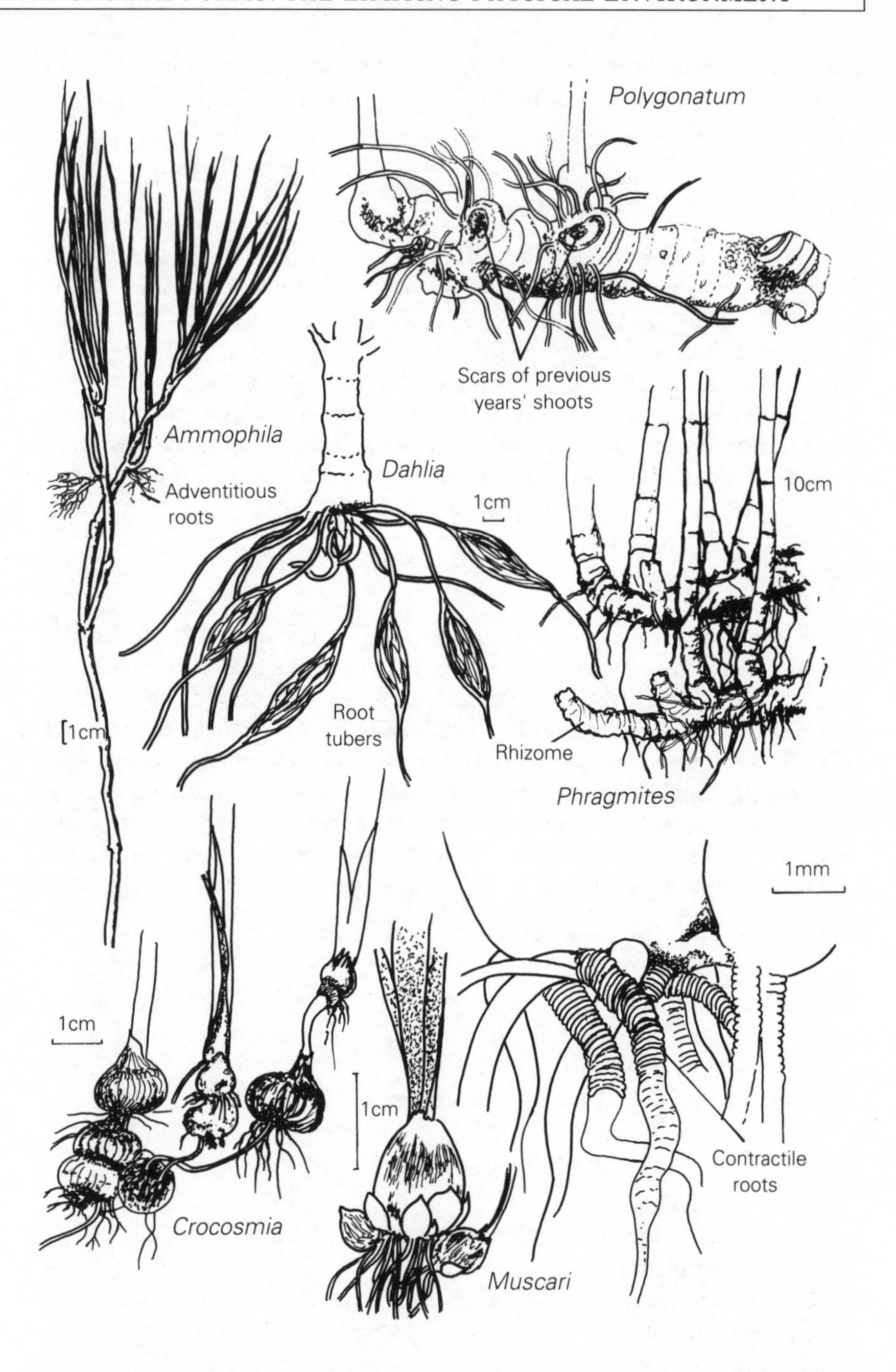

Fig. 7.13. Geophytes with different underground organs: some are specialized for storage like the rhizomes of *Polygonatum*, the tubers of *Dahlia*, the corms of *Crocosmia* and the bulbs of *Muscari*; the rhizomes of *Ammophila* and *Phragmites* are designed for vegetative spread vertically and horizontally respectively. The contractile roots of *Muscari* pull the bulb into the soil.

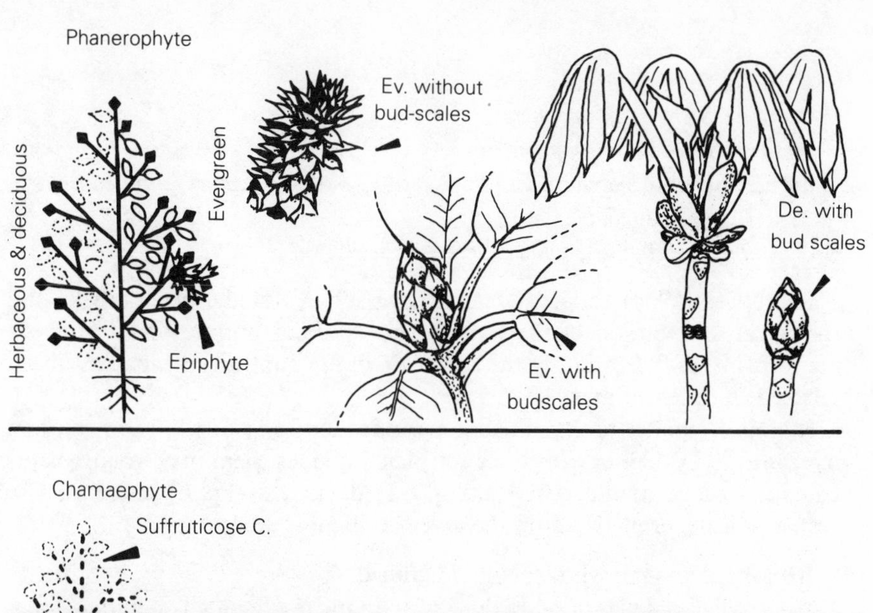

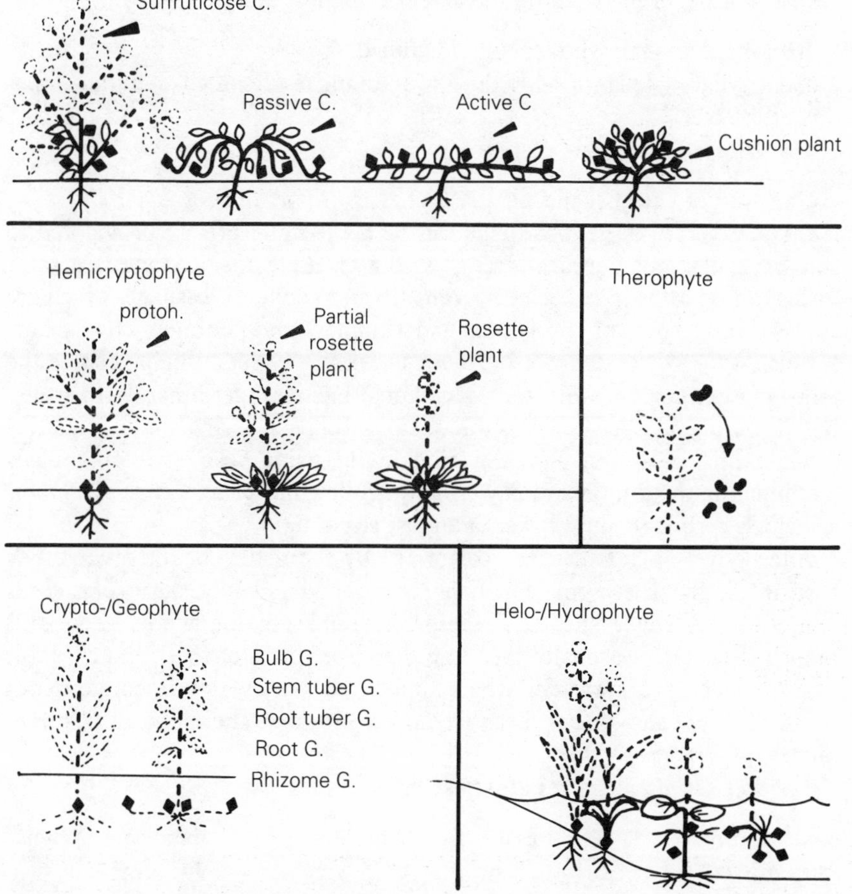

Fig. 7.14. Life forms (after Raunkiaer (1934)) which are based on the position of buds in the harsh season for growth except for therophytes which survive as seed.

(i) Rhizome geophyte
(ii) Bulb geophyte
(iii) Stem tuber geophyte
(iv) Root tuber geophyte
(m) Helophyte (marsh plants with resting buds in water saturated soil)
(n) Hydrophyte (buds in water)
5. *Therophytes (annuals, which survive as seed)*

Raunkiaer (1934) carried out extensive surveys of the geographical and ecological distribution of life forms. The different proportion of different life forms is an indication of the spectrum of life strategies which have been selected in different areas (Fig. 7.15).

Running parallel to Raunkiaer's classification, one of the simplest, and most useful, system of categories for plants divides them into woody plants and herbaceous plants. Du Rietz (1931) devised a slightly more precise system of categories based on the extent of lignification:

1. Holoxylales – the whole plant is lignified
2. Semixylales – plants with the lower branches lignified and the upper herbaceous
3. Axylales – herbaceous plants

One way of categorizing plants, included in part in Raunkiaer's system, is the seasonality of growth. This can be made more precise or adapted to suit particular geographical areas, so for example mediterranean growth forms can be categorized as evergreens, winter-active, biseasonals which are active during parts of the winter and summer, and summer-active plants (Orshan 1983). A system for northern temperate Europe might include the following categories: evergreen; deciduous; biennial (germinating in spring but flowering in the second subsequent summer); spring/summer annual (germinating in spring and flowering in the same year); winter annuals (germinating in autumn and flowering the following spring); ruderals (annuals which germinate and flower at almost any time).

Alternatively plants can be categorized by the nature of the organs they shed in the harsh season: (1) whole plant shedders which survive as seeds (annuals); (2) shoot shedders, seasonally renewing the whole shoot; (3) branch shedders, seasonally shedding the upper parts of their branches; (4) leaf shedders. This system when applied to Negev desert communities (Orshan, 1983) shows that there are no leaf shedders there but all phanerophytes are branch shedders.

Another set of categories are those based on leaf size:

subleptophyll/	leptophyll/	nanophyll/	nano-microphyll/
<0.1cm^2	0.1–0.25cm^2	0.25–2cm^2	2–12cm^2

microphyll/	micro-mesophyll/	mesophyll/	macrophyll/	megaphyll
12–20cm^2	20–56cm^2	56–180cm^2	180–1640^2	>1640cm^2

Leaf consistency characters are malocophyll (soft), sclerophyll (hard),

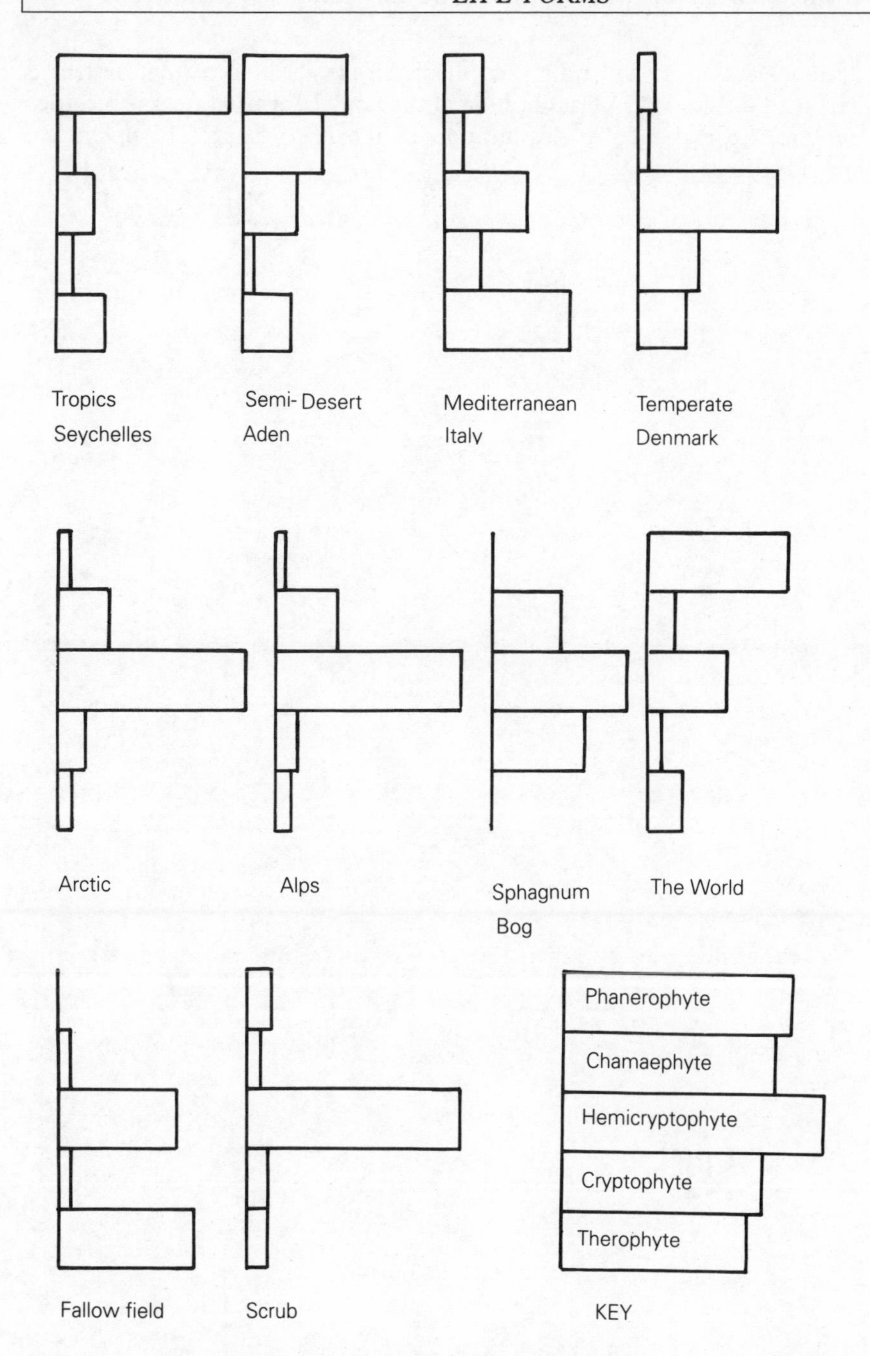

Fig. 7.15. Geographical distribution of life forms (data from Raunkiaer (1934)): phanerophytes (trees and epiphytes) dominate in the humid tropics, therophytes (annuals) in the Mediterranean region where there is a regular summer drought; the fallow field and scrub show changes in the distribution as the succession proceeds from an open habitat to a closed and highly competitive one.

resinous succulent and water succulent. A classification of root system is recorded in Fig. 7.8. A classification of the growth forms of trees is recorded in Figs 2.6 and 6.5. A classification of the life forms of bryophytes is provided in section 7.2.

Competition, herbivory and dispersal: the limiting biotic environment

8

8.1 REPRODUCTIVE STRATEGIES
8.2 HERBIVORY
8.3 DISPERSAL
8.4 ESTABLISHMENT
8.5 A BEHAVIOURAL CLASSIFICATION OF PLANTS

SUMMARY

In this chapter adaptations relating to the multiplication of the individuals of a species, the competition between species and the interactions between different species are described. This kind of biology is called population biology. It is because of interactions between species that evolution is not static. An individual, a species, is adapted relative to its biotic environment. Other species surround it compete for the same light, water and nutrients. As its competitors evolve to exploit the environment more effectively each species must evolve in turn to compete more effectively by growth or multiplication. Or it must evolve mechanisms to escape the competition. These may involve the kind of specializations which have been described in Chapter 7, or adaptations for dispersal to, and establishment in, less competitive areas. An important element of the biotic environment are the animals which exploit the plants. Plants have many adaptations relating to herbivory. The close relationship between animals and flowers has been described in Chapter 5.

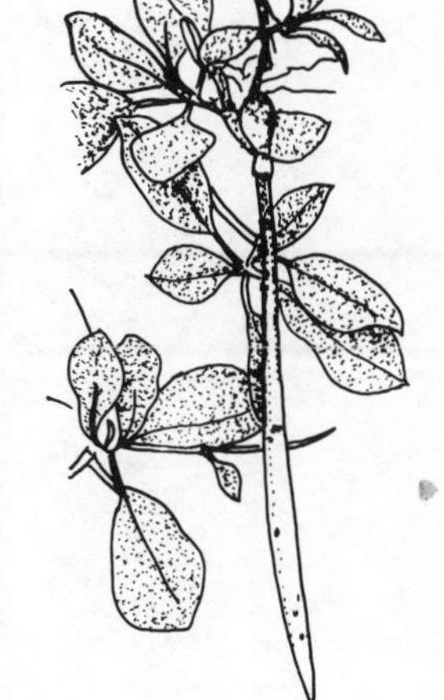

8.1 REPRODUCTIVE STRATEGIES

Two extreme reproductive strategies can be caricatured. **r-strategists** are adapted to reproduce rapidly and colonize new open and perhaps transitory habitats where competition is low. They tend to be short lived and monocarpic. They include ruderals, weedy plants. They multiply by producing disseminules (diaspores) either spores, seeds or fruits adapted for dispersal. They may be adapted for wind, water or animal dispersal. **K-strategists** are long lived and tend to be polycarpic. They grow tall or spread laterally vigorously. Competition occurs by the growth and reproduction of the plant without or with limited dispersal. The most important K-strategists, the trees, have been descibed in Chapter 6. Other important K-strategists are some grasses.

8.1.1 Propagules

Plants have a remarkable ability for vegetative reproduction because they grow by the replication of modules. The modules have varying degrees of independence. **Adventitious roots** are readily produced and the plant can fragment so that each part can grow into an independent plant. We make extensive use of this ability in the propagation of cultivated plants. Some plants have specially designed propagules. Vegetatively produced propagules are not usually not very well adapted for dispersal since they lack the protective coat of spores, seeds and fruits. In dry areas they are poor disseminules but in wet areas they are not. The propagules may be several times larger than normal disseminules and may establish themselves quicker which may be important in areas with a short growing season. Propagules often develop precociously while still attached to the parent plant. Propagules are generally not adapted as resting organs like bulbs, tubers or seeds. Nevertheless the production of propagules is an effective means of population expansion.

Vegetative propagation is particularly common in the bryophytes, many of which only reproduce sexually very occasionally. **Gemmae** are the asexual propagules of some liverworts and mosses. The thalloid liverworts *Lunularia* and *Marchantia* have gemma cups in which small disc shaped gemmae are produced (Fig. 8.1). Each gemma has a characteristic notch at the apical meristem. *Blasia* has a flask-shaped organ inside which gemmae are budded off. They are extruded from the flask when mucilage in the flask takes up water. Following release each gemma produces some rhizoids and grows into a new thallus. *Riccardia* produces two-celled gemmae within thallus cells. Each gemma is released when the containing cell wall breaks down. Some leafy bryophytes also produce gemmae. They arise on their leaves, at the apex or costa, or on the stems or rhizoids. Some species have specialized gemmae producing structures. The moss *Aulocomnium androgynum* produces a stalk with a spherical mass of gemmae at its tip. Some pteridophytes also produce gemmae. They arise on the developing prothallus of *Psilotum*, *Tmesipteris* and epiphytic *Lycopodium*.

Vegetative reproduction can also occur just by the fragmentation of the plant. Each fragment may grow into a new plant. Leafy bryophytes with a delicate structure fragment easily. In the moss *Dicranum flagellare* the shoot is modified to be especially fragile. Fragmentation of the plant as a regular reproductive strategy is quite common in aquatic plants of all sorts.

Many ferns, some lycopodiums and flowering plants, produce bulbils. The bulbil starts as a swelling. They may arise in different positions, on the margin or lamina of the leaf as in *Kalanchoe* and many ferns, on the axil of the leaves as in *Saxifraga granulata*. The bulbils may remain like small tubers or gemmae until after they are shed, or they may develop precociously into

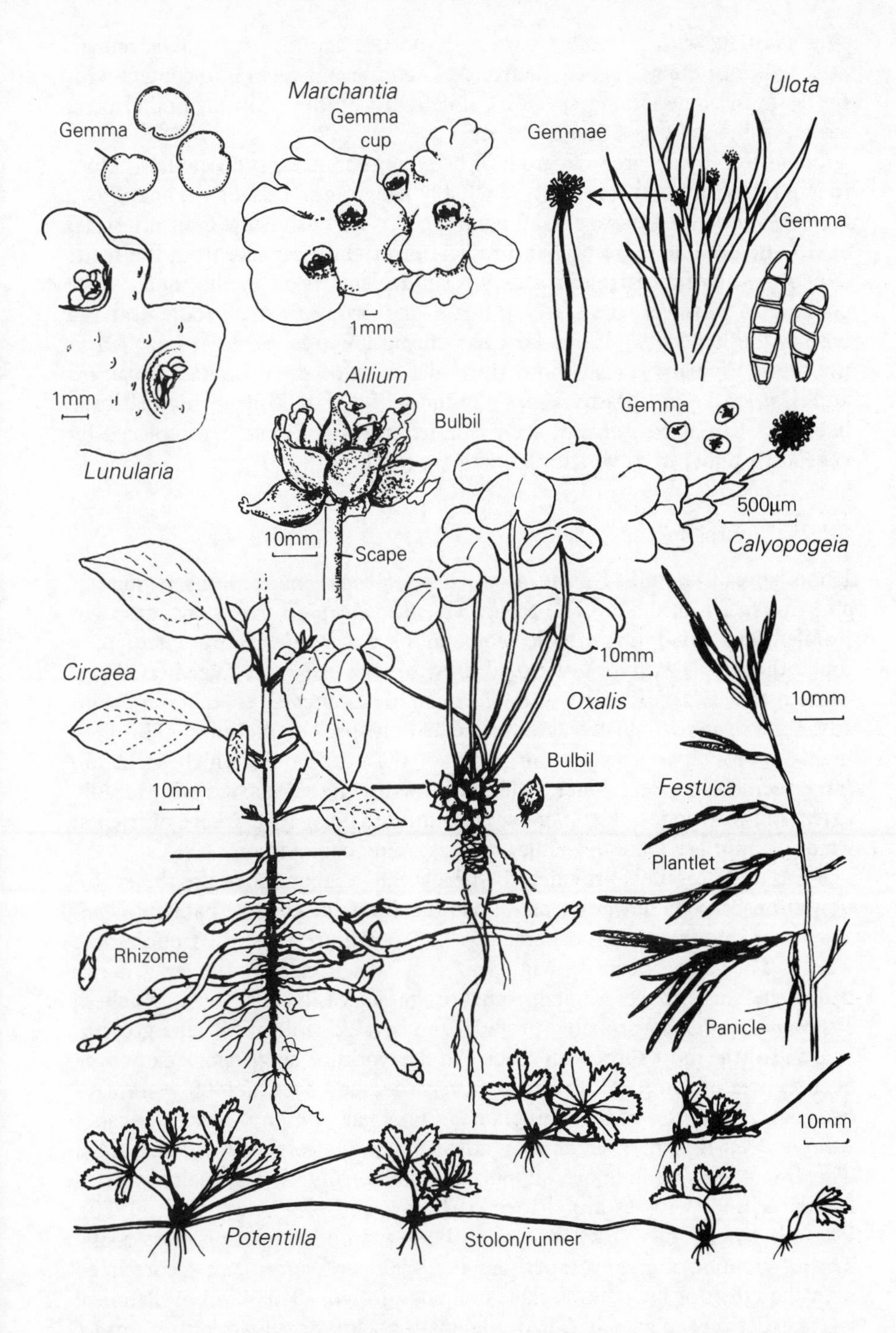

Fig. 8.1. Vegetative propagation by the production of gemmae, bulbils, plantlets, stolons and rhizomes: the *Festuca* species illustrated is viviparous, producing plantlets where other species produce seeds.

tiny plantlets with a swollen base and roots. They may fall off or remain attached until the wilting of the frond or stem brings them into contact with the soil. In some *Allium* species bulblets are produced in an umbel on a scape, in the same position as florets.

Some seed plants are viviparous. The seed germinates precociously before it is released from the parent. A notable example is that of the mangrove *Rhizophora* where vivipary is an adaption for establishment in an intertidal environment. The hypocotyl and radicle protrude spear-like from the fruit. When the seedling is released it falls, and plants itself in the mud. More commonly, however, the seedling floats, and produces small roots near the base of the hypocotyl. When beached during low tide these get attached to the mud. Vivipary is also shown by other mangroves *Avicennia*, *Bruguiera* and *Aegiceras*. Some grasses especially those of high altitude or high latitude habitats, such as *Poa alpina*, are pseudoviviparous. Florets are replaced by vegetative buds, from which plantlets are produced.

8.1.2 Lateral spread

Plants spread laterally by the production of **rhizomes**, **stolons** (runners), and invasive roots. The fern *Asplenium rhizophyllum* spreads by means of proliferating frond tips which root and produce new plants where they touch the soil. Two types of clonal growth have been recognized (Bell and Tomlinson, 1980). Phalanx growth is highly branched so that the plant advances relatively slowly as a set of closely packed modules. This is a highly competitive growth form. Most of the interactions which occur are intraspecific even intraclonal. Guerilla growth is less branched with rapidly extending internodes. The plant seeks out more open areas where nutrients, water or light are more available, thus escaping competition.

Ferns and horsetails provided the herbaceous vegetation before the evolution of herbaceous flowering plants. In the Jurassic, in some habitats *Equisetum* was dominant. Some horsetails like *E.arvense*, are still troublesome weeds. They have deep growing rhizomes which can grow over a metre below the soil surface. The rhizomes fragment easily. The fern, bracken, *Pteridium aquilinum*, probably provides the best example of phalanx growth. It is one of the most successful species in the world. A single clone extending over an area of 474 x 292 m has been recorded and ages of 1400 years have been suggested for individual plants (Oinoinen, 1967). Bracken invades using a deep rhizome, which may advance one metre ahead of the fronds. The front guard fronds are supported nutritionally by the main wave of fronds behind. Grasses are shaded or buried in frond litter (Watt, 1970). Phenolic compounds, leachates, slow the growth of competing grass roots. The distribution of grass competitors like *Agrostis* v *Deschampsia* is correlated with the effect of bracken leachates on root growth. This kind of chemical warfare between plants is called allelopathy. Other well-worked examples are the poisoning of the underflora beneath the American walnut, *Juglans nigra*. The desert shrub from California *Encelia farinosa* inhibits the growth of annuals within a circle of 1 m. There is a toxin, a benzene derivative, in the leaves which is released as fallen leaves decompose (Gray and Bonner, 1948). In the Californian chapparal *Salvia leucophylla* and *Artemisia californica* have zones of bare soil around them. Further away grasses grow. It has been suggested that volatile terpenes are responsible for this distribution (Muller, 1970).

Grasses are now the overall champion as lateral spreaders. They are highly competitive sward formers. They produce a fibrous shallow adventitious root system. They spread by rhizomes and stolons at the soil surface or rhizomes in the soil. These readily produce new shoots called tillers. The grass colonizers of dune systems have different patterns of growth. The sand couch grass *Elymus farctus* and the sea lyme grass *Leymus arenarius*, which colonize embryo dunes, have remarkable abilities to spread laterally but restricted abilities to grow vertically if buried by sand, unlike the marram grass *Ammophila arenaria*, which can grow vertically and horizontally up to 1 m a year in growing dunes. The deeply buried parts of the plant die.

Grass leaves are divided into a sheath and blade as in other monocots. The sheaths are tightly wrapped together and combined with the stem. The basal meristems of the youngest leaves are protected within this structure. This allows tillers to recover when their tips have been grazed. Young leaves are extruded like the sections of a telescope (Fig. 8.2). The leaf blades are arranged closely and held more or less vertically. The depth of the sward that results intercepts a large proportion of the incident light in a way analogous to the deep canopy of fast growing trees.

The clover, *Trifolium repens*, has a guerilla kind of growth form. It produces stolons which can quickly ramble through the sward to find openings where it can root, and most importantly where it can get light. It is commonly overtopped by grasses. It is not a very strong competitor but neither is it much affected by the species of the competitor. Its interactions are not intraspecific but with a wide range of species. It responds very well in comparison to other grassland species to grazing. It is protected from predation by the presence of cyanogenic glycosides and can rapidly regrow (many refs. in Harper, 1977).

8.2 HERBIVORY

Some of the earliest land plants were being eaten by herbivorous arthropods. The Rhynie chert contains fossil mites and mite-like creatures as well as collembolans ('springtails'). Living collembolans are litter feeders. The fossil arthropod *Rhyniella* may have lived on dead fungal remains. However several *Rhynia* plants show lesions on the surface which may indicate areas where a herbivore was tapping into the vascular strand. Some lesions show a wound response growth which would indicate living herbivory rather than saprophytism (Chaloner and Macdonald, 1980).

The earliest tetrapods have jaws and teeth which are not adapted for herbivory (Milner, 1980). Some important plant groups which are important in the fossil floras could not have been very palatable or provided much energy. The leaves of cycads, ginkgos and conifers are thick, tough and

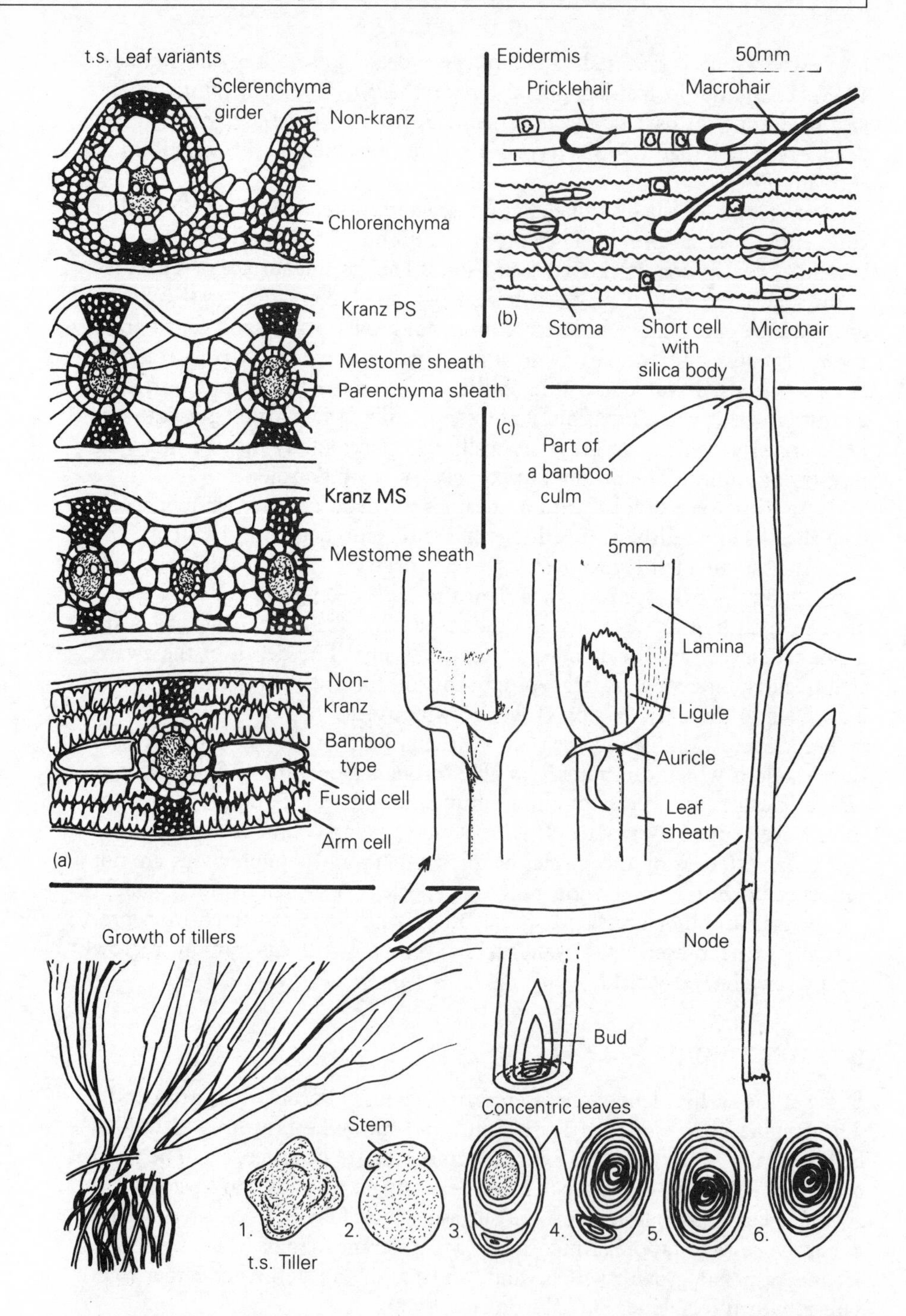

Fig. 8.2. Vegetative characteristics of the grasses (Poaceae): (a) the diverse leaf anatomies associated with different patterns of photosynthesis; (b) the epidermis with trichomes and silica bodies; (c) stem and leaf morphology showing sheath lamina transition with auricles and ligule, concentric arrangement of sheaths in the culm and production of tillers and adventitious roots.

leathery. Ferns contain high levels of toxic, especially phenolic compounds. These morphologies and 'biochemistries' may represent antifeeding strategies evolved over the many millions of years when these groups dominated the flora.

A very important trend in the evolution of herbivory has been the increase in body size of the herbivore (Wing and Tiffney, 1987). Small animals have high metabolic rates and require high energy food. They concentrate on seeds and fruits. If in the past small animals concentrated on reproductive structures this may have been one of the evolutionary pressures which encouraged the protection of the female gametophyte and embryonic sporophyte leading to the development of the seed and fruit.

In the late Carboniferous insects diversified. The first flying insects are found. Insect groups include surviving ones like the cockroaches but also others which have since become extinct. Orthopterids (grasshoppers) may have eaten plant material. Many of the insect groups are likely to have been detritus feeders, eating rotting plant material or fungal hyphae rather than predating living plants. Spore eating was particularly prevalent; evidence for insect herbivory comes from abundant coprolites (fossilized droppings) containing spores (Scott *et al.*, 1985). Some fossils show damage of different sorts including bore and chew holes. The Megasecoptera and Palaeodictyoptera had probosces which some have interpreted as piercing mouthparts for probing for sap, though it is possible they only probed sporangia for spores. The Hemiptera, bugs with piercing and sucking mouth parts, evolved in the Permian. Possibly they were able to take advantage of the newer kinds of plants which evolved after the Carboniferous.

Woodiness and growing as a tall single-stemmed tree are protection against large ground based herbivores, especially if the bark is tough and unpalatable. The adaptations of some herbivores to overcome these defences are remarkable. The neck of the giraffe and the trunk of the elephant are two living examples. Many dinosaurs had very long necks. The late Jurassic herbivore fauna was dominated by huge dinosaurs. Of the biomass 95% of the tetrapod communities were large sauropods and stegosaurs. There was probably a large consumption of low quality vegetable material with slow passage times through the gut so that it could ferment. Fermentation would break it down into energy providing compounds and also possibly detoxify it. The plant matter was not chewed but ground by ingested stones in a gastric mill.The maximum browse height was not much greater than at present but there were at times many more large species and so the average browse line of dinosaur faunas was much higher (Coe *et al.*, 1987). These huge animals must have created vast open spaces just because of their ability to knock trees down.

In the early Cretaceous there was a shift to smaller but still very large, ornithopods. They had a limited ability to chew vegetable matter. The angiosperms enter the fossil record in the Mid Cretaceous. Possibly related to their arrival on the scene there was an increased diversity of dinosaur families in the late Cretaceous (Coe *et al.*, 1987) and there was a decrease in dinosaur average body size. Although there were still some very large dinosaurs, small dinosaur herbivores, with a body weight 10–100 kg became more important. These smaller herbivores were specialized feeders. *Hypsilophodon* had a narrow muzzle and probably browsed selectively on shoots. *Edmontosaurus* had a broad beak and probably cropped low swards of vegetation. The earliest flowering plants were semi-woody shrubs. The decrease in dinosaur size may record a shift from browsing trees to browsing flowering plant shrubs. The diffuse bushy growth forms of primitive flowering plants probably resisted browsing better than the rather simply branched forms of gymnosperms (Bakker, 1978). However, even after angiosperms appeared, the herbaceous flora was at first still dominated by ferns and horsetails. In wet and boggy areas the evolution of the monocots provided diverse new herbaceous forms which may have provided a richer, improved quality food source.

At the transition from the Cretaceous to the Tertiary all herbivores larger than 10 kg were eliminated. In some places the boundary clay is marked by a temporary superabundance of ferns, which may reflect some global destructive event which gave an opportunity to fern colonizers. It was not for another 30 million years that larger, but now mammalian herbivores (>100 kg) predominated again. In the early Tertiary angiosperms may have formed dense forests restricting the evolution of terrestial herbivores. Birds (Olson, 1985) radiated in the Early Tertiary, perhaps recording the diversity of fruits becoming available. It was not until the Early Tertiary that seeds and fruits show adaptations to animal dispersal. Climatic cooling, and probably increasing aridity, seems to have opened the forests encouraging the development of grasslands. In the Middle to Late Tertiary there was a great diversity of herbivores of all sizes, including larger long-limbed forms similar to those of today. This kind of diversity is characteristic today of open savanna. Grasses which may have first arisen in the Palaeocene became abundant in the Oligocene (Muller, 1981). Hypsodont teeth of herbivores evolved to grind the silica rich leaves of grasses.

8.2.1 Physical defences

Physical defences against herbivory are widespread in living plants. The existence of thorns, hairs and hard integuments and bark, in plants from the Triassic and Jurassic are taken as evidence of extensive herbivory. *Equisetum*, which was a dominant herbaceous species in the Jurassic, has a hard silicified epidermis and abundant sclerenchyma which still today provides an effective defence against herbivores. The rough surface has given them another name, that of the 'scouring rushes'.

Many different families have species with thorns or spines (Fig. 8.3). In different groups they are derived from lateral branches, leaves, petioles or are newly formed enations. Prickles may form at the margins of leaves or along the midrib, or on ribs of the stem or anywhere. *Zizyphus spinus-christi* has particularly effective thorns, hooked in pairs in opposite directions. The production of mechanical defences can be encouraged by predation. One

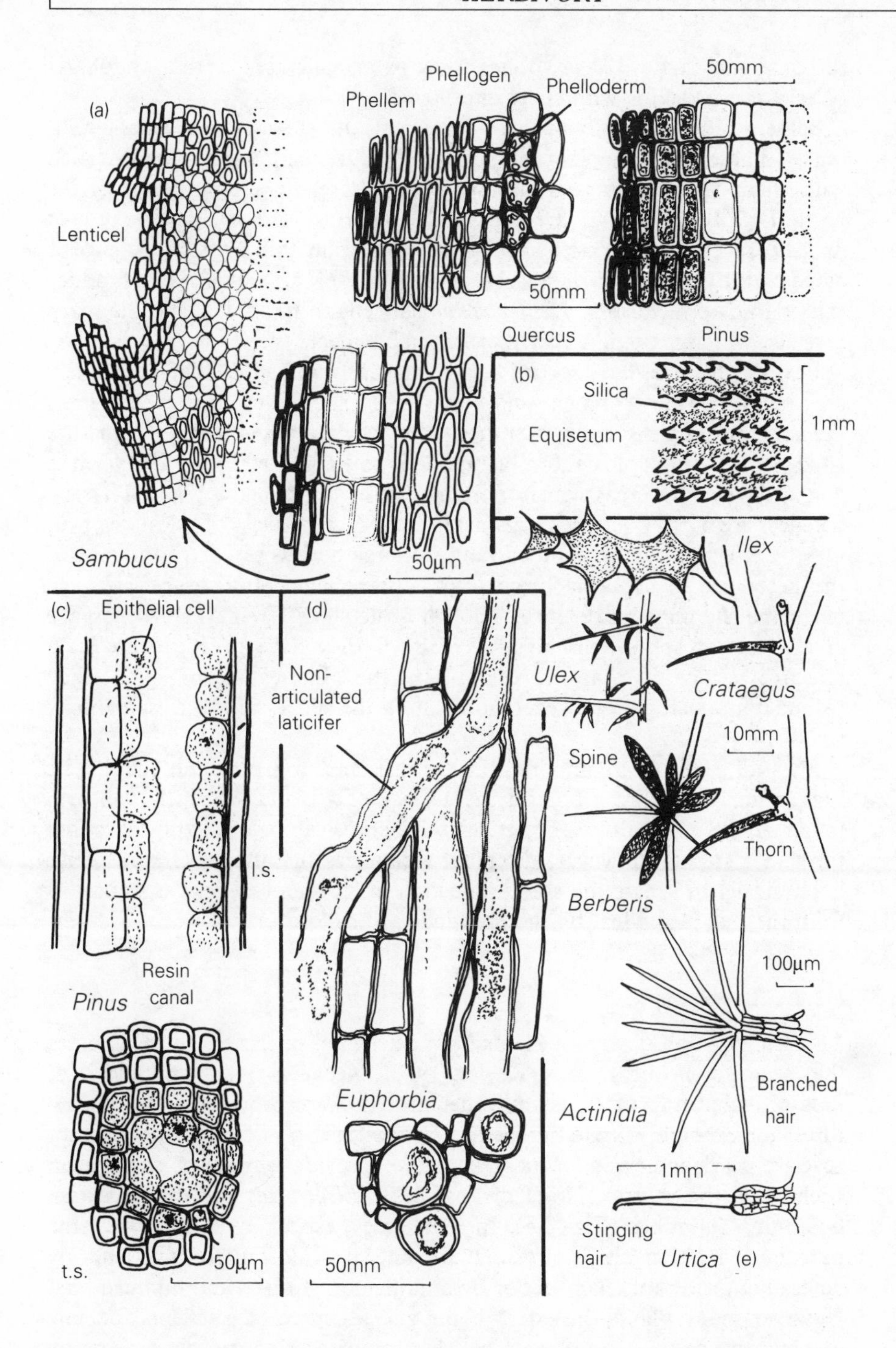

Fig. 8.3. Morphological and anatomical defences: (a) three variant periderms; (b) external appearance of branch of *Equisetum* showing protective silica; (c) resin canal; (d) laticifer; (e) thorns, spines and trichomes in a range of species.

response that some *Acacia* species show to being grazed is the production of very thorny stems with short internodes.

Some groups especially are associated with the possession of hairs (trichomes). These make the plant less palatable and frequently they are associated with chemical repellants. A familiar example are the stinging hairs of the nettle *Urtica dioica* (Fig. 8.3). A tough epidermis with a thick waxy coat may be an effective barrier against insects. Again this physical defence is aided by the presence of chemical repellants in the wax. Some plants produce cells with hard inclusions. *Dieffenbachia* has cells with raphides, needle sharp crystals, which make it a very uncomfortable meal.

Physical defences have several disadvantages over chemical defences. They are complex in development and costly in energy. They may even restrict the amount of light for photosynthesis and they are only effective in mature condition. A developing thorn or prickle is soft and ineffective and yet it is young developing areas which most need protection in the plant. Physical defences are not very effective against insects. Nevertheless physical defences may be effective against a broad range of large herbivores. It is difficult for the herbivore to evolve a mechanism, either behavioural or physical, to overcome the physical defence. Though goats show a remarkable tolerance of all sorts of spines, prickles and hairs if they have to. They are such effective herbivores that, for example in the Mediterranean region, they maintain an open garigue vegetation halting the succession to closed climax woodland.

Physical defences are important in the resistance to pathogenic fungi. The fungus may puncture a hole in the surface of a plant but thick wax or a tough cuticle may help restrict attacks. Defences can be evaded by invading through a stoma. However, the plant recognizes the attack. One response is physically to isolate the site of infection in an island of dead cells cut off from the rest of the leaf by the production of callose.

8.2.2 Ant plants

At least 1% of all flowering plants have extrafloral nectaries. In many cases they seem to encourage ants (Keeler, 1989). Nectar is not the only kind of food provided for ants: oils and other nutrients are provided in leaf structures: for example the Beltian bodies on the leaf tip of some American ant acacias; pearl bodies and Mullerian bodies provided by hairy pads called trichilia on the ventral side of the leaf base in *Cecropia*; and the Beccarian bodies in *Macaranga* (Fig. 8.4). In some cases, especially in epiphytes, the nutritional relationship is two-way: the ants provide mineral nutrients by defecating inside ant chambers or by building ant gardens and humus nests. However many plants provide a home and a source of nutrients for ants without any obvious nutritional benefit. In these cases the plant gains by being protected by the aggressive ants: from attack by herbivores and from encroachment by other plant species including epiphytes. The latter is particularly common in vines.

The effectiveness and costs of ants as herbivore deterrants is illustrated

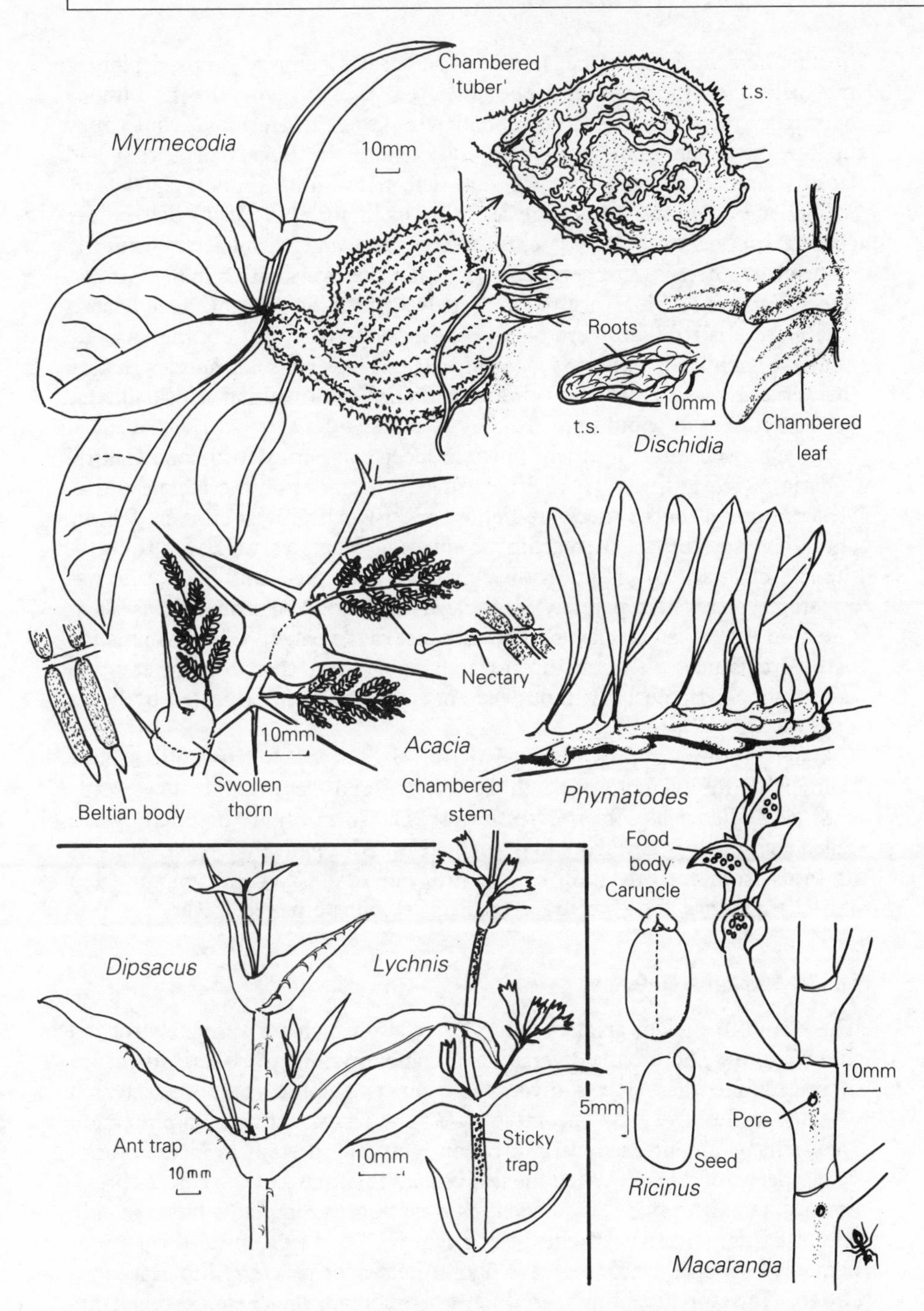

Fig. 8.4. Ant plants providing brood chambers or food, nectar and oil or starch bodies. *Myrmecodia*, *Dischidia* and *Phymatodes* are epiphytes. *Ricinus* is one example of the ant dispersal of seed, where ants are attracted by the provision of a caruncle. *Dipsacus* and *Lychnis* protect their flowers and fruits from robbing ants by the provision of traps.

in the *Cecropia/Aztecha* association. *Cecropia* is a genus of tropical pioneer tree with a chambered pith. The ant, *Aztecha*, gain entry to the pith by cutting a hole at the top end of a shallow groove, the prostoma, which runs up the internode from the axillary bud. Here the stem wall is thin and lacks sclerenchyma and collenchyma. The trichilium produces 2500–8000 Mullerian bodies, packages of glycogen, in its lifetime, 20–25 per day, before ceasing production. The presence of prostoma and trichilia is genetically controlled. There is both between and within species variation in these traits. Species and individual plants from high altitude on West Indian Islands where there is less herbivore pressure, especially from leaf cutting ants, do not have trichilia or prostoma (Janzen, 1973). However some species lacking the *Aztecha* association have a waxy, slippery stem making it difficult for the leaf-cutters to climb up.

An ant association is present in some species of both African and Central American *Acacia* trees. There are extrafloral nectaries on the petiole and as well as normal thorns there are hollow ones which provide a home for the ants. The ants prune encroaching competing plants as well as rushing to the site of attack to discourage herbivores. The acacia–ant association has apparently evolved separately in the African and American species. It is absent in Australian acacias, perhaps because of the paucity of its mammalian herbivore fauna. The structure of the swollen ant thorns is different in each case. Some species without the ant association are protected by being cyanogenic.

Many species use ants for seed dispersal. The seeds have oily arils or elaiosomes; for example the orchids *Acriopsis* and *Dendrochilum* have seeds with oil bodies, elaiosomes (Dressler, 1981). An aril near the micropyle is called a caruncle in *Ricinus* and other species. Alternatively *Lychnis* protects its seeds from ant predation by the presence of sticky bands on the stem and *Dipsacus* has ant water traps formed by connate paired leaves.

8.2.3 Chemical defences

There are three major areas of natural diversity which are intimately related to each other. There is the diversity of the angiosperms, there is the diversity of insects and there is the diversity of plant chemical compounds which discourage herbivory by the insects. Where other methods of protection appear to be absent chemical protection is usually present as in the cyanogenic species of *Acacia*. When the leaf is chewed cyanogenic glycosides break down to release cyanide. Cyanogenesis is widespread in plants but it is only one of a huge diversity of chemical defences. The biochemical diversity of plants is beyond the scope of this book: the reader is referred to Harborne (1988). However, it is important to at least mention the chemical repellants produced by plants because chemical diversity is fundamental to an understanding of the success and great diversity of the angiosperms (Fig. 8.5).

Chemical repellants and defences have evolved along three main biosynthetic routes, to produce terpenoids and terpenoid-like compounds, phenolic compounds and nitrogen containing compounds respectively. Some of these

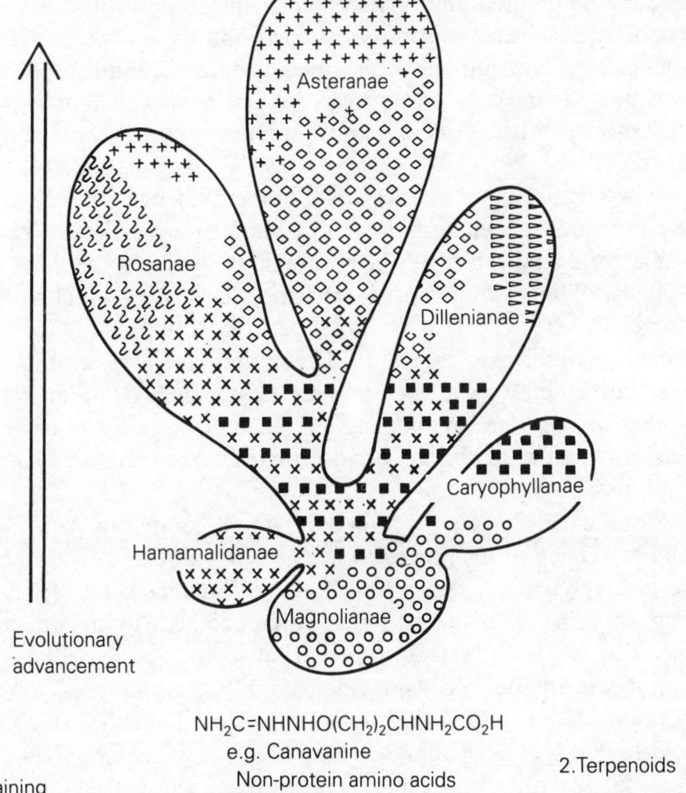

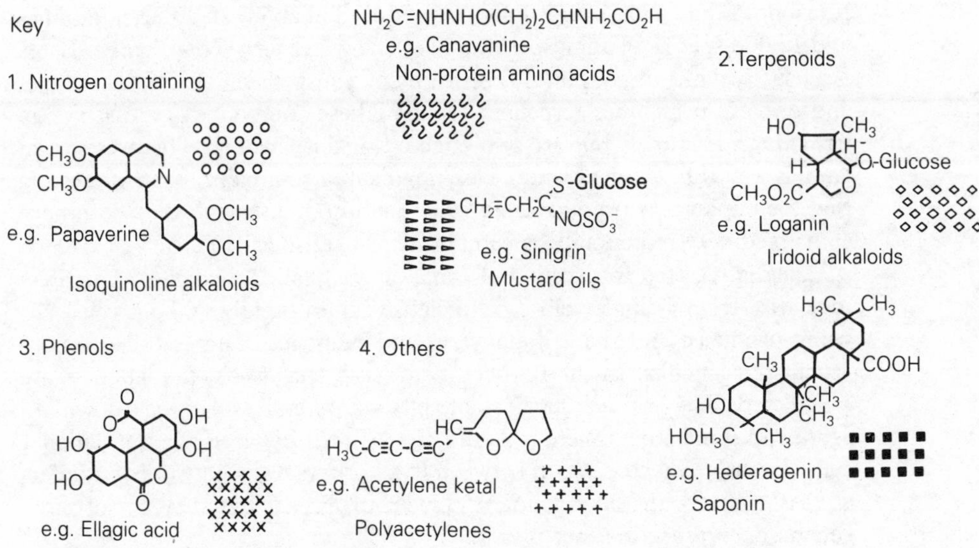

Fig. 8.5. Trends in the evolution of chemical defences in the dicotyledons showing the evolutionary replacement of compounds in derived taxa (data from Cronquist, 1981, and Harborne, 1988). Only a few classes of compound are illustrated.

compounds are toxic and others reduce palatability. The most complex kinds of compounds are found in herbaceous flowering plants. The simplest kinds, condensed tannins, are found in ferns and gymnosperms. There has been gradually increasing chemical sophistication. Cronquist (1981) ascribes the remarkable diversity and success of the advanced family of flowers, the Asteraceae, to their possession of sesquiterpene lactones and pyrrolizidine alkaloids.

One peculiarity of conifer wood is the presence of resin canals or ducts. Resin canals are elongated cavities lined by special resin canal cells. The canal forms by schizogeny, the breaking apart of the cells at the middle lamella. The resin secreted into the duct is a complex of compounds especially terpenes. It is possible that resin canals evolved as a response to dinosaur herbivory (Fig. 8.3). They may have formed initially as a direct response to injury and then become established. Resin decreases the palatability of twigs and buds. The existence of large quantities of fossil resin, amber, testifies to the importance of resin production in some plant communities.

Resin canals have evolved in several non-coniferous plants. Frankincense and myrrh are resins produced by *Boswellia* spp. and *Commiphora abyssinica* respectively, in the family Burseraceae. Lacquer comes from *Rhus verniciflua*. In some cases the resin can be tapped. It hardens on exposure to the air. The soft resin, mastic, comes from several species of *Pistacia*.

Latex is another repellant produced in large quantities. Latex production is taxonomically widespread. Latex is produced in about 20 different families and 12 000 species of angiosperm, and in the water fern *Regnellidium*. Latex is a usually white viscous liquid consisting of a suspension of rubber particles and often also includes terpenoid toxins. These compounds are nasty skin irritants in mammals but are also effective against insects. The presence of latex is a useful characteristic for distinguishing some taxa. It is interesting that the euphorbias produce latex while cacti do not though they both have converged very remarkably as cereiform, prickly succulents.

Latex is secreted into a kind of canal or duct called a laticifer. Laticifers may arise from a single cell. The nucleus divides rapidly without cell walls being produced and the cell elongates and branches. This is called a non-articulated laticifer. In the rubber tree, *Hevea brasiliensis*, the laticifers are articulated. They arise from lines of cells whose end walls break down. In *Musa* and some others there is a kind of trap door between adjacent laticifer cells. Later lateral connection between laticifers creates a network. Laticifers are associated with the secondary phloem in *Hevea*, but they are of more general occurrence in other plants.

Latex and resin production constitute a final line of defence. Most repellants are located in the external parts of the plant, or with particularly vulnerable parts of the plant like the young and tender buds. Trichomes constitute a first line of defence. For example the leaf trichomes of the tomato have ketohydrocarbon 2-tridecanone which is toxic to several potential pests (Williams *et al.*, 1980). The wild ancestors of the cultivated tomato have higher levels of the toxin, and more trichomes. The stinging hairs of *Urtica dioica* the stinging nettle contain the amine 5-hydroxytryptamine.

A second line of defence is the leaf surface waxes or resins. Frequently mixed with the wax are toxic constituents. The apple leaf has wax containing dihydrochalcone phloridzin which repels some species of aphid (Klingauf, 1971). The next line of defence is within the epidermal cells where toxins or repellants may be concentrated in the vacuoles.

The evolution of the flowering plants has been marked by an increasing sophistication and diversity of their chemical defences. Some large groups are marked by particular compounds. There has been extensive co-evolution of herbivore and plant. If an animal evolves a mechanism allowing it to detoxify a plant's toxins, it then has an advantage over other herbivores because it now has its own reserved food source. The situation of the plant moves from being protected to being extensively and selectively predated. The once toxic or repellant compound may even become an attractant to the herbivore enabling it to identify its reserved food plant. For example the apple leaf wax constituent repellant to some aphids is actually attractive to the apple aphid providing a signal to feed. There is very strong selection for the evolution of a more sophisticated defence by the plant, usually a new toxic compound, based on a modification of the original defensive compound. Once embarked upon, sophisticated chemical defence becomes an arms race.

An association between the rate of speciation and the evolution of chemical defences has been described in the Apiaceae which have coumarin toxins based on the hydroxycoumarin, umbelliferone (Fig. 8.6) (Harborne, 1988). Some have a phototoxic effect whereby light greatly magnifies the toxicity. Some insects have overcome this problem by leaf rolling behaviour, shielding themselves from light before eating. There are two advanced classes of toxin; linear and angular furanocoumarins. Those genera which have just the basic hydroxycoumarins have fewer species than those with the more advanced furanocoumarins. Genera with both linear and angular furanocoumarins have many more species. The evolution of more advanced compounds has restricted the range of possible herbivores and a close relationship with a particular herbivore has resulted. Coevolution of plant and herbivore has resulted in greater rates of speciation.

One large and successful advanced group, the grasses, have few chemical defences. Their growth form, having basal meristem and mechanisms of rapid lateral growth, allows them to survive herbivory while not so adapted competitors are eliminated by the herbivores. The grass leaf can be strongly sclerified. The epidermis contains silica cells which make them less palatable. Some herbivores do show preferences for particular grasses. Sheep favour soft species like *Lolium* over the hard-leaved *Nardus*. However, there is generally no close herbivore/grass relationship. Graminivorous herbivores like cattle and horses are not very choosy and no sophisticated chemical defences have evolved. Even in *Sorghum*, where some species are distasteful to the locust, because of the presence of a range of alkanes and acid esters

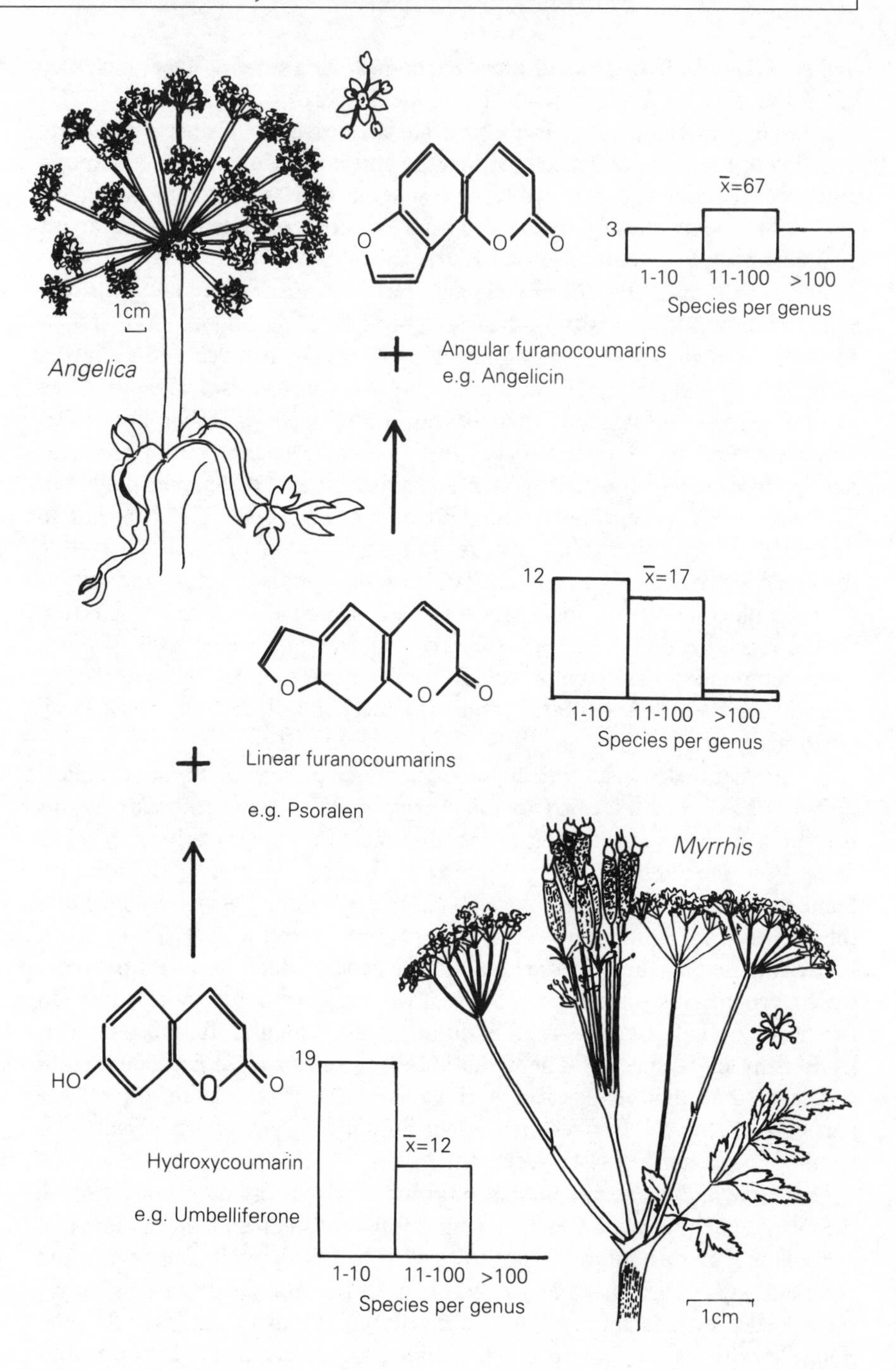

Fig. 8.6. Chemical defences and rates of speciation in the Apiaceae (Umbelliferae) (after Harborne, 1988). Histograms show genera divided into three size classes, showing that genera are larger in the group with the most sophisticated chemical defences.

in the wax these are rather general repellants (Woodhead and Chapman, 1986).

Herbivores are not the only source of biotic damage. Antimicrobial compounds are widespread in many plant tissues but are particularly associated with surface layers. One example is the presence of large quantities of catechol and protocatechuic acid in the dead outer scales of especially coloured onions. These compounds prevent the germination of spores of the onion smudge disease organism.

One class of compounds, called phytoalexins, is synthesized after a fungal attack. They inhibit the development of the fungus in hypersensitive tissues. Several hundred novel compounds have been isolated and characterized as phytoalexins. They are widely ranging in chemical structure, with different kinds associated with particular taxonomic groups. The Solanaceae have sesquiterpenoids (Harborne, 1988). The advanced families Asteraceae and Fabaceae, and advanced species within these families have diverse arrays of compounds. In evolution the phytoalexin response has become increasingly sophisticated.

The heartwood of trees is packed with terpenoids and phenolic compounds. Many have been demonstrated to have an antimicrobial, especially antifungal action but their value in the living plant is uncertain. However, since the heartwood is largely dead tissue its destruction by wood-rotting fungi would greatly damage the physical strength of the tree. Some woods like teak and elm are noted for their resistance to rot in wet conditions. Some compounds may limit the destructive potential of termites not by a direct toxicity to the termite host but by being toxic to their flagellate gut fauna which actually digests the wood. Certainly leaf cutting ants reject leaves which have terpenoids which would be damaging to their fungal associate (Hubbell *et al.*, 1983).

8.3 DISPERSAL

The disseminules of lower plants are spores but only in homosporous forms are spores effective in aiding the colonization of new geographical areas. In heterosporous pteridophytes and in higher plants the microspores and pollen can only promote cross-fertilization. Colonization occurs either by the dispersal of vegetative disseminules or in seed plants by the dispersal of seeds and fruits. Spores and pollen grains $<200\ \mu m$ in diameter are called collectively miospores. All disseminules are also called diaspores.

8.3.1 Passive release of wind-dispersed disseminules

The same physical rules influence the release and dispersal in the air and deposition of all disseminules. The aerial motion of diaspores is related to the Reynolds number (R_e) which is related to the velocity of the object (V_p), the length of the object and the kinematic coefficient of viscosity of the air (V); $R_e = V_p l/V$. Pollen spores and dust seeds are so tiny that they are well within the range for which the viscous forces of air are dominant. They have terminal velocities less than 100 mm/s (Burrows, 1986).

The means by which the disseminules get into the air stream is very important in wind dispersal. There is a change in the speed of airflow from zero at the surface of an object to ambient wind speed at a certain, quite small distance. The thickness of this boundary layer depends upon the smoothness of the surface; the smoother the surface the narrower the distance. A small difference in the elevation of the disseminule into the boundary layer will substantially increase the local airflow conditions that the spore experiences. There are also edge effects, whereby air speed increases the closer to the leading edge of the object; the boundary layer decreases in proportion to the square root of the distance from the margin.

The majority of plants have a passive release of spores. In many species, spores are released above or outside the boundary-layer effects of leaves, bracts or perianth segments, by being located on modified branch systems where the leaves are absent or reduced. For example moss and liverwort capsules are elevated on setae above the gametophyte. Anthers in flowering plants are exposed on long filaments in *Plantago*, *Thalictrum*, grasses and some wind-pollinated trees. The position and arrangement of sporangia and anthers allows the wind to shake spores and pollen from them. The vibration of these organs may be very regular. Even under moderate airflow of 2–3 m/s grass inflorescences have been observed to shake harmonically. Many species have a male cone, strobilus or catkin. In this case the shape of the associated sporophylls and bracts is important as the airflow is either channelled between them or eddies are set up which extract pollen and spores from dehiscing sporangia and anthers (Niklas, 1985).

8.3.2 Mechanical and explosive discharge of spores and seeds

Mechanical methods of spore release are powered by changes in water pressure as cells dry. **Elaters** which move hygroscopically within the spore mass are found in three groups: in liverworts, hornworts and *Equisetum*. As the elater dries and rehydrates it twists and turns to fluff up the spore mass and allow slight air movements to catch the spores. The shrinkage of the moss **capsule** pushes the **operculum** off because it shrinks less than the rest of the capsule. Shrinkage also pushes spores to the opening where the peristome teeth lift the spores by flexing and reflexing hygroscopically into the airstream. In some species there is a double peristome (Fig. 4.6). The outer ring interdigitates between the inner and collects spores on its toothed segment tips, flicking them out.

Explosive methods of spore discharge are found in many groups (Fig. 8.7). In *Sphagnum* the capsule is only slightly elevated on a short pseudopodium (part of the gametophyte). The spherical capsule shrinks as it dries, putting the air it contains under pressure. The cells below the **annulus** collapse and the lid (**operculum**) of the capsule bursts off. A cloud of spores are released. In the liverworts the spring-like elaters can either be compressed or stretched until the tensions are released explosively. In one,

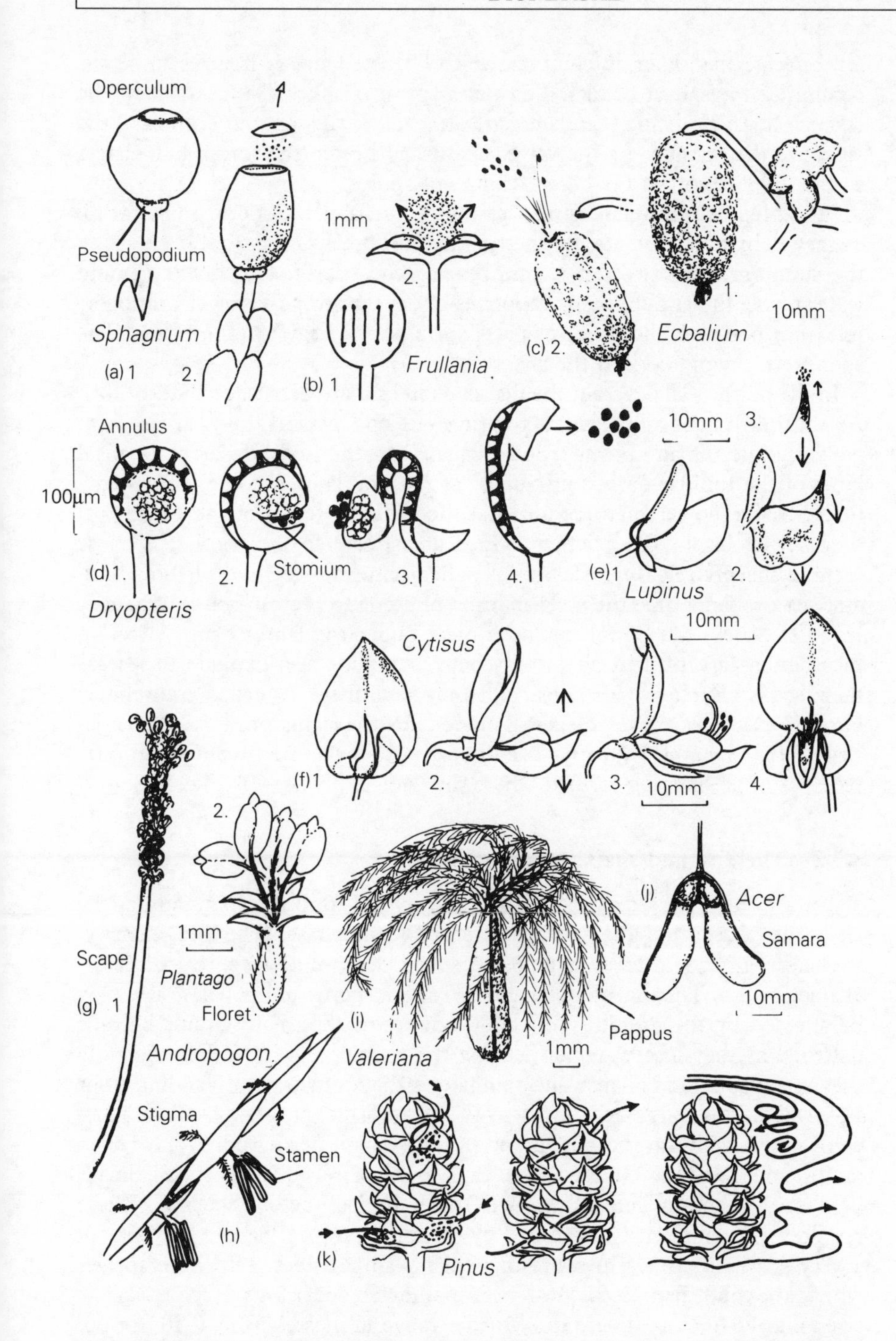

Fig. 8.7. Dispersal of spores, seeds and fruits: (a–f) Mechanical and explosive release; (g) and (h) wind pollinated herbs showing exserted stamens and stigmas; (i) and (j) wind dispersed diaspores; (k) air flow patterns around a female *Pinus* cone (after Niklas, 1985).

Cephalozia, the elater is compressed and twisted up as it dries until the column of water in it breaks. The elater springs back, twisting to shake the spores off. In *Frullania* the elaters are attached to the top and bottom of the valves of the capsule. As the valves open they stretch the spring-like elaters so that they rip off at the base flicking out spores.

In the **leptosporangiate** ferns the annulus cells of the sporangium shrink as they dehydrate. Eventually the sporangium breaks catastrophically along the stomium, into two parts, one bearing the spores. Quickly the liquid water inside the annulus cells vaporizes under tension, causing cavitation so releasing pressure and the two parts spring back near to each other. The spores are catapulted into the air.

In the nettles, Urticaceae, the filaments in bud are bent back like springs. As the bud opens the filaments spring out and pollen grains are thrown violently into the air. In the tropical mistletoes, the Loranthaceae, the petal lobes of the tubular corolla are joined at the tip. They become increasingly turgid as the flower matures until, at the slightest touch of the pollinator, they spring back. The stamens which are adnate to the corolla inflex or recurve sharply releasing a cloud of pollen onto the forehead of the pollinator. At the same time the style snaps to one side to prevent self-pollination. In gorse, *Ulex*, the flower explodes open under the landing insect so that the stamens are pushed against its belly. Its pods also explode to release their seeds. Explosive dispersal of seeds and fruits is called **autochory**. Notable examples are the dispersal of seeds by the jaculator in the Bignoniaceae. *Ecbalium*, the squirting cucumber, is well named because it squirts seed out of an opening formed where the pedicel breaks off (Fig. 8.7).

8.3.3 The size and shape of disseminules

There are three phases of spore dispersal; release from the sporangium, the air-borne phase, and the settling or entrapment phase. The size, density, stickiness and to a lesser extent the shape determine a spore's dispersal characteristics. The morphology and density of spores and pollen grains may be affected by the humidity of the air. It seems that many grains become flattened as they lose water on leaving the sporangium or anther. Possibly they are adapted to collapse and reinflate as they rehydrate on landing prior to germination. Large and low density spores or pollen grains will be more easily lifted from the sporangium or anther. Large, dense spores and pollen grains will settle or collide more easily with a receptive surface. Small, light spores and pollen grains will travel further before settling. These requirements conflict. Constrained by the same physical factors there is not a very great difference in spore or pollen grain size between those species which are wind dispersed. Most wind pollinated species have spores in the size range 20–60 μm. There is a broader range in mosses from 7–10 μm up. Larger spores have been described but generally they are not designed for long distance wind dispersal.

Wind dispersed pollen in higher plants is dry and tends to have a rather simple external morphology. There is a smooth or very finely roughened exine. In some conifers like *Pinus* the exine is formed into two large air-bladders which help the pollen grain to stay air-borne.

Wind dispersed seed, which is of course much larger, tends to have a rough surface. The tiny seeds of parasitic plants like *Orobanche*, which are produced in huge numbers, 2000 in each capsule, may be only 200 μm long. They have a testa with ridged reticulations. Orchid seeds range in weight from 0.3 to 14 μg. They are produced in huge numbers, up to several million per capsule. However, this seed may only be dispersed relatively short distances. It has also been suggested that the reticulations are an adaptation for water dispersal, trapping a film of water over the seed.

An alternative strategy which has been adopted by many plants is to have a large seed in a complex wind-dispersed fruit. There are the pappus parachutes of dandelions, *Valeriana* and *Clematis*. Many trees have winged fruits, samaras, which roll or spin. The spinning slows the fall to the ground so extending the scope of wind transport. In *Tilia* and *Carpinus* there are bracts connected to the fruit which act as parachutes. Most of these fruits serve only to disperse the seed relatively short distances away from the parent plant. Wind dispersal of seeds and fruits is not effective over long distances so that the floras of oceanic islands have a very low proportion of wind-dispersed species (Carlquist, 1967). Nevertheless within continental areas regular dispersal by only very short distances can be very effective. This is amply demonstrated by the very rapid recolonization of northern Europe after the last glacial period by a wide range of species, many with few obvious dispersal mechanisms.

8.3.4 Water dispersal

The flowers of water-pollinated angiosperms are reduced and often unisexual. The marine families, Posidoniaceae and Zosteraceae have thread-like pollen grains which increase the chances of lodgement on the stigma. The marine genera of the Hydrocharitaceae have a similar adaptation, but this time provided by the dispersal of a chain tetrad.

Water-dispersed fruits and seeds are relatively large or even very large. The largest fruit is that of the double coconut *Lodoicea maldavica* which weighs up to 90 kg and contains a two-lobed seed 50 cm long. It floats but apparently has lost its ability to be dispersed in this way because it is killed by prolonged exposure to sea water. Coconuts of *Cocos nucifera* are one seeded drupes. The fruit wall, the pericarp, has a skin-like exocarp, a fibrous mesocarp, which can be used for matting and cord, and a hard endocarp with three pores. The seed testa, a thin brown layer adheres to the endocarp. The endosperm is abundant with a white outer zone, the 'meat', and a liquid centre, the coconut juice. The embryo is at the base.

Water-dispersed seeds and fruits have bouyancy because of the presence of air chambers either as part of the fruit as in *Xanthium*, or pseudofruit in *Atriplex*, or in the testa as in *Menyanthes*. Interesting examples are the three species of cycad which are potentially water dispersed, *Cycas circinalis*, *C.rumphii* and *C.thouarsii*. They can survive floating in sea-water for at least

5 weeks. The outer sarcotesta is covered in wax and has resin- and wax-filled cells as well as starch cells. In other cycads it is this starch which is eaten by dispersing animals. Inside it there is a very tough sarcotesta which in the three named species has an extra inner spongy layer of large thin-walled parenchyma with intercellular air spaces which gives bouyancy (Murray, 1986).

8.3.5 Animal dispersal

Insect pollination represents a kind of spore dispersal though it does not disperse the plant. Insect dispersal of spores is found in the moss *Splachnum rubrum*, which has a capsule with a swollen, bright red, fouly smelling apophysis which attracts flies (Schofield, 1985). The importance of spore eating in the Carboniferous flora must have made insect dispersal important. Insect dispersed spores and pollen grains may be sculptured and sticky to help them adhere to the bodies of insects. The presence of pollinia in the orchids, asclepiads and *Acacia* lead to the dispersal of many pollen grains at once.

Animal dispersal of seed and fruits is called zoochory. Generally the taller the plant or the higher the growing position of an epiphyte the more likely animal dispersal is to be replaced by wind dispersal. Zoochory is very common and predominates in some communities. It appears to be especially common in moist and wet tropical forests (Gentry, 1982) and in heathlands or grasslands. This predominance of zoochory is probably only a relatively recent feature, that is of the last 70 million years. The evolution of fruits may have provided an important new kind of high energy food especially for small multituberculate and marsupial mammals which were becoming more diverse in the period after the origin of the angiosperms. Dispersal prior to that was probably more generalized (Tiffney, 1986).

Dispersal by mammals and birds is either external, ectozoochory, or internal, endozoochory. Different fruits or seeds have characteristic retention rates on or in the animal. Ectozoochorous fruits and seeds are adapted by having hooks, burrs or being sticky. There is some specialization for particular kinds of animals though some are simple and dispersed, for example, in the mud clinging to the animals feet.

A special example of zoochory is the dispersal of seed by ants, myrmechory. Myrmechory has been recorded in over 80 families of plants (Fenner, 1985). In some communities myrmechorous species account for 35% of all plants. It is especially common in dry heathlands. In some cases there is a close mutualistic relationship between the plant and the species of ant. Ant dispersed seeds may possess a special oil-body, an elaiosome, or an aril, as a food supply for the ants. The elaiosome is clipped off in the nest. However, seeds lacking these structures may also be dispersed by ants. One limitation of seed dispersed by ants is that the seed has to be relatively small. Ant dispersed seed has the advantage of being abandoned/planted in or near the fine nutrient rich tilth of an ant nest (Beattie and Culver, 1982).

Endozoochorous agents include mainly mammals and birds but also fish,

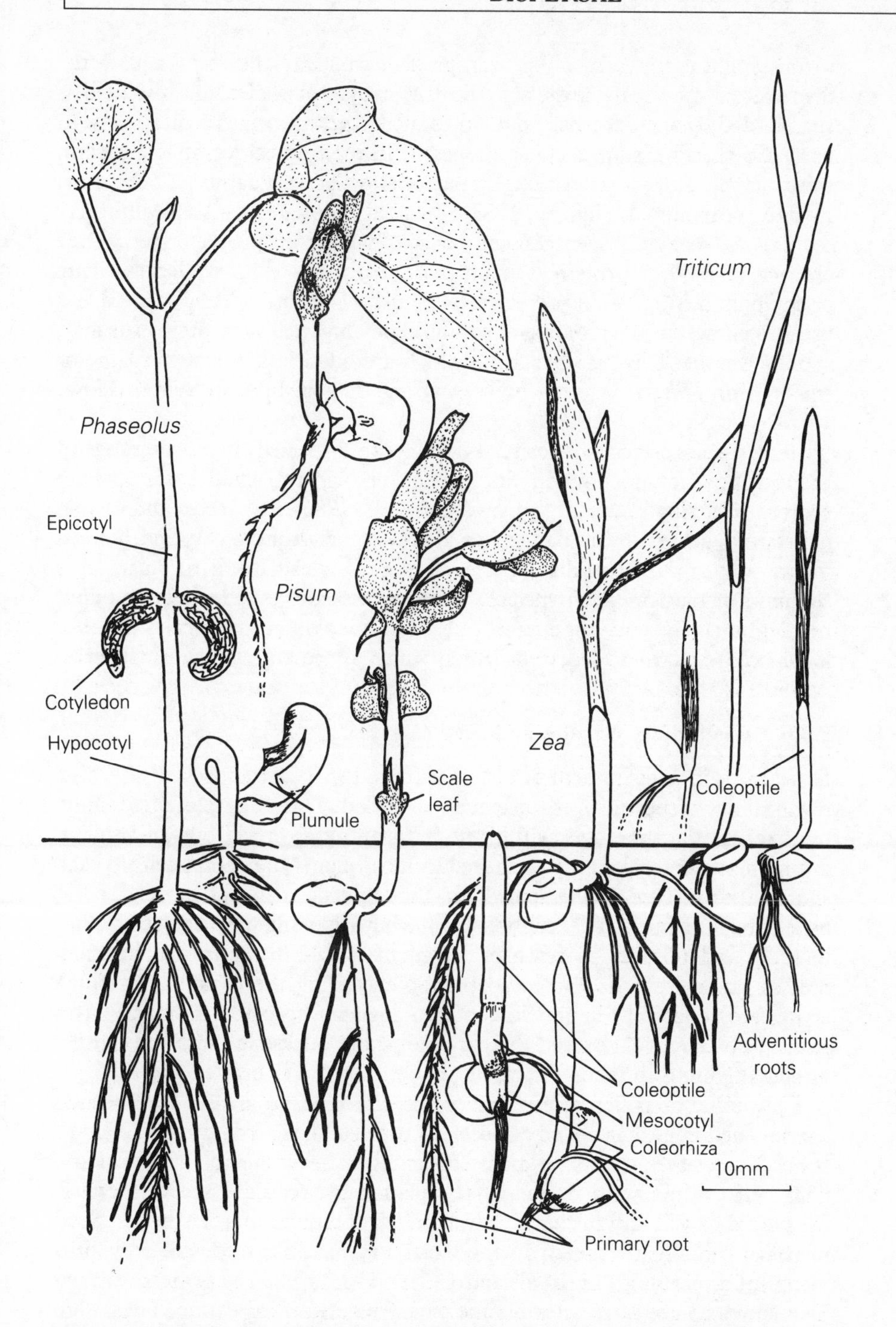

Fig. 8.8. Seedling germination and establishment in some crop plants. Epigeal germination in *Phaseolus*, hypogeal in the others. The grasses *Triticum* and *Zea* have a coleoptile, *Zea* a coleorhiza, and both show rapid replacement of the primary roots by an adventitious root system.

tortoises and earthworms. For example fish are especially important in the dispersal of species of river bank figs (Murray, 1986). Usually the seed is surrounded by an attractive, sometimes succulent, nutritious fruit. The seed has a digestion resistant testa or endocarp. In the tropical forest of Peru two thirds of the fruit species belong to one of two classes, adapted for dispersal by birds or mammals (Janson, 1983). Bird diaspores are small and unprotected drupes. They are scentless and black, blue, green, purple or red. They are rich in lipid or protein. Mammal (i.e. monkey) fruits are large arillate or compound fruits often protected by a husk. They are heavily scented and green, yellow, brown, orange or white. They are rich in protein, sugar or starch. Another class are bat fruits which are odourless or musky; large or small; white, whitish, green or yellow, and rich in lipid or starch (Howe 1986).

Some species have fruits or seeds which are dispersed by a wide range of agents. It is not uncommon for a range of dispersal syndromes to have evolved in a single genus (O'Dowd and Gill, 1986). In *Acacia* there is an interesting geographical pattern of dispersal syndromes. Myrmechory is found only in the Australian species. The seed is arillate. Bird dispersal is common in the American species. They have seeds embedded in and contrasting in colour to a bright fruit pulp. In Africa the many species produce indehiscent or tardily dehiscent fruits, which are eaten by large herbivores.

8.3.6 The timing of spore and seed release

Differences in development of different reproductive structures on the same plant allows dispersal over an extended period. One example is the filmy ferns where the sporangia within the indusium are carried outwards by an extending receptacle as they mature. In most plants the spores within each sporangium mature at the same time though they may not be released at once. The hornworts are exceptional in having a sporangium which continues to grow and releases spores over a long period. The peristome in many mosses can be regarded as an adaptation extending the period over which spores are released. The indehiscent sporangia and capsules of many lower plants also effectively extend the period of spore release because the capsule or sporangium wall takes a variable amount of time to break down.

The timing of pollen release is very precise in some grasses with related species shedding their pollen at different times of day. For example, *Agrostis stolonifera* releases its pollen in the morning and *A. tenuis* in the afternoon. This helps to maintain reproductive isolation between the species because the period of viability of the air-borne pollen is limited to a few hours. Seed dispersal can also be timed. In the violet *Viola nutallii* seeds are only shed in the mid-morning (Turnbull and Culver 1983). The seeds are dispersed by ants which are active at this time of day whereas seed-eating rodents are active at night.

8.4 ESTABLISHMENT

Spores are at a disadvantage over seeds in the establishment phase of growth. They are smaller and do not have large food reserves to support the young plantlet for an extended period. The seeds of plants of competitive habitats can be large allowing the rapid establishment of the plant and its survival for a longer period. Endozoochorous seed can be quite large. Large seeds may be retained longer in the animals gut and therefore dispersed further. The large size of the seed gives the seedling a large energy source to enable it to establish itself rapidly. However one disadvantage that these seeds suffer is that they are concentrated in the animals gut. Then there is very strong competition between seedlings growing up from a single faecal deposit.

Some species have adaptations for planting the seed. The awns of grasses flex with changes in humidity so that the 'fruit' moves around searching for a small crevice or hole in which it can lodge. The single cotyledon of monocots may have evolved as a adaptation for pushing the radicle into the soil. In some seedlings, the embryo breaks out of the testa (epigeal germination) (Fig. 8.8). In others the cotyledons remain inside the testa and only the radicle, and the stem above the cotyledons, called the epicotyl, bearing the first true leaves, breaks out of the seed (hypogeal germination). The latter have large cotyledons containing food reserves which can be rapidly utilized.

8.5 A BEHAVIOURAL CLASSIFICATION OF PLANTS

Grime (1979) has described the ecological strategies of herbaceous plants on a system of three different kinds of characters, each an axis so that the behaviour of the plant can be represented graphically. One axis represented the ability of a plant to compete, another the ability of a plant to colonize. This is very similar to the categorization of plants into 'r' and 'K' strategists which has proved so very useful in studying plant behaviour. The third of Grime's axes recorded differences in adaptation to the physical environment, coping with environmental stress (Fig. 8.9).

A limited number of marker species were scored so that they could be located to one of seven points within the triangular system of axes (C, C-R, C-S, C-S-R, S-R, R and S) using the key in Table 8.1.

Vegetation quadrats were then scored for the presence and absence of these marker species. Each quadrat was then located to one of 91 hexagonal zones within the triangular system on the positions and frequencies of marker species. In this way the detailed distribution of a species, including non-marker species, based on the distribution of all the quadrats in which it was found, could be plotted. The distributions of three very different species are illustrated in Fig. 8.9. They summarize the behavioural limits of the species within the vegetation that was in the Sheffield region of northern England.

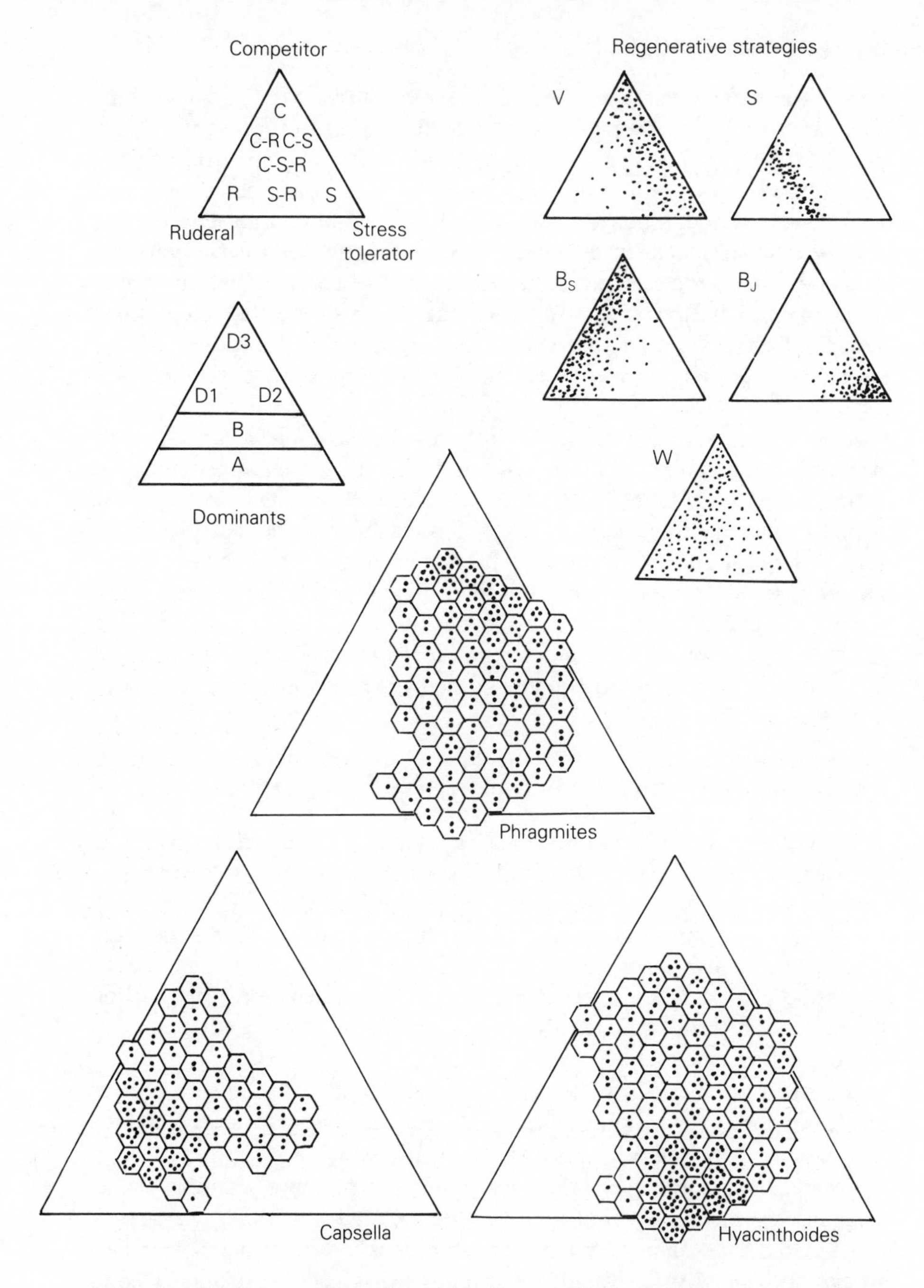

Fig. 8.9. Plant strategies adapted from Grime *et al.*, 1988. Having V = vegetative expansion, S = seasonal regeneration, B_s = a persistent seed bank, B_j = persistent juveniles, W = numerous widely dispersed seeds. A = adapted to extremely disturbed or unproductive habitats, B = subordinates, D = potential dominants, D1 = ruderal dominants, D2 = competitive dominants, D3 = stress-tolerant dominants.

Table 8.1 A key to strategies in herbaceous plants (adapted from Grime, 1984)

Key	Result
1. Perennial (polycarpic)	2
Biennial/annual (monocarpic)	7
2. Vernal geophyte	S-R
Not vernal geophyte	3
3. Slow leaf turnover	4
Rapid leaf turnover	5
4. Shoots tall and/or laterally extensive	S-C
Shoots short and not laterally extensive	S
5. Rapid proliferation and shoot fragmentation	C-R
Shoots not fragmenting rapidly	6
6. Shoots tall and laterally extensive	C
Shoots short and creeping	C-S-R
7. Small and slow growing	S-R
Potentially fast growing (and large)	8
8. Flowering precocious	R
Flowering delayed	C-R

Abbreviations: C = competitor, R = ruderal, S = stress tolerator

If this kind of analysis was carried out elsewhere the list of marker species used and possibly even the key characteristics used would have to be modified. This limits the value of this kind of analysis for comparing different ecosystems and the behaviour of one species in different ecosystems. Nevertheless, the method does provide a good method of comparing the strategic composition of closely related vegetations such as two species rich calcareous grasslands. The distribution of species characterizes a vegetation.

Dominants (D1–D3) make a major contribution to that vegetation in biomass and also by modifying the environment for other species. Highly competitive species (D3) are potentially dominants only in highly productive undisturbed conditions. Ruderal-dominants (D1) are potentially dominant where there is a single predictable annual major disturbance like flooding or ploughing. Stress-tolerant competitors (D2) are relatively slow growing but potentially dominant in moderately productive conditions such as nutrient-poor soils or in shade. Area B is occupied subordinate species which can coexist with dominants. They either have strategies which allow them to evade the stresses imposed by the dominants or are relatively resistant to those stresses. They may be a conspicuous element in the vegetation though they do not modify the environment or contribute a large amount to the total biomass. Area A is occupied by species which are excluded by dominants. They grow only in highly stressed environments where the dominants cannot exist or in very disturbed habitats where the dominants cannot become established.

Species with very different life forms can appear in the same position on the graph. This is especially true of stress-adapted species so that species adapted to shade, a woodland herb, a spring-flowering geophyte like *Hyacinthoides non-scripta* (Fig. 8.9) for example, or xerophytes, such as a leaf succulent *Sedum album*, growing on exposed rock, may have overlapping graphical distributions. This kind of analysis emphasizes that such very different species can share some important characteristics in the way they grow, propagate and disperse. Most species have a fairly broad graphical distribution though they favour a particular kind of niche. The tall grass, *Phragmites communis*, which is a strong competitor, actually overlaps a large part of the graphical distribution of *Hyacinthoides*.

Alternatively species with very similar basic morphologies can have very different graphical distributions. For example the grasses *Phragmites communis*, *Poa annua* and *Nardus stricta* are competitor, ruderal and stress-tolerator, respectively. Some of the most important aspects of a species behaviour are not related directly to their morphology.

One of the most important aspects of behaviour is the regenerative strategy. Species with characteristic regenerative strategies have different predicted distributions on the triangular diagram. Species which expand primarily vegetatively (V) tend to be competitors or stress tolerators. Species with seasonal regeneration (S) share some features of the most extreme ruderals. Species with a persistant seed bank (B_s) are either competitors or ruderals. Seed remains dormant until it can germinate into a gap in the vegetation. The fast-growing ruderals are dominant at first but are replaced later in the succession by slower growing competitors. In contrast stress-tolerators generally have persistant juveniles (B_j). They grow very slowly. Species which produce widely dispersed diaspores (W) are found in all categories of plant but are concentrated among the ruderals and competitors.

Cultivated plants: conclusion

9

9.1 EXPLOITED DIVERSITY, THE USES OF PLANTS
9.2 THE GENETIC HISTORY OF CROPS
9.3 FUTURE PROSPECTS
9.4 CONCLUSION

SUMMARY

All parts of at least some plants have been directly exploited for food, for nutrients or flavourings: roots (root-tubers, tap-roots), stems (tubers, sugar cane), leaves, flowers (nectar and pollen in honey), seeds and fruits. They have also been cultivated for food for domesticated animals. They have been exploited as a source of dyes, fibres, medicines and poisons. They have provided materials for crafts and construction.

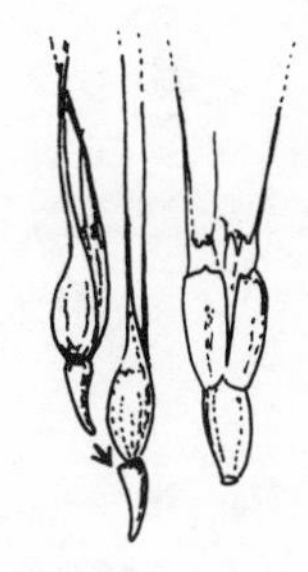

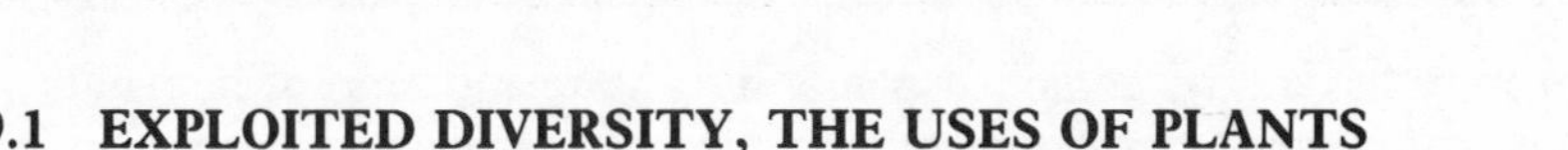

9.1 EXPLOITED DIVERSITY, THE USES OF PLANTS

Some families have provided a large proportion of species exploited by mankind. In the forefront are the cereals grasses (Poaceae) which provide the world with most of its carbohydrate either directly in the grain crops, wheat, rice, maize, barley, or as feed to animals. Over 70% of all the world's cropland is occupied by cereals (Tivy, 1990). Other important families are the Fabaceae (legumes and beans of all sorts), the Brassicaceae (leaf, roots and oil-seed) and the Solanaceae (fruits and the potato). Many crop species have multiple uses. They started being used for one purpose and then a different part of the plant was exploited. The brassicas are outstanding example of a kind of plant in which almost all parts of the plant have been exploited in one way or another. *Brassica oleracea*, *B.campestris*, and *B.napus* are closely related species which exist as an astonishing range of crops: cabbages, kales, chinese mustards and cabbages (wong nga baak, baak choi, choi sum, etc.), broccolis, cauliflowers, turnips, swedes, kohlrabi, sarson, oil-seed rapes.

Plants provide one quarter of all prescribed medicines. A few are listed in Table 9.1. Some of the uses of plants have been superseded by advances in chemical engineering and construction, but not many. Plants are the earth's great renewable resource. There is great unrealized potential. For example only 5000 species have been tested exhaustively for medicinal value.

Table 9.1 Drugs from plants

Source	*Drug*	*Ailment (use)*
Agave sisalina	steroids	(hormonal)
Catharanthus roseus	vincristine	childhood leukaemia
Cephaelis ipecacuanha		amoebic dysentry
Chondrodendron tomentosum	D-tubocurarine	surgery muscle relaxant
Cimifuga racemosa snakeroot	reserpine	high blood pressure
Cinchona calisaya	quinine	malaria
Colchicum colchicine		gout, arthritis
Datura stramonium	stramonium (hyoscamine)	asthma
Digitalis digoxin		irregular heartbeat
Dioscorea spp.	diosgenin	(contraceptive pill)
Glycorrhiza liquorice	carbenoxalone	peptic ulcers
Hydnocarpus kurzii	chaulmoogra oil	leprosy
Papaver somniferum	morphine etc.	(analgesics)
Physostigma venenosum	physostigmine	myasthenia gravis
Podophyllum peltatum	etoposide	warts, cancer
Posoquiera Amazonian oak		coagulate proteins
Salix bark	aspirin (now synthesized)	(analgesic)
Tinospora crispa		diabetes

9.1.1 Directed evolution

Humanity has always exploited plants but the relationship of humanity to plants is different from that of other herbivores. Almost uniquely man is a gardener, a cultivator. Actually if primitive cultures tell us anything about early man the first gardeners were likely to be women. What women may have done in the use of plants was to promote the chosen plant species. From the first domestications of plants 6000–7000 years ago some species have come under intense evolutionary pressure. They have evolved rapidly. Their diversity has been exploited and their genetic pool fragmented into thousands of local cultivars or land races.

Early selection of improved varieties may have been more or less accidental. Until this century we have not had a scientific theory which explains the response to selection. Nevertheless the naive promotion of favoured varieties by the saving of seed or clones generation after generation has produced remarkable results.

In relation to the total number of plant species only a tiny handful have been domesticated by man; less than 100 important ones from a possible 250 000 (<0.04%). Just the four species wheat, rice, maize and potatoes provide over 50% of all the world's production by weight. Why these species? Primitive cultures exploit a much wider range of wild species. Obviously the chosen crops had some features which favoured them. However, there must have been a large element of chance in their selection. It is startling to think we might have ended up with a very different array of crops.

9.1.2 The features of crops

Simmonds (1979) has listed the features in which crops differ from their wild relatives. Yield has been increased markedly. The size of the harvested organs and in some cases the number of harvested organs per plant has increased. There have been changes in branching pattern associated with changes in yield. Increased yield has been achieved in some cases by selecting a more favourable partition towards the utilized part of the plant; the amount and size of grain produced relative to the foliage in cereals for example. In this way smaller plants have been favoured.

A whole set of changes has improved the quality of the crop especially as a food; reduced spininess and toxic constituents, increased sugar or starch content, increased attractiveness. Important adaptations have been related to agricultural practice. Inflorescences which do not break apart allow easier harvesting. One of the most important changes in this century has been the selection of varieties which are able to make use of the high levels of fertilizers made available to them. Cultivars have become adapted to different climatic regimes and different photoperiods (that is, day-lengths in different latitudes).

In some respects crop plants have been pushed towards 'r' on the r–K continuum (Section 8.5). They have faced less competition, and stress from the environment has been managed, by irrigation and the application of fertilizers, for example. A rapid or annual life cycle has been selected. Smaller plants have been selected sometimes because they are associated with a shorter life cycle. Seeds and tubers lacking dormancy and with rapid and uniform germination have been selected. Synchronous development and ripening has been selected. They differ from ruderals in their diminished dispersability.

9.2. THE GENETIC HISTORY OF CROPS

Crop plants have a complex genetic history which nevertheless has been worked out in some cases. There have been mutations with a profound effect on the morphology of the plant and also the gradual accumulation of genes with minor but cumulative effect. Hybridization has been an important source of new variation. Polyploidy is especially frequent. Some crops are self-pollinating derivatives of outbreeding ancestors. Inbreeding is widespread because it creates a homogeneous highly homozygous cultivar amenable to agriculture. However the crop may also be very vulnerable to disease. Some crops have become adapted for clonal reproduction at the expense of sexual reproduction. In other crops like maize techniques have been developed to produce hybrid varieties which are grown as highly heterozygous but homogeneous crops.

Many of these genetic changes occur naturally in wild plants and have resulted in the patterns of diversity seen in nature. An interesting example is the gamoheterotopic mutation in maize which converted a male inflorescence into a female cob. The same kind of major mutation has been suggested as being the one which gave rise to flowering plants from something like the Bennettites (Meyen, 1988). The evolutionary history of one of the major crops of the world, wheat (Feldman, 1976), is illustrated in Fig. 9.1.

9.3 FUTURE PROSPECTS

It is clear that the 'genetic engineering' of plants has been going on for thousands of years through the process of hybridization and selection, first by farmers and then by plant breeders. What has become possible more recently is the insertion of completely foreign, even animal, genes into plants and to get them expressed there (Jones and Jones, 1989). This new technology has yet to prove itself but there are exciting prospects ahead.

Arising from the new technology is the ability to identify and target particular genes. Since it seems that many even profoundly different looking species differ only in a few genes, some of which are only expressed at critical moments in development, this ability may lead us to an understanding of the process of plant evolution in a new way. At long last a great missing chapter in our understanding of plants may be revealed to us; the control of development. We live in interesting times.

9.4 CONCLUSION

There can be little doubt that the land plants are going through more rapid changes now than they have ever before. Within a few thousand years huge areas of the world's vegetation has become dominated by the influence of human beings. First, organized pastoralism and then civilization with its reliance on agriculture, have created new plant communities and even new kinds of plants. The most diverse and ancient vegetations, the tropical forests managed to escape until this century. They have taken many millions of years to evolve and we have scarcely begun to understand them, and yet almost incredibly they are likely to be lost or changed out of all recognition within the space of our own lifetime.

Even the tropical forests have come to be regarded primarily as a resource to be exploited for the benefit of mankind. One sees this, even in arguments put forward for the preservation of the forests. 10 km^2 of Amazonian forest can contain 2200 different species of plant. They might be useful as new types of crops or by providing new drugs. All true, but how depressing it is to see the forests only as a possible commodity to be exploited for profit. Scientific research is becoming a tool of the market place; it is to be concentrated only on the exploitable, the potentially profitable. What a poverty of the imagination and spirit this represents!

The effect of acid rain has concentrated our attention on the environmental damage that industrialization has caused in under 200 years. However, we can be sure the plants will respond to the changes, including the climatic change which now seems inevitable. They have responded in innumerable

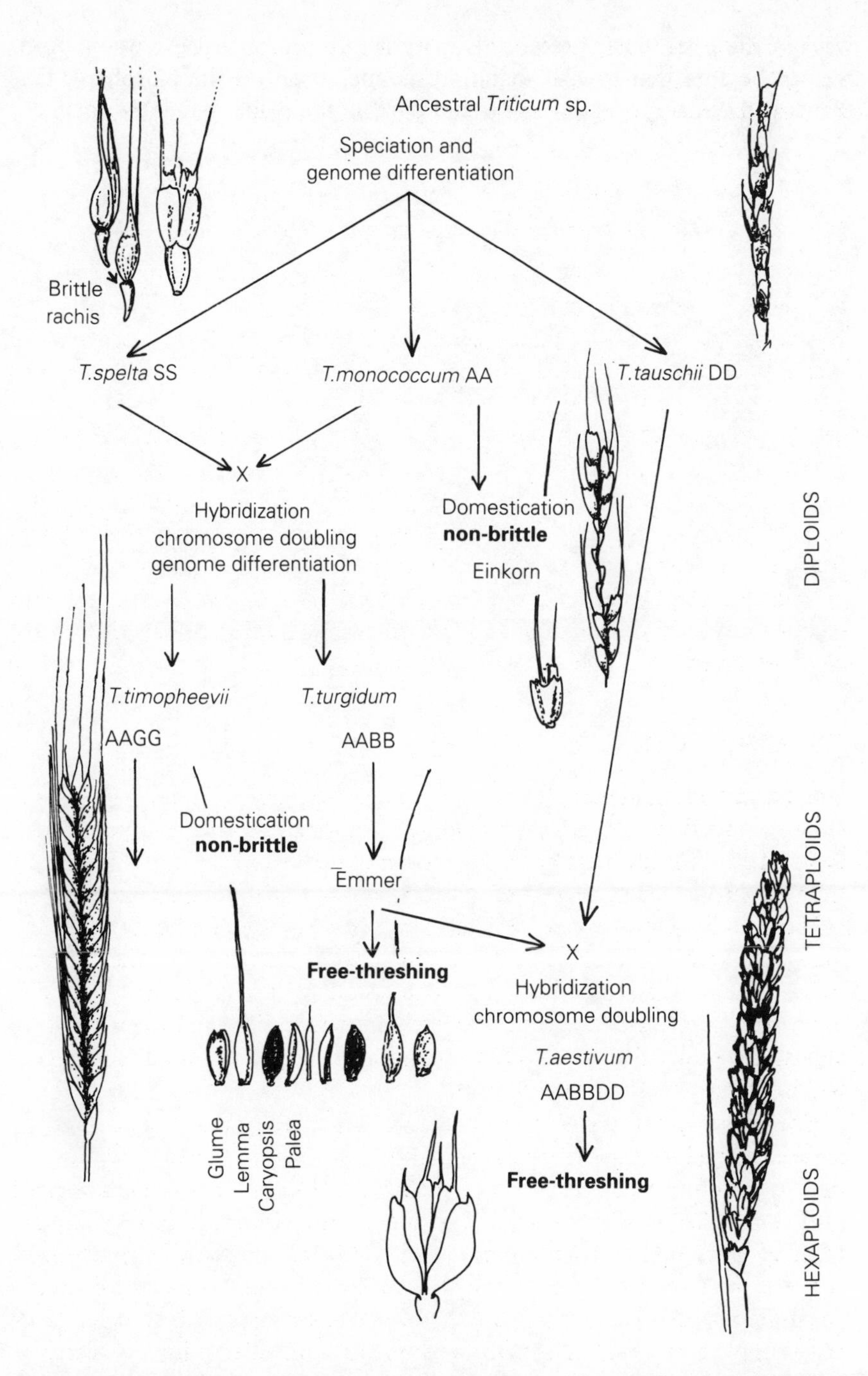

Fig. 9.1. The evolution of wheat showing patterns of mutation, hybridization and chromosome doubling (adapted from Feldman, 1976). A brittle rachis makes harvesting difficult and non-free-threshing grain produces poor quality flour. Another important change has been in the number of grains in each spikelet.

ways in the past. Their present diversity is a record of that evolution. And we can be sure that even if mankind pollutes itself off the biosphere, the plants will remain, evolving anew and making the planet habitable again.

Glossary

a- ab- Prefix meaning away from.
a- an- Prefix meaning lacking, without.
abscission The formation of a layer of cells which become separated from each other thereby cutting off the organ so that it falls.
abaxial On the side facing away from the axis.
achene A single seeded dry **indehiscent** fruit.
achlamydeous Lacking a **perianth**.
acro- Prefix meaning apically, at the tip
acrocarpous In mosses where the **sporophyte** emerges apically from the **gametophyte** (c.f. **pleurocarpous**).
acropetal Towards the apex (c.f. **basipetal**).
actinomorphic Radially symmetrical.
acyclic Applied to flowers with parts arranged in a spiral and not whorls.
ad- Prefix meaning to, towards, near.
adaptation A characteristic which improves the probability of survival or reproduction (as a noun) or a process of responding physiologically to the environment.
adaxial On the side facing upwards or towards the axis.
adnation The fusion of two unlike parts together.
adventitious Applied to roots arising not in the primary root system but from the stem or of buds arising from the roots.
aerenchyma A tissue with many large intercellular air spaces.
aerophyte An epiphyte without roots.
aestivation The way petals and sepals overlap.
albuminous Applied to seeds having an endosperm.
alternation of generations The life cycle of all land plants, where a **gametophyte** gives rise to a **sporophyte** after fertilization and the latter gives rise to the gametophyte after **meiosis** and **sporogenesis**.
ament A catkin.
amphi- Prefix meaning on both sides, double, both
amphigastria The underleaves of leafy liverworts.
amphithecium The embryonic outer part of the developing **sporangium** which becomes the **jacket**.

amphitropous An orientation of the **ovule** so that it is bent, with the **micropyle** near the **funiculus**.
anatropous An arrangement of the **ovule** where the ovule is straight but the **micropyle** is near the **funiculus**.
androdioecious Having separate male and **hermaphrodite** plants.
androecium The male reproductive area or organ.
anemophily Wind pollination.
anisophylly Having leaves of clearly unequal sizes.
angiosperms Flowering plants.
anisospory A condition where spores of different sizes are produced within the same **sporangium**.
annual A plant growing from seed and reproducing by seed within a single year.
annulus A line of cells distinguished from others in the wall of a **sporangium** (**capsule** in bryophytes) and having a function in dehiscence.
anther The part of a stamen containing pollen.
antheridiophore A stalked outgrowth of the **thallus** bearing **antheridia** in some liverworts.
antheridium A male sex organ, usually sac like, containing many sperm.
anthesis The flowering period from bud opening to seed setting.
anthophytes Plants with flowers or flower-like reproductive structures, includes **angiosperms**, gnetophytes and some fossil groups.
ant- anti- Prefix meaning against, opposite to.
anticlinal At right angles to the surface.
apetalous Lacking petals.
apo- ap- Prefix meaning downwards, below, from, away.
apocarpous Having free, separate **carpels**.
apophysis The basal part of the moss **capsule**, sometimes photosynthetic and having stomata.
apoplasm (apoplast) The part of a plant exterior to the **plasmalemma**, made up of intercellular spaces and cell walls.
apposition growth Extension where the apical meristem continues to grow but is replaced as the main apex by the growth of an **axillary** meristem.
archegoniophore (carpocephalum) A stalk (=phore) bearing **archegonia**, and **sporophytes** after **fertilization**, in some liverworts.
archegonium A female sex organ, bottle or flask shaped and containing a single egg.
areole A part of a cactus, a modified short shoot, the area occupied by a group of spines; the area in a leaf bounded by the finest terminal veins.
aril The succulent part of some seeds, sometimes bright, an outgrowth of the **funicle**.
arista A long fine point to a leaf.
auricle A small ear-like appendage at the base of some leaf lamina.
awn A bristle-like point to a leaf or bract.
axil The angle between the adaxial part of a leaf and a stem.
axillary Lateral, growing out of the **axil**.
axis The main stem to which other organs are attached as appendages.

bark The tissue in the secondary stem outside the **vascular cambium** made of the **phloem**, remnants of **cortex**, **phelloderm**, **phellogen** and **phellem**.
basipetal Towards the base.
berry A succulent fruit, usually many seeded.
bicollateral Having phloem on both inner and outer sides of the xylem in vascular bundles.
bisexual Having both sexes in the same plant, either **monoecious** or **hermaphrodite**.
blade The leaf **lamina** of plants with leaf sheaths.
bordered pit A pit where the secondary wall over-arches the primary to give it the appearance of having a halo.
bract A leaf associated with a reproductive structure.
bracteate With bracts.
bryophytes A group containing mosses, liverworts and hornworts.
bulbil A kind of **propagule**, small bulb or tuber, usually arises in an **axil**.
bundle sheath A layer(s) of different cells surrounding the vascular bundle in leaves.

caducous Falling off early, especially of sepals and stipules which fall off as the bud opens.
caespitose Tufted.
calyptra The cap covering the capsule in mosses, derived from the top of the archegonium.
calyx The sepaloid part of the perianth, usually green and protecting the flower in bud.
cambium A lateral meristem which gives rise to secondary tissues such as the wood or bark, e.g. vascular cambium and phellogen, frequently applied just to the vascular cambium which gives rise to the **xylem** and **phloem**.
campylotropus An arrangement of the **ovule** where it is curved, the **funiculus** is in a median position.
capitulum A kind of inflorescence, a head, characteristic of the **Asteraceae**.
capsule The spore containing part of the sporophyte in bryophytes, or dry **dehiscent** multi-seeded fruit in angiosperms.
carpel A leaf-like unit of a pistil bearing ovules near its margin. A pistil may be formed from one or more carpels.
caryopsis A single seeded dry indehiscent fruit where the seed is fused to the fruit wall, especially in grasses.
casparian strip The **suberized** area or strip running around cells, especially in the **endodermis** which is impervious to water.
cataphyll A scale leaf or bud scale.
catkin (ament) A spike of unisexual flowers, usually pendulous.
caudex The main axis of stem and roots together.
caudiciform Having a swollen caudex.
caulonema A part of the young **gametophyte** of mosses, a **uniseriate** filament with oblique cross walls.

chalaza The basal region of the **ovule**.
chamaephyte One of Raunkiaer's life forms, a small shrub.
character A feature.
characteristic A defining feature of a group.
chimera A plant with cells derived from two kinds of plants, contrived by grafting or tissue culture.
chlamydeous Having a **perianth**.
chlorenchyma Parenchymatous cells with chloroplasts.
chloronema The filamentous green photosynthetic part of the young moss **gametophyte**.
choripetalous **Polypetalous**.
cincinnus A kind of cymose inflorescence (see Fig. 5.8).
clade An evolutionary lineage.
cladistics Systematic studies which attempt to discover clades.
cladode A flattened branch which looks like a leaf.
claw The narrow lower basal part of some petals and sepals (c.f. limb).
cleistogamy Pollination within an unopened floral bud.
cline A continuous gradation in the measurement or frequency of a character with either small or large geographical distance, sometimes correlating with an environmental variable.
clone A vegetative offshoot of a plant, not genetically different from the parent.
coenocarpium (sorosis) A kind of fleshy compound fruit, e.g. pineapple.
coenocytic Having a multinucleate mass of protoplasm.
coleoptile The protective sheath around the **plumule** in seedlings.
coleorhiza The protective sheath around the **radicle** in seedlings.
collenchyma A cell or tissue type where the cell walls are primary and are thickened sometimes asymmetrically.
columella The sterile central rod or dome of tissue in some sporangia or a rod like part of the **exine** below the **tectum**.
commissural bundle A leaf vein connecting the main, usually parallel veins, of a leaf **lamina**.
companion cell A cell developmentally related and adjacent to sieve tube element.
cone (strobilus) A group of **sporophylls** usually arranged spirally on an axis.
conifer One group of **gymnosperms**, cone-bearing seed plants.
connation The fusion of similar parts together.
connective Part of a **stamen** between the thecae.
convergent evolution (convergence) Evolution of similar, but usually not homologous, organs in different only distantly related groups.
coprolite Fossilized faeces.
corm Basal storage organ of the stem.
corolla The petaloid part of the **perianth**, often showy and attractive or modified to promote pollination.
corona Frill of the '**perianth**', either derived from the **stamens** or the **tepals**.

cortex The tissue of the primary stem which is not pith, **stele** or **epidermis**.
corymb A flat-topped racemose inflorescence.
costa Midrib of a leaf.
cotyledon A seed leaf, part of the **embryo** in the seed.
cryptophyte A geophyte.
culm The upright shoot of grasses and similar plants.
cultivar A cultivated variant.
cupule The cup-shaped organ surrounding the seed or fruit in some seed plants.
cyathium The kind of inflorescence characteristic of the **Euphorbiaceae**.
cycads A group of gymnosperms, mostly small, **pachycaul**, dioecious trees.
cyclic Having floral parts arranged in whorls.
cyme (cymose) A kind of determinate inflorescence (see Fig. 5.8).
cypsela A single seeded dry **indehiscent** fruit where the testa is fused to the **pericarp**.

de- Prefix meaning downwards, from.
deciduous Shedding leaves at the end of the growing season.
decumbent Prostrate with the apex pointing upwards.
decussate An arrangement on a stem where at each successive node the leaves are at 90° to those at the previous node.
dehiscent Opening to release seeds, pollen or spores.
deme A variant, a population, an evolutionary unit within the species.
determinate A kind of growth which is strictly limited and decided early in development.
diadelphous With two groups of **stamens** united by their **filaments** or one group and a single free stamen.
diaspore (disseminule) A dispersal unit, either a spore, seed or one-seeded fruit.
dichasium A kind of cymose inflorescence (see Fig. 5.8).
dichotomous Applied to branching where there is more or less equal division of the apical meristem and to venation which branches equally.
dicliny The separation of male and female reproductive organs, either into different regions of the same plant (**monoecy**) or on different plants (**dioecy**).
didynamous Having 4 stamens, 2 short and 2 long.
dioecy (dioecious) Having separate male and female plants.
diploid Having nuclei where there are pairs of homologous, chromosomes, the normal condition in the sporophyte of land plants.
disk The part of the **receptacle** around the base of the **ovary**, often nectiferous.
disseminule A part of a plant adapted for dispersal, a diaspore; either seed or fruit or spore or a vegetative part of the plant.
distichous An arrangement on two opposite sides of an axis.
double fertilization Fertilization in angiosperms where one male gamete

fuses with the egg to form the **zygote** and another fuses with the polar nuclei to form the triploid **endosperm** nucleus.
drapanium A kind of cymose inflorescence (see Fig. 5.8).
drupe A fleshy, usually single seeded fruit, in which the **endocarp** is stony.
drupecetum An aggregation of drupelets, commonly and mistakenly called a berry, e.g. blackberry.
drupelet A small **drupe**.

ecocline A **cline** in which the continuously varying character correlates to an ecological gradient.
ecodeme A variant associated with a particular ecological situation.
ecotype A genetic variant associated with a particular ecological situation.
ectexine The outer region of the **exine** of a pollen grain.
ectohydric Applied to a plant which obtains water by absorbing it through the surface of the shoot.
elaiosome (elaiophore) An oily organ which attracts insects, especially ants.
elater An elongated accessory organ found with the spores in some sporangia; unicellular, multicellular or part or the spore wall.
elaterophore The column or dome to which the **elaters** are attached in some liverworts.
embryo A young plant developing from the zygote after fertilization.
embryogenesis The development of the **embryo**.
embryosac The female **gametophyte** of angiosperms.
enation A superficial outgrowth of a plant.
endarch Centrifugal development of the primary **xylem** (c.f. **exarch**).
endexine The inner part of the **exine** of a pollen grain.
endo- Prefix meaning inside, inwards.
endocarp The inner part of the **pericarp**.
endodermis A layer (sheath) of cells, usually **suberized** and irregularly thickened, surrounding the **stele** in many roots and some stems.
endogenous Arising within the plant.
endohydric Applied to plants which obtain water through their roots or **rhizoids** and transport it internally in a **stele**.
endosperm The triploid nutritive tissue in the seed of **angiosperms** (c.f. **perisperm**).
endosporic Developing within the spore wall.
endothecium The inner part of the developing **sporangium** which becomes the sporogenous cells and **columella** in some groups.
entomophily Insect pollination.
ep- epi- Prefix meaning on, upon or attached to.
epicalyx A whorl of bracts outside the calyx.
epicormic Applied to a dormant bud or growth from such a bud which may have become hidden within the bark by subsequent growth.
epicortical Arising within the stem **cortex**.
epicotyl The first **internode** above the **cotyledons** in seedlings.
epidermis The outer layer of the primary plant.

epigyny The condition where the **ovary** is inferior, the **perianth** and **stamens** arising around the base of the **style**.
epipetalous Inserted on the petals.
epiphyll A plant living **epiphytically** on the leaf of another plant.
epiphyllous Inserted on the leaves.
epiphyte A plant living on another plant.
eusporangium A kind of **sporangium** with a thick **multistratose** wall.
e- ect- ecto- ex- Prefix meaning from out, without, on the outside.
exarch Centripetal development of the primary xylem (c.f. **endarch**).
exine The outer part of the wall of the pollen grain, **ectexine** and **endexine**.
exocarp The outer part of the **pericarp**.
exserted Protruding beyond associated structures.
extrorse Facing outwards.

fascicle (fascicular) A bundle (pertaining to the bundle).
fastigiate Having branches erect and running more or less in parallel.
filament The stalk part of a **stamen**.
floret A small flower, one of several in an inflorescence.
foliose Leafy.
folivore A leaf eater.
follicle A dry dehiscent fruit of a single **carpel**, which opens along one edge.
form (a) A taxonomic category below **variety** (varietas).
free central placentation Attachment of the ovules to a central column of the ovary.
free-nuclear phase Nuclear division, mitosis, without cell division giving rise to a **coenocytic** mass.
frond The leaf of a fern.
frugivore A fruit eater.
fruticose Shrubby, bushy.
funiculus (funicle) The stalk of an ovule or seed attaching it to the placenta.
fusiform Elongated and spindle shaped.
fusoid Rounded in section, wide in the middle but not markedly elongated.
gametogenesis Gamete formation, occurs by mitotic divisions in plants.
gametophore A stalk bearing the sex organs.
gametophyte A plant producing the gametes, the **haploid** phase of the **alternation of generations**.
gamopetalous Sympetalous.
gemma An asexual **propagule** especially in mosses and liverworts.
genecology A study which relates the genetic variation of populations and species to their evolutionary and ecological situation.
generative (genetic) spiral The sequence of development of primordia in the apical meristem.
genetic drift The accidental change in the spectrum of genetic variants present due to sampling which occurs especially in small populations.
genotype The genetic constitution of an organism (c.f. phenotype).

geophyte (cryptophyte) One of Raunkiaer's life forms, plants with underground survival organs such as a bulb, **corm**, tuber or rhizome.
geotropic Growing towards the ground or downwards.
glabrous Hairless, smooth, lacking **trichomes**.
glaucous Covered with a white or grey waxy bloom.
glume The lower **bract** in a grass spikelet.
gnetophytes A group containing three genera of peculiar **gymnosperms**, *Gnetum*, *Ephedra*, *Welwitschia*.
graminovore A grass eater.
ground meristem A general and poorly localized meristem.
guard cell One of a pair of cells surrounding the pore in a stoma.
gymnosperms A group of seed plants, all seed plants except **angiosperms**, living representatives include **cycads**, *Ginkgo*, **conifers** and **gnetophytes**.
gynobasic Arising from the base of the **pistil**.
gynodioecious Having separate female and **hermaphrodite** plants.
gynoecium The female reproductive area or organ.
gynostemium Part of orchid flower, the column.

habit The characteristic way of growing of a species.
habitat The place where a plant grows, including all the environmental and biotic factors which impinge on it (c.f. niche).
hadrom Conductive tissue analogous to the **xylem** in some mosses.
halophyte A plant adapted to saline conditions.
hapaxanthy Transformation of a shoot into a reproductive axis.
haploid Having nuclei with a single chromosome of each type present (c.f. **diploid**), the normal condition in the gametophyte phase of the life cycle.
haustorium The organ by which a parasite attaches and absorbs nutrients from its host, combines parasite and host tissue
head A dense crowded inflorescence of florets, of which **capitulum** is a special kind.
hemicryptophyte One of Raunkiaer's life forms, a plant which dies back to buds at the soil surface.
hemiparasite A plant parasite which is photosynthetic (c.f. **holoparasite**).
hep A kind of fruit of where achenes are surrounded by a fleshy receptacle or transformed **hypanthium**.
herbaceous Non-woody.
herbivore A plant eater.
hermaphrodite Having both male and female within the same reproductive structure.
hesperidium A kind of berry like an orange.
heteroblastic Development where the organs are produced in sequence but the later ones are markedly different from the early ones.
heterochlamydeous Having a **perianth** differentiated into **calyx** and **corolla** (c.f. **monochlamydeous**).
heteromorphic Having alternate plants with different morphologies e.g. a heteromorphic **alternation of generations** has alternating **gametophyte**

and **sporophyte** with different morphologies (c.f. **homomorphic**, **heterothallic**).

heterospory (**heterosporous**) A condition where the **sporangia** are of two sorts, **megasporangia** containing few large **megaspores** and **microsporangia** containing many small **microspores**.

heterostyly Having floral morphs with styles of different lengths, associated with **self-incompatibility**.

heterothallic Having gametophytes of different morphologies, sometimes associated with the different distribution of sexes and arising from different sizes of spore.

hilum Point of attachment of seed by **funiculus** to **placenta**, or its scar on the seed.

holoparasite Plant parasite wholly nutritionally dependent on its host (c.f. **hemiparasite**).

homoeohydric Being able to maintain a particular water content (c.f. **poikilohydric**).

homologous Having arisen from the same ancestral organ.

homomorphic Having alternate plants with the same morphology (c.f. **heteromorphic**).

homoplasy Resemblance due to **convergent** or **parallel evolution**.

homospory Having a spore of a single type and size.

homothallic Having only one kind of gametophyte, either bisexual or dioecious.

humus epiphyte An **epiphyte** which traps leaf litter.

hyaline Transparent, applied to some cells which lack a protoplasm at maturity.

hydathode An organ secreting water.

hydroid A cell of the **hadrom**.

hydrophyte A water plant.

hypanthium The floral tube surrounding the **gynoecium**, especially obvious in perigynous flowers.

hypo- Prefix meaning below.

hypocotyl The region of the stem below the **cotyledons** in seedlings.

hypodermis A layer or layer of cells inside the **epidermis** which are different from the **palisade layer** or **mesophyll**.

hypoepigyny An intermediate condition between **hypogyny** and **epigyny**.

hypogyny A condition where the **ovary** is superior, with the **perianth** and **stamens** originating below the **gynoecium**.

idioblast An isolated cell clearly distinguished from the others around it.

imbricate Overlapping, like the tiles on a roof.

incubous In liverworts where the anterior margin of a leaf overlaps the posterior margin of the next **acropetal** leaf on the same side of the shoot (c.f. **succubous**).

indehiscent Applied to a fruit or **sporangium** which does not open to release its **diaspores**.

indeterminate Having the potential for continued growth.

indusium An outgrowth of the underside of the **frond** present in some **ferns**, a flap covering each **sorus**.
inflorescence A group of flowers on the same **axis**.
insectivorous Trapping and gaining nutrients from dead insects.
inserted Growing out of another organ.
integument An outer layer of the **ovule**, surrounding the **nucellus** and **megaspore**/female **gametophyte**, becomes the **testa** in the seed.
internode A section of the stem between **nodes**.
intine The inner part of the pollen wall.
introrse Facing inwards.
involucre Protective structures surrounding a reproductive area, upgrowths of a **thallus** or whorl(s) of **bracts**.
isomerous Having an equal number of floral parts in each whorl.
jacket The wall of a **sporangium**.

'K' strategist A plant which commits a large proportion of its energy to vegetative growth, generally long lived and a good competitor.
kranz a kind of leaf anatomy where the vascular bundles are no more than four mesophyll cells apart, and where there may be an obvious sheath around the vascular bundles.

lamina (blade) Expanded flat part of a leaf (c.f. petiole and sheath).
laticifer A latex canal.
legume A kind of dry fruit, derived from a single **carpel** but dehiscing by splitting along two margins.
lemma Lower bract in grass **floret** (c.f. **palea, glume**).
leptocaul Thin stemmed and more or less profusely branched (c.f. **pachycaul**).
leptomorph A plant with a slender shoot system where branches or **culms** arise dispersed.
leptosporangium A kind of **sporangium** in some **ferns**, small with a **unistratox** wall, and containing relatively few spores (c.f. **eusporangium**).
liane (liana) A climber with a long woody stem.
libriform fibre An elongated thick-walled non-**tracheary** cell of the **xylem**.
ligulate Having a strap- or tongue-like outgrowth.
ligule A membranous or delicate outgrowth at the base of the leaf lamina especially in grasses and heterosporous lycopods.
limb The upper expanded part of a **petal** or **sepal** (c.f. **claw, tube**).
locule A compartment of an **ovary**, where it may represent a **carpel**, or of a **pollen sac**.
loculicidal Dehiscing by openings in the wall of each **locule** of a **capsule** (c.f. **septicidal**).
lodicule A small sterile part of a grass **floret**, which opens the floret by swelling.
loment A kind of legume which is constricted between the seeds.
lycopods Lycophytina, the group of clubmosses (*Lycopodium sensu lato*, *Phylloglossum*, *Selaginella*, *Isoetes* and *Stylites*)

male nuclei Male gametes in **siphonogamous** plants.
marcescent Withered but remaining attached to the plant.
megaphyll A large leaf, supposedly a modified lateral branch system, usually identified by being **determinate** and having an axillary bud.
megasporangium A **sporangium** containing **megaspores**.
megaspore A large spore in **heterosporous** plants, usually gives rise to a female gametophyte.
megasporophylls (megasporangiophyll) Leaf (bract) having a **megasporangium** in its **axil** or bearing megasporangia.
meiosis (**meiotic**) A cell division where one of each pair of 'homologous chromosomes' ends up in different cell so that the number of chromosomes per cell is halved.
mericarp A one seeded fragment of a **schizocarp**.
meristem A region of actively dividing cells.
mesocarp The middle part of a **pericarp**.
mesomorphic Having characteristics which adapt a plant to areas without climatic extremes i.e. moist, warm or temperate conditions.
mesophyll The tissue in the centre of the leaf and also a leaf size category.
metamer A leaf, **axillary** bud, stem unit.
metaxylem The part of the primary **xylem** which differentiates in the part of the **axis** which is ceasing to elongate.
microclimate The local atmospheric conditions of a plant.
microhabitat The local area and conditions which impinge on an individual plant (c.f. **habitat**).
microphyll A small leaf, said to have a different evolutionary origin from **megaphylls** as an **enation**, and lacking an **axillary** bud.
micropyle The opening between the **integuments** of the **ovule**, can be seen in the seed **testa**.
microsporangiophore A stalk bearing **microsporangia**.
microspore The small spore of **heterosporous** plants and giving rise to the male **gametophyte**.
microsporophyll (microsporangiophyll) A leaf(**bract**) with a **microsporangium** in its axil or bearing a microsporangium.
middle lamella The boundary between the cell walls of two different cells made of pectic substances.
midrib The largest vein of a leaf, a continuation of the **petiole** into the leaf **lamina**.
mitosis (mitotic) The normal kind of cell division which produces two daughter cells identical genetically and chromosomally to the parent cell.
module The basic unit of plant architecture.
monadelphous Having the **stamens** fused by their **filaments** into a group.
monocarpic (semelparous) Reproducing just once and then dying (c.f. **polycarpic**).
monocaul Single stemmed.
monochlamydeous Having a **perianth** of undifferentiated **tepals** (c.f. **heterochlamydeous**).

monoecy (**monoecious**) Bisexual but **diclinous** (having separate male and female regions) (c.f. **dioecy, hermaphrodite**).
monophyletic A taxonomic group derived from a single common ancestor and including all the derivatives of that ancestor.
monopodial Growth where the apical meristems have continued activity (c.f. **sympodial**).
monosporic Producing a single viable spore in a **sporangium**.
monotypic Applied to a supraspecific **taxon** which has a single species.
multistratose Several cell layers thick.
mycorrhiza The combination of plant tissue and a fungus, a type of root, indicative of a nutritional relationship between the plant and its fungal associate.

neck The top part of an **archegonium**.
neck canal cells Cells within the neck of the **archegonium** which disintegrate at maturity.
nectary Organ secreting nectar.
niche The role of an organism in a community or the set of physical and biotic relationships with that organism with the environment and other organisms.
node The part of the stem where there is a leaf and a bud.
nodule Swelling on the root of a plant usually indicative of an association with a symbiotic nitrogen fixing organism.
nucellus The tissue in the ovule between the integuments and the **megaspore**, homologous to the megasporangium wall of heterosporous pteridophytes.
nut A fruit, usually single seeded, with a hard **pericarp**.

ochrea A fringe around the stem at the base of the leaves of some plants.
oligomerous Having a few floral parts, usually with a constant number in each species.
oogonium A female sex organ in fungi and algae.
operculum The lid of the capsule in mosses.
orthotropous An arrangement of the **ovule** where it is straight and the **micropyle** is at the other end from the **funiculus**.
ovule A female organ in seed plants, made up of **integument**(s), **nucellus** and female **gametophyte** or **embryosac**, becomes the seed after fertilization.
ovuliferous Bearing an ovule.

pachycaul Thick stemmed and weakly branched.
pachymorph A plant with a thick congested shoot system so that branches or in bamboos, the culms, arise crowded together.
paedomorphic The retention of a juvenile characteristic.
palea The upper **bract** in a grass **floret** (c.f. **lemma, glume**).
palisade layer Layer(s) of photosynthetic columnar or obconical cells below the epidermis in leaves.

panicle A loose compound **racemose** or mixed inflorescence.
parallel evolution The separate evolution of similar, often homologous, organs in different groups. It is particularly common in plants.
parallel venation Venation where the main veins run in parallel from the base of the leaf (c.f. reticulate venation).
paraphyllia Finely divided leaves.
paraphysis A sterile hair, simple or branched, unicellular or multicellular, associated with a reproductive structure.
parenchyma A living cell or tissue type where the cell walls are primary and unthickened.
pedicel A stalk of an individual flower or floret in an inflorescence.
peduncle The stalk of a reproductive structure, flower or inflorescence.
perennial Growing for more then 2 years.
peri- Prefix meaning around.
perianth The sterile floral parts below or outside the fertile ones, made up of **tepals**, or if differentiated into **calyx** and **corolla**, of **sepals** and **petals** respectively: sometimes used for accessory structures in reproductive axes which are not flowers.
pericarp The fruit wall, made up of 3 layers; **exocarp**, **mesocarp** and **endocarp**.
perichaetium (perigynium) A rosette of leaves (sometimes called the perianth), cup-like or tubular sheath surrounding the **gynoecium**.
periclinal Parallel to the surface.
periderm The outer protective layer of plants, replacing the epidermis in plants which have undergone secondary growth.
perigonium A rosette of leaves surrounding the antheridia in mosses.
perigynium A perichaetium.
perigyny (perigynous) A condition of flowers where there is a cup like structure, the **hypanthium**, around the **gynoecium**, from the rim of which sepals, petal, and stamens arise.
perisperm The nutritive tissue in some gymnosperm seeds derived from the **nucellus** (c.f. **endosperm**).
peristome The ring(s) of teeth around the opening in a moss capsule.
petal A part of the corolla.
petiole The stalk of a leaf.
phanerophyte One of Raunkiaer's life forms, a tree, tall shrub or **epiphyte**.
phellem The outer part of the **periderm**, made of cork cells.
phelloderm The inner part of the **periderm**.
phellogen The cork cambium, the lateral meristem of the **periderm**, which gives rise to the **phelloderm** centripetally and the phellem centrifugally.
phenotype The physical expression of the **genotype**.
phloem The tissue which conducts dissolved carbohydrates and minerals around the plant.
phreatophyte Deep rooted plant which taps the water table.
phylloclades A flattened branch system which looks like a leaf.
phyllode A flattened **petiole** which functions as a leaf.
phyllotaxis The arrangement of leaves.

pinna A leaflet of a compound leaf.
pinnule A leaflet of a compound pinna.
pistil The **gynoecium**, female part of a flower, made up of **ovary**, **style** and **stigma** and containing **ovules**.
pit A thin area of the primary cell wall where the secondary wall does not develop.
placenta The part of the **ovary** or fruit to which an **ovule** or seed is attached by its **funiculus**.
placentation The arrangement of ovules in the ovary.
plagiotropic Growing horizontally.
plasmalemma The outer membrane of the cell.
plasmodesmata The cytoplasmic connections between cells.
plastic Applied to development which is potentially modified by the environment.
plastochron The interval between the production of two **metamers** in the apical meristem.
platyspermic Applied to seeds with bilateral symmetry (c.f. **radiospermic**).
pleonanthy The production of reproductive structures laterally as the main axis continues to grow vegetatively.
pleur- pleuro- Prefix meaning lateral or ribbed.
pleurocarpous **Sporophyte**(s) emerging on short lateral branches in mosses.
plumule The shoot in the **embryo**.
pneumatode Air cell or region in a **velamen**.
pneumatophore An aerating negatively geotropic root of swamp plants.
poikilohydric Having very limited ability to maintain a different water content than the exterior.
pollen A **microspore** of seed plants, produces a pollen tube.
pollen chamber The area above the **nucellus** below the **micropyle**.
pollen sac A microsporangium in seed plants.
pollinium A group of pollen grains connected together and shed as a unit.
polycarpic (iteroparous) Reproducing several times (c.f. **monocarpic**).
polymerous Having numerous floral parts, sometimes of variable number in different individuals of the same species (c.f. **oligomerous**).
polymorphism The existence of several, usually obviously distinct, variants within a single population
polypetalous (choripetalous) Having a corolla of several separate petals (c.f. **gamopetalous**).
polystele Having several separate steles in an axis. Not applied to the vascular bundles of a primary stem.
pome A kind of fleshy fruit with a papery core, derived from an inferior ovary e.g. apple.
population A group of individuals of a species in a single location, which may potentially interbreed.
poricidal Opening by pores, in fruits and **sporangia**.

primordium The group of cells at the edge of a **meristem** which will give rise to a differentiated lateral appendage.
pro- Prefix meaning before or in front of.
procambium Tissue derived from the apical meristem which differentiates as the stele.
procumbent Lying along the ground.
proembryo The cells or **coenocytic** mass derived mitotically from the **zygote** before the organization of the **embryo**.
prolepsis Growth which is rhythmic.
promeristem A group of cells which later become meristematic.
propagule A fragment or part of a plant designed for multiplication.
protandry (protandrous) A condition where the male matures before the female.
prothallus The **gametophyte** of **pteridophytes**.
protoderm The tissue which differentiates as the epidermis.
protogyny A condition where the female matures before the male.
protonema The filamentous first stages of the moss **gametophyte**, produced on germination of the spore.
protoxylem The kind of primary **xylem** which differentiates in a part of an **axis** which is elongating, just behind the **meristem**.
pseudocarp 'False fruit', structure associated with seeds or fruits functioning as a fruit but not derived from the **pericarp**.
pseudoelater Multicellular **elater** produced by hornworts.
pseudoparenchyma The main kind of plant tissue in some multicellular algae, found below the epidermis. In section the densely packed filaments look a little like the **parenchyma** of land plants.
pteridophytes The group of all non-seedland plants which are not **bryophytes**, includes living ferns, horsetails, clubmosses, and whisk-ferns.
pteridosperm A group of extinct plants, the seed-ferns.

'r' strategist A plant which commits a large proportion of its energy in the production of **diaspores**, generally one with a rapid life cycle and low competitive ability.
raceme (racemose) A kind of **indeterminate inflorescence** (Fig. 5.8).
rachis (rhachis) The main axis of an **inflorescence**, or the **axis** of a pinnately compound leaf to which **pinnae** are attached (c.f. midrib).
radicle The root in the **embryo**.
radiospermic Applied to seeds with radial symmetry (c.f. **platyspermic**).
ray File of parenchymatous cells reaching transversely across a stele.
reaction wood A kind of wood produced in branches as a result of weight stresses.
receptacle The part of the plant which bears reproductive organs such as **sporangia**, the area of the stem to which floral parts are attached.
reductionism Philosophy of trying to understand everything by reducing it to its component parts.
repent Lying on the soil and rooting.

reproductive isolation The inability of two different species to breed, important in the process of speciation.
resupination The 180° twisting of the flower seen in some groups.
reticulate venation Complex branching venation where the veins branch unequally and there are many connections between different veins.
rhipidium A kind of cymose inflorescence (Fig. 5.8).
rhizoid A non-photosynthetic filament, unicellular or multicellular, which penetrates the substrate, produced by bryophytes, whisk-ferns and the gametophytes of pteridophytes.
rhizome An underground stem.
rhizophore A kind of aerial adventitious root in *Selaginella*.
rhizosphere The region immediately around the roots.
rhytidome Protective part of the outer bark, made up of alternating **periderm** and dead **phloem**.
root cap A group of cells with its own **meristem** at the tip of the root, which protects the root tip as it grows.
root hair A cell of the root epidermis which is elongated to penetrate the soil.
rostellum Part of the orchid flower.

saccate Bag-like.
samara A winged single-seeded fruit.
saprophyte A plant obtaining nutrition from dead and decaying material with the help of a fungal symbiont.
scalariform Ladder-like, applied to a kind of secondary cell wall.
scape A leafless stem usually with a terminal inflorescence and arising from a basal rosette.
scarious Membranous and dried-up.
schizocarp A dry fruit which splits at maturity into usually one-seeded fragments called **mericarps**.
sclereid A cell with a thick secondary cell wall and often lignified, a type of **sclerenchyma** cell, which is not a **fibre**.
sclerenchyma A type of cell and tissue where there is a thick secondary cell wall which is lignified.
sclerophyll A hard, stiff fibrous leaf.
seed A fertilized **ovule**, made up of **testa**, and **embryo** and sometimes including a nutritive tissue such as **endosperm** or **perisperm**.
self-incompatibility Self-sterility.
sepal A part of the **calyx**.
septicidal A capsule which dehisces by pores or slits along the line of the **septum**.
septum The wall between the **locules** of a multilocular **ovary**.
seta The stalk of the **sporophyte** in mosses and liverworts.
sheath The basal part of the leaf of plants like grasses which clasps or runs cylindrically down the stem.
short shoot A lateral shoot with limited growth.
sieve cell Part of the **phloem** in **gymnosperms**.

sieve tube A tube in the **phloem** made up of sieve tube elements, cells connected by protoplasm through sieve areas.
sieve tube element (member) The main translocating cell type of **sieve tubes** in **angiosperms**.
silicula A short bicarpellary fruit characteristic of the **Brassicaceae**.
siliqua A long bicarpellary fruit characteristic of the **Brassicaceae**.
siphonogamy (siphonogamous) Fertilization effected by the transfer of the male gamete to the egg by a pollen tube.
sorosis A **coenocarpium**.
sorus A group of **sporangia** in ferns.
spadix A kind of **inflorescence** with **florets** crowded on a swollen **axis** (c.f. **spike**).
spathe The **bract** enclosing an **inflorescence**.
spermatocyte A stage in **gametogenesis** of the male gamete.
spike A kind of **inflorescence** having flowers without **pedicels** on an extended **axis** (Fig. 5.8).
spikelet A group of **florets** in grasses, usually with a pair of basal **glumes**.
sporangiophore A stalk bearing **sporangia**.
sporangium The organ in which spores are produced.
sporophyll (sporangiophyll) A leaf with a **sporangium** in its **axil** or bearing a sporangium.
sporophyte A plant producing spores, diploid phase of the **alternation of generations**.
stamen The **microsporangiophore** of **angiosperms**, made up of **filament**, and **anther**.
staminode A sterile **stamen**.
stele The vascular tissues of the plant.
stigma The receptive terminal part of the **pistil**.
stipule An appendage at the base of some leaves, often leaf-like and in pairs.
stoma(ta) A pore(s) and the **guard cells** around it.
stomium A line of weakness along which a **sporangium** or **anther** splits to release spores or pollen.
strategy The characteristic behaviour of a species.
strobilus A **cone** or **catkin**.
style The part of the **pistil** connecting **stigma** and **ovary**.
suberized Made impermeable to water with suberin.
substitution growth Growth where the apical **meristem** ceases growth and continued growth is carried on by the activity of a lateral meristem.
succubous An arrangement in liverworts where the anterior margin of a leaf is overlapped by the posterior margin of the next **acropetal** leaf on the same side of the shoot (c.f. **incubous**).
succulent Having fleshy leaves or stems.
suckering Reproducing vegetatively by producing underground lateral shoots.
suffruticose Semi-woody, lower part shrubby.
supine Lying back.

suspensor A group of cells arising from the **zygote** along with the **embryo** which push it into the nutritive tissue.
sy- syl- sym- syn- syr- or sys- Prefix meaning together or joined.
syconium An aggregate fruit, e.g. fig.
syllepsis Continuous growth.
sympetalous (gamopetalous) Having a **corolla** of fused **petals**.
symplasm (symplast) The interconnected protoplasts of the plant.
sympodial Growth where the apical **meristem** ceases activity and forward growth is continued by the activity of a meristem of a lateral.
synangium A group of **sporangia** fused together.
syncarpous Having a compound **ovary** with **carpels** fused together.
synsepalous Having **sepals** fused together.

tank plant An **epiphyte** with a rosette of leaves or other structure which catches rainwater.
tapetum The nutritive region of an anther.
taxon (pl. taxa) A taxonomic group of any rank.
tectum The outer part of the **ectexine**.
telome A fundamental lateral shoot.
tension wood A kind of **reaction wood** produced in dicotyledons.
tepals The parts of a **perianth** when it cannot be differentiated into **calyx** and **corolla**.
terete Elongated cylindrical-conical tapering to a point.
testa The outer part of a **seed**, derived from the **integuments** of the **ovule**.
tetrad A group of four spores, the products of **meiosis**.
tetradynamous (tetradidynamous) Having six stamens, two shorter than the others.
thallus A simple flattened green **gametophyte** of hornworts and some liverworts, not differentiated into an axis and leaves.
theca A part of an **anther**, usually with two **pollen sacs**, joined in pairs to the **filament** by the **connective**.
thyrse A **racemose** inflorescence with lateral cymes.
tiller A shoot produced at the base of the stem, especially in grasses.
tonoplast The membrane which surrounds the vacuole.
topocline A cline expressed over a geographical distance
tracheary Part of the water conducting system of the **xylem**.
tracheid An imperforate **tracheary** cell.
tracheophytes Vascular plants, i.e. those with xylem, a group which includes all land plants except bryophytes though some of these do in fact have a kind of vascular system.
transfer cell A special cell involved in the short distance transfer of solutes.
transpiration The evaporation of water from a plant through its **stomata**.
trash basket plants **Humus epiphytes** with a rosette of leaves which traps leaf litter.
trichome An epidermal appendage which may take many forms; hairs or scales.

trichosclereid A kind of **sclereid** with long thin processes.
tropism Movement in plants.
tube The basal fused part of a **sympetalous corolla** or **synsepalous calyx** (c.f. limb).
tubercle (tuberculate) A bump on the surface.

uniseriate A filament with a single line of cells.
unistratose Having a single cell layer.
utricle A kind of **disseminule** where the fruit is surrounded by an inflated membranous bract or bracts.

valvate Having the **sepal** or **petal** margins tightly abutting each other, a kind of **aestivation**.
variant A kind of plant (c.f. variety).
variety (var.) A taxonomic group of sub subspecific rank (c.f. **variant**).
vascular bundle A part of the **stele** in the primary stem.
vasicentric An arrangement around the **xylem vessel elements**.
velamen Sheath of dead cells around the aerial roots of some **epiphytes**.
ventral canal cell Cell in the **archegonium** above the egg.
verticel A **whorl**.
verticillaster An **inflorescence** with **whorls** of flowers around the stem.
viviparous Applied to plants which either have precocious seed germination, while the seed is still attached to the parent, or where plantlets arise in unusual parts of the plant, such as on the margins of the leaves or in the flower.

whorl Arrangement of parts in an encircling ring.

xeromorphic Having characteristics which adapt a plant to a dry habitat.
xerophyte A plant adapted to a dry habitat.
xylem The vascular or wood tissue, the conducting system for water.
xylem vessel element A perforate **tracheary cell**, dead at maturity.
zygomorphic Bilaterally symmetrical.
zygote The product of the fusion of two gametes.

References

Albersheim, P. and Darvill, A. G. (1985) Oligosaccharins. *Scientific American*, **253**, 44–50.

Andrews, H. N. (1963) Early seed plants. *Science*, **142**, 925–31.

Andrews, H. N. and Murphy, W. H. (1958) *Lepidophloios* – and ontogeny in arborescent lycopods. *American Journal of Botany*, **45**, 552–60.

Andrews, H. N., Gensel, P. G. and Forbes, W. H. (1974) An apparently heterosporous plant from the Middle Devonian of New Brunswick. *Palaeontology*, **17**, 387–408.

Argent, G. C. G. (1976) The wild bananas of Papua New Guinea. *Notes Royal Botanical Gardens Edinburgh*, **35**, 77–114.

Atsatt, P. R. (1983) Mistletoes leaf shape: a host morphogen hypothesis, in *Biology of Mistletoes* (eds. M.Calder and P.Bernhardt), Academic Press, Sydney, Australia, pp. 259–75.

Baas, P. (1986) Ecological patterns in xylem anatomy, in *The Economy of Form and Function in Plants* (ed. T.J.Givnish), Cambridge University Press, Cambridge, pp. 327–52.

Baas, P. Werker, E. and Fahn, A. (1983) Some ecological trends in vessel characters. *IAWA Bulletin*, **4**, 141–159.

Bailey, I. W. and Tupper, W. W. (1918) Size variation in tracheary cells 1. A comparison between the secondary xylem of vascular cryptograms, gymnosperms and angiosperms. *Proceedings American Acadamy of Arts and Sciences*, **54**, 149–204.

Bakker, R. T. (1978) Dinosaur feeding behaviour and the origin of flowering plants. *Nature*, **274**, 661–3.

Barlow, P. W. (1987) The cellular organization of roots and its response to the physical environment, in *Root Development and Function* (ed. P. J. Gregory), Cambridge University Press, Cambridge, pp. 1–26.

Barlow, P. W. (1989) Meristems, metamers and modules and the development of shoot and root systems. *Botanical Journal of the Linnean Society, London*, **100**, 255–79.

Barlow, B. A. and Wiens, D. (1977) Host parasite resemblance in Australian mistletoes: the case for cryptic mimicry. *Evolution*, **31**, 69–84.

Baylis, G. T. (1975) The magnolioid and mycotrophy in root systems derived from it, in *Endomycorrhizas* (eds. F. E. Saunders, B. Moss and Tinker, P.B.), Academic Press, London, pp. 373–89.

Beattie, A.J. and Culver, D.C. (1982) Inhumation: how ants and other invertebrates help seeds. *Nature*, **297**, 627.

Beck, C. B. (1962) Reconstruction of *Archaeopteris* and further consideration of its phylogenetic position. *American Journal of Botany*, **49**, 373–82.

Beck, C. B. (1971) On the anatomy and morphology of lateral branch systems of *Archaeopteris*. *American Journal of Botany*, **58**, 758–84.

Beck, C. B., Schmid, R. and Rothwell, G. W. (1982) Stelar morphology and the primary vascular system of seed plants. *Botanical Reviews*, **48**, 691–815.

Bell, A. D. (1980) Adaptive architecture in rhizomatous plants. *Botanical Journal of the Linnean Society*, **85**, 125–160.

Bell, A. D. (1991) *Plant Form: An Illustrated Guide to Flowering Plant Morphology*, Oxford University Press, Oxford.

Benzing, D. H. (1976) Bromeliad trichomes: structure function and ecological significance. *Selbyana*, **1**, 330–348.

Benzing, D. H. (1989) The Evolution of Epiphytism, in *Vascular Plants as Epiphytes: Evolution and Ecophysiology* (ed. U. Luttge), Springer Verlag, Berlin, pp. 15–42.

Bold, H. C., Alexopoulos, C. J. and Delevoryas, T. (1987) *Morphology of Plants and Fungi*. 5th edn, Harper and Row, New York.

Boureau, E. (1964) *Stylocalamites, Calamatina, Crucicalamites. Traite Paleobotanique, 3. Sphenophyta, Noeggerathiophyta*. Masson, Paris.

Bower, F. O. (1935) *Primitive Land Plants*. Macmillan, London.

Braam, J. and Davis, R. W. (1989) Rain–, wind–, and touch–induced expression of calmodulin and calmodulin–related genes in *Arabidopsis*. *Cell*, **60**, 357–364.

Bradshaw, A.D. (1965) Evolutionary significance of phenotypic plasticity in plants. *Advances in Genetics*, **13**, 115–55.

Briggs, B. G. and Johnson, L. A. S. (1979) Evolution in the Myrtaceae – evidence from inflorescence structure. *Proceedings of the Linnean Society of New South Wales*, **102**, 157–256.

Bruce, J. G. (1979a) Gametophyte and young sporophyte of *Lycopodium carolinianum*. *American Journal of Botany*, **66**, 1138–1150.

Bruce, J. G. (1979b) Gametophyte and young sporophyte of *Lycopodium carolinianum*. *American Journal of Botany*, **66**, 1156–63.

Bruchman, H. (1910) Die keimung der sporen und die entwicklung der prothallien von *Lycopodium clavatum* L., *L.annotinum* L. und *L.selago* L. *Flora*, **101**, 220–267.

Burgess, N.D. and Edwards, D. (1988) A new Palaeozoic plant closely allied to *Protoaxites* Dawson. *Botanical Journal of the Linnean Society, London*, **97**, 189–203.

Burrows, F.M. (1986) The aerial motion of fruits, spores and pollen, in *Seed Dispersal* (ed. Murray, D.R.), Academic Press, Australia, Sydney, pp. 1–48.

Caldwell, M. M. (1987) Competition beteween root systems in natural communities. in *Root development and function* (eds. P.J. Gregory, J.V. Lake and D.A. Rose), Cambridge University Press, Cambridge.

Cannon, W.A. (1949) A tentative classification of root systems. *Ecology*, **30**, 542–8.

Carlquist, S. (1967) The biota of long distance dispersal V. Plant dispersal to Pacific Islands. *Bulletin of the Torrey Botanical Club*, **94**, 129–62.

Carlquist, S. (1975) *Ecological Strategies of Xylem Evolution*. University of California Press, Berkeley.

Carlquist, S. (1988) *Comparative Wood Anatomy*. Springer Verlag, Berlin.

Chaloner, W.G. and Macdonald, P. (1980) *Plants Invade the Land*. The Royal Scottish Museum/HMSO, Edinburgh.

Chapman, J. M. (1988) Leaf patterns in plants; leaf positioning at the plant shoot apex. *Plants Today*, **1**, 59–66.

Charlton, W. A. (1983) Patterns of distribution of lateral root primordia. *Annals of Botany*, **51**, 417–27.

Clayton, W. D. and Renvoize, S. A. (1986) *Genera Graminum, Grass of the World*. Kew Bulletin Additional Series XIII. Royal Botanic Gardens, Kew/HMSO, London.

Coe, M. J., Dilcher, D. L., Farlow, J.O., Jarzen, D. M. and Russel, D. A. (1987) Dinosaurs and land plants, in *The Origins of Angiosperms and their Biological Consequences* (eds. E. M. Friis, W. G. Chaloner and P.R. Crane), Cambridge University Press, Cambridge, pp. 225–58.

Cope, M.J. and Chaloner W.G. (1980) Fossil charcoal as evidence of past atmospheric composition. *Nature* (London), **283**, 647–9.

Coulter, J. M. and Chamberlain, C. J. (1910) *Morphology of the Gymnosperms*. Cambridge University Press, Cambridge.

Crane, P. R. (1985) Phylogenetic analysis of seed plants and the origin of angiosperms. *Annals of the Missouri Botanical Garden*, **72**, 716–93.

Crane, P. R. (1987) Vegetational consequences of angiosperm diversification, in *The Origins of Angiosperms and their Biological Consequences* (eds. E. M. Friis, W. G. Chaloner and P.R. Crane), Cambridge University Press, Cambridge, pp. 107–144.

Crawford, R. M. M. (1989) *Studies in plant survival: ecological case histories of plant adaptation to adversity*. Blackwell Scientific Publications, Oxford.

Crepet, W. C. and Friis, E. M. (1987) The evolution of insect pollination in angiosperms, in *The Origins of Angiosperms and their Biological Consequences* (eds. E. M. Friis, W. G. Chaloner and P.R. Crane), Cambridge University Press, Cambridge, pp. 145–180.

Cridland, A. A. (1964) *Amyelon* in American coal balls. *Palaeontology*, **7**, 186–209.

Cronquist, A. (1981) *An Integrated System of Classification of Flowering Plants*, Columbia University Press, New York.

Cutler, D. F. (1978) *Applied Plant Anatomy*. Longman, London.

Daniels, R. E. (1989) Adaptation and variation in bog mosses. *Plants Today*, **2**, 139–144.

Dawes, C. J. (1981) *Marine botany*. John Wiley, New York.

De Soó R. and Webb, D. A. (1972) *Melampyrum* L. in *Flora Europaea* vol. 3, (eds. T. G. Tutin, A. O. Chater, R. A. De Filipps, I. K. Ferguson and I. B. K. Richardson), Cambridge University Press, Cambridge.

Delvoryas, T. (1971) Biotic provinces and the Jurassic–Cretaceous floral transition. *Proceedings of the American Paleontological Convention*, **1**, 1660–74.

Dobbins, D. R. and Fisher, J. B. (1986) Wound response in girdled stems of lianas. *Botanical Gazette*, **147**, 278–89.

Doyle, J. A. (1978) Origin of angiosperms. *Annual Review of Ecology and Systematics*, **9**, 365–92.

Doyle, J. A. and M. J. Donoghue (1987) The origin of the angiosperms: a cladistic approach, in *The Origins of Angiosperms and their Biological Consequences* (eds. E. M. Friis, W. G. Chaloner and P.R. Crane), Cambridge University Press, Cambridge, pp. 17–50.

Dressler, R. L. (1981) *The Orchids, Natural History and Classification*. Harvard University Press, Cambridge, Massachussets.

Du Rietz, G. E. (1931) Life forms of terrestrial plants. *Acta Phytogeographica Suecica*, **3**, 1–95.

Duckett, J. G. (1973) Comparative morphology of the gametophytes of *Equisetum*, subgenus *Equisetum*. *Botanical Journal of the Linnean Society, London*, **66**, 1–22.

Duckett, J. G. and Pang, W. C. (1984) The origins of heterospory: a comparative study of sexual behaviour in the fern *Platyzoma microphyllum* R.Br. and the horsetail *Equisetum giganteum* L. *Botanical Journal of the Linnean Society, London*, **88**, 11–34.

East, E. M. (1940) The distribution of self–sterility in flowering plants. *Proceedings of the American Philosophical Society*, **82**, 449–518.

Edwards, D. (1986) *Aglaophyton major*, a non–vascular land plant from the Devonian Rhynie Chert. *Botanical Journal of the Linnean Society, London*, **93**, 173–204.

Eggert, D. A. (1974) The sporangium of *Horneophyton lignieri*. *American Journal of Botany*, **61**, 405–13.

El–Saadaway, W. E. and Lacey W. S. (1979a) Observations on *Nothia aphylla* Lyon ex Hoeg. *Review of Palaeobotany and Palynology*, **27**, 119–147.

El–Saadaway, W. E. and Lacey W. S. (1979b) The sporangia of *Horneophyton lignieri* (Kidston and Lang) Barghorn and Darrah. *Review of Palaeobotany and Palynology*, **28**, 137–44.

Engler, A. and Prantl K. (1887–1915) *Die naturlichen Pflanzenfamilien*. Wilhelm Engelman, Leipzig.

Esau, K. (1965) *Plant anatomy*. John Wiley and Sons, New York.
Faegri, K. and van der Pijl, L. (1979) *The Principles of Pollination Ecology*, 3rd edn, Pergamon, Oxford.
Fahn, A. (1982) *Plant Anatomy* 3rd edn, Pergamon Press, Oxford.
Feldman, M. (1976) Wheats. in *Evolution of Crop Plants* (ed. N. W. Simmonds). Longman, London, pp. 120–8.
Fenner, M. (1985) *Seed Ecology*. Chapman and Hall, London.
Fisher, D. G. and Evert, R. F. (1982) Studies on the leaf of *Amaranthus retroflexus* (Amaranthaceae): ultrastructure, plasmodesmatal frequency, and solute concentrations in relation to phloem loading. *Planta.*, **155**, 377–87.
Forsaith, C. C. (1926) The technology of New York State timbers. Technical Publication 18, vol.26. New York State College of Forestry, Syracuse University, Syracuse.
Frei, J.K. (1973) Effect of bark substrate on germination and early growth of *Encyclia tampensis* seeds. *American Orchid Society Bulletin*, **42**, 701–08.
Friis, E. M. and Crepet, W. L. (1986) Time of appearance of floral features, in *The Origins of Angiosperms and their Biological Consequences* (eds. E. M. Friis, W. G. Chaloner and P.R. Crane), Cambridge University Press, Cambridge, pp. 145–80.
Friis, E.M., Chaloner, W.G. and Crane, P. R. (1987) Introduction to angiosperms, in *The Origins of Angiosperms and their Biological Consequences* (eds. E. M. Friis, W. G. Chaloner and P.R. Crane), Cambridge University Press, Cambridge, pp. 1–15.
Fry, S. C. (1989) Dissecting the complexities of the plant cell wall. *Plants Today*, **2**, 126–32.
Gentry, A. H. (1982) Patterns of neotropical species diversity. *Ecological Biology*, **15**, 1–84.
Gibson, A. C. (1973) Comparative anatomy of secondary xylem in Cactoideae (Cactaceae). *Biotropica*, **5**, 29–65.
Gibson, A. C. and P. S. Nobel (1986) *The Cactus Primer*. Harvard University Press, Cambridge, Massachussetts.
Giddings, T. H., Brower, D. L. and Staehelin L. A. (1987) Visualisation of particle complexes in the plasma membrane of *Micrasterias denticulata* associated with the formation of cellulose fibrils in primary and secondary cell walls. *Journal of Cell Biology*, **84**, 327–39.
Gifford, E. M. and Foster, A. S. (1990) *Comparative Morphology of Vascular Plants*, 3rd edn, W. H. Freeman, San Francisco.
Givnish, T. J. (1986) Biomechanical constraints on crown geometry in forest herbs. in *The economy of form and function in plants* (ed. T.J.Givnish), Cambridge University Press, Cambridge, pp. 525–84.
Godley, E. J. and Smith, D. H. (1981) Breeding systems in New Zealand plants 5. *Pseudowinteri colerata* (Winteraceae). *New Zealand Journal of Botany*, **19**, 151–6.
Goh, C. J. and J. Kluge (1989) Gas Exchange and Water Relations in Epiphytic Orchids. in *Vascular Plants as Epiphytes: Evolution and Ecophysiology* (ed. U. Luttge) Springer Verlag, Berlin, pp. 137–66.
Goldring, W. (1924) The Upper Devonian forest seed ferns of eastern New York State. *New York State Museum Bulletin*, **251**, 49–72.
Gornall, R. J. (1988) The coastal ecodeme of *Parnassia palustris* L. *Watsonia*, **17**, 139–43.
Gould, S. J. and Lewontin, R. C. (1979) The spandrels of San Marco and the Panglossian paradigm: a critique of the adaptationist programme. *Proceedings of the Royal Society of London, series B*, **205**, 581–98.
Gray, R. and Bonner, J. (1948) *Encelia*. *American Journal Botany*, **34**, 52–7.
Grime, J. P. (1979) *Plant Strategies and Vegetation Processes*. John Wiley and Sons, Chichester.
Grime, J. P. (1984) The ecology of species, families and communities of the contemporary British flora. *New Phytologist*, **98**, 15–33.

Grime, J. P., Hodgson, J. G. and R. Hunt (1988) *Comparative Plant Ecology*. Unwin Hyman, London.

Hadley, G. and Williamson, B. (1972) Features of mycorrhizal infection in some Malayan orchids. *New Phytologist*, **71**, 1111–8.

Hagman, M. (1975) Incompatibility in forest trees. *Proceedings of the Royal Society B*, **188**, 313–26.

Haig, D. and Westoby, M. (1989) Selective forces in the emergence of the seed habit. *Biological Journal of the Linnean Society*, **38**, 215–38.

Hallé, F. R., Oldeman, A. A. and P. B. Tomlinson, (1978) *Tropical trees and forests: an architectural analysis*. Springer Verlag, Berlin.

Harborne, J. B. (1988) *Introduction to Ecological Biochemistry*, 3rd edn. Academic Press, London.

Harper, J. L. (1977) *Population Biology of Plants*. Academic Press, London.

Harris, T. M. (1973) What use are fossil ferns? in *The Phylogeny and Classification of Ferns*, *Botanical Journal of the Linnean Society* suppl. 1 (eds. A.C.Jermy, J.A.Crabbe and B.A.Thomas), pp. 41–4.

Henderson, A. (1986) A review of pollination studies in the Palmae. *Botanical Review*, **52**, 221–59.

Hickey, L. J. (1977) Stratigraphy and palaeobotany of the Golden Valley Formation (early Tertiary) of western Dakota. *Memoirs of the Geological Society of America* **150**, 1–183.

Hickey, L. J. and Doyle J. A. (1977) Early Cretaceous fossil evidence for angiosperm evolution. *Botanical Review*, **43**, 3–104.

Hirmer, M. (1927) *Handbuch de Paläobotanik*. Von R. Oldenbourg, München, Berlin.

Hirmer, M. (1933) Rekonstruktion von *Pleuromeia sternbergi* Corda nebst bemerkungen zur Morphologie der Lycopodiales. *Palaeontographica*, **78B**, 47–56.

Hodgson, J. (1989) Are families of flowering plants ecologically specialized? *Plants Today*, **2**, 132–8.

Horn, H. S. (1975) Forest succession. *Scientific American*, **232**, 90–8.

Hoster, H. R. and Liese, W. (1966) Über das Vorkommen von Reaktionsgewebe in Wurzeln und Ästen der Dikotylodnen. *Holzforschung*, **20**, 80–90.

Howe, H. F. (1986) Seed dispersal by fruit–eating birds and mammals, in *Seed Dispersal* (ed. D.R. Murray), Academic Press, Australia, Sydney, pp. 123–90.

Hubbell, S. P., Wiemer, D. F. and Adejore, A. (1983) An antifungal terpenoid defends a neotropical tree (*Hymenaea*) aganst attack by fungus–growing ants (*Atta*). *Oecologia*, **60**, 321–27.

Jackson, R. M. and Mason P. A. (1984) *Mycorrhiza. Studies in Biology*, no. 159, Edward Arnold, London.

Janson, C. H. (1983) Adaptation of fruit morphology to dispersal agents in a neotropical forest. *Science*, **219**, 187–9.

Janzen D. H. (1973) Dissolution of mutualism between *Cecropia* and its *Azteca* ants. *Biotropica*, **5**, 15–28.

Janzen, D. H. (1975) *Ecology of Plants in the Tropics*. Edward Arnold, London.

Jeffrey, D. W. (1986) *Soil Plant Relationships, an Ecological Approach*. Croom Helm, London.

Johansson, D. R. (1975) The ecology of epiphytic orchids in West African rain forests. *American Orchid Society Bulletin*, **44**, 125–36.

Jones, H. and Jones, M. G. K. (1989) Direct gene transfer into plant protoplasts, *Plants Today*, **2**, 175–78.

Kalkman, C. (1988) The phylogeny of the Rosaceae. *Botanical Journal of the Linnean Society, London*, **98**, 37–59.

Kaplan, D.R. (1975) Comparative developmental evaluation of the morphology of unifacial leaves in the monocotyledons. *Botanisch Jahrbuch Systematisch*, **95**, 1–105.

Kay, Q. (1988) More than the eye can see: the unexpected complexity of petal structure. *Plants Today*, **1**, 109–14.

Keeler, K. H. (1989) Ant–Plant interactions, in *Plant–Animal interactions* (ed. W.G. Abrahamson), McGraw Hill, New York, pp. 207–42.
Kidston, R. and Lang, W. H. (1917) On Old Red Sandstone plants showing structure from the Rhynie Chert Bed, Aberdeenshire I. *Rhynia Gwynne vaughanii* Kidston and Lang. *Transactions of the Royal Society of Edinburgh*, **51**, 761–84.
Kidston, R. and Lang, W. H. (1920a) On Old Red Sandstone plants showing structure from the Rhynie Chert Bed, Aberdeenshire II. Additional notes on *Rhynia major* n.sp. and *Hornea lignieri* n.g.,n.sp. *Transactions of the Royal Society of Edinburgh*, **52**, 603–27.
Kidston, R. and Lang, W. H. (1920b) On Old Red Sandstone plants showing structure from the Rhynie Chert Bed, Aberdeenshire III. *Asteroxylon mackiei* Kidston and Lang. *Transactions of the Royal Society of Edinburgh*, **52**, 643–80.
Kidston, R. and Lang, W. H. (1920c) On Old Red Sandstone plants showing structure from the Rhynie Chert Bed, Aberdeenshire IV. Restorations of the vascular cryptogams and discussion of their bearing on the general morphology of Pteridophyta and the origin and organization of land plants. *Transactions of the Royal Society of Edinburgh*, **52**, 831–54.
Klepper, B. (1987) Origin, branching and distribution of root systems, in *Root development and Function* (eds. P. J. Gregory, J. V. Lake and D. A. Rose), Cambridge University Press, Cambridge, pp. 104–124.
Klingauf, F. (1971) Die Wirkung des Glucosids Phlorizin auf das Wirtswahlverhalten von *Rhopalosiphum insertum* (Walk.) und *Aphis pomi* de Geer (Homoptera: Aphididae). *Zeitschrift angewandte Entomologie*, **68**, 41–55.
Kress, W. J. (1989) The Systematic Distribution of Vascular Epiphytes, in *Vascular Plants as Epiphytes: Evolution and Ecophysiology* (ed. U. Luttge), Springer Verlag, Berlin, pp. 234–61.
Kubíková, V. A. (1967) Contribution to the classification of root systems of woody plants. *Preslia*, **39**, 236–43.
Kujit, J. (1969) *The Biology of Parasitic Flowering Plants*. University of California Press, Berkeley.
Kwak, M. M. (1988) Pollination ecology and seed-set in the rare annual species *Melampyrum arvense* L. Scrophulariaceae. *Acta Botanica Neerlandica*, **37**, 153–63.
Lamont, B. B. (1983) Strategies for maximising nutrient uptake in two mediterranean ecosystems of low nutrient status, in *Mediterranean Type Ecosystems: The Role of Nutrients* (eds. F.J.Kruger, D.T.Mitchell and J.U.M.Jarvis), Springer Verlag, Berlin, pp. 246–73.
Leclercq, S. and Banks, H. P. (1962) *Pseudosporochnus nodosus* sp.nov. a Middle Devonian plant with cladoxylean affinities. *Palaeontographica*, **110B**, 1–34.
Lee, D. W. (1986) Unusual strategies of light absorption in rain-forest herbs, in *The Economy of Form and Function in Plants* (ed. T.J.Givnish), Cambridge University Press, Cambridge, 105–32.
Lewontin, R. C. (1977) Adaptation. *Scientific American*, **239**, 156–69.
Long, H. G. (1975) Further observations on some lower Carboniferous seeds and cupules. *Transactions of the Royal Society of Edinburgh*, **69**, 267–93.
Longman, K. A. and Jenik, J. (1987) *Tropical Forest and its Environment*. 2nd edn, Longman, London.
Longton, R. E. (1988) Adaptations and strategies of polar bryophytes. *Botanical Journal of the Linnean Society, London*, **98**, 253–68.
Louda, S. M. (1982) Limitations of the recruitment of the shrub *Haplopappus squarrosus* Asteraceae by flower and seed feeding insects. *Journal of Ecology*, **70**, 43–53.
Lovtrup, S. (1987) *Darwinism: the Refutation of a Myth*. Croom Helm, London.
Mabberley, D. J. (1987) *The Plant Book*. Cambridge University Press, Cambridge.
Madison, M. (1977) Vascular epiphytes: their systematic occurrence and salient features. *Selbyana*, **2**, 1–13.
Magdefrau, K. (1956) *Paläobiologie der Pflanzen*. 3rd edn, Veb: Gastav Fischer.

Marks, G. C. and Foster, R. C. (1973) Structure, morphogenesis and ultrastructure of ectomycorrhizae, in *Ectomycorrhizae: their ecology and physiology* (eds. G.C.Marks and T.T.Kozlowski), Academic Press, London, pp. 2–36.

Martin, E. S., Donkin, M. E. and Stevens, R. A. (1983) *Stomata. Studies in Biology*, no. 155, Edward Arnold, London.

Mauseth, J. D. (1988) *Plant Anatomy*. Benjamin/Cummings, Menlo Park, California.

Meinzer, F. and Goldstein, G. (1986) Adaptations for water and thermal balance in Andean giant rosette plants, in *The economy of form and function in plants* (ed. T.J.Givnish), Cambridge University Press, Cambridge.

Metcalfe, C. R. and Chalk, L. (1950) *Anatomy of the Dicotyledons*. 4 vols. Clarendon Press, Oxford.

Meyen, S. V. (1987) *Fundamentals of Palaeobotany*. Chapman and Hall, London.

Meyen, S. V. (1988) Origin of the angiosperm gynoecium by gamoheterotropy. *Botanical Journal of the Linnean Society, London*, **97**, 171–8.

Meyer, F.H. (1973) Distribution of ectomycorrhizae in nature and man–made forests, in *Ectomycorrhizae: their ecology and physiology* (eds. Marks, G. C. and T. T. Kozlowski), Academic Press, London, pp. 79–106.

Milner, A. R. (1980) The Tetrapod assemblage from Nyrany Czechoslovakia, in *The Terrestrial Environment and the Origin of Land Vertebrates* (ed. Panchen A. C.), Academic Press, London, pp. 439–96.

Mody, N. W., Henson, R., Hedlin, P. A., Kokpol, U., and Miles, D. H. (1976) Isolation of the insect paralyzing agent Coniine from *Sarracenia*. *Experientia*, **32**, 829–30.

Morgan Jeanne (1959) The morphology and anatomy of American species of the genus *Psaronius*. *Illinois Biological Monographs*, **27**, 1–108.

Mosse, B. (1963) Vesicular–arbuscular mycorrhiza: an extreme form of fungal adaptation, in *Symbiotic Associations* (eds. P.S. Nutman and B. Mosse), Cambridge University Press, Cambridge, pp. 146–70.

Muller, C. H. (1970) Phytotoxins as plant habitat variables. *Recent Advances in Phytochemistry*, **3**, 106–21.

Muller, J. (1981) Fossil pollen records of extant Angiosperms. *Botanical Reviews*, **47**, 1–145.

Murray, D. R. (1986) Seed dispersal by water, in *Seed Dispersal* (ed. D.R. Murray), Academic Press, Australia, Sydney, pp. 49–85.

Nambiar, E. K. (1977) The effects of drying on the topsoil and of micronutrients in the subsoil on micronutrient uptake by an intermittently defoliated ryegrass. *Plant and Soil*, **46**, 185–93.

Niklas, K. J. (1985) The aerodynamics of wind pollination. *Botanical Review*, **51**, 328–86.

Niklas, K. J., Tiffney, B. H. and Knoll, A. H. (1985) Patterns in vascular plant diversification: an analysis at the species level, in *Phanerozoic Diversity Patterns: Profiles in Macroevolution* (ed. J. W. Valentine), Princeton University Press Princeton, pp. 97–128.

Nobel, P. S. (1987) Photosynthesis and productivity of desert plants, in *Progress in Desert Research* (eds. L. Berkofsky and M. G. Wurtele), Rowman and Littlefield, New Jersey, pp. 41–66.

O'Dowd, D. J. and Gill, A. M. (1986) Seed dispersal syndromes in Australian *Acacia*, in *Seed Dispersal* (ed. D. R. Murray) Academic Press, Australia, Sydney, pp. 87–121.

Oinoinen, E. (1967) The correlation between the size of Finnish bracken (*Pteridium aquilinum* (L.)Kuhn) clones and certain periods of site history. *Acta forestalia Fennica*, **83**, 1–51.

Olson S. L. (1985) The fossil record of birds. *Avian Biology*, **8**, 79–237.

Orshan, G. (1983) Approaches to the definition of mediterranean growth forms, in *Mediterranean Type Ecosystems: the Role of Nutrients* (eds. F.J.Kruger, D.T.Mitchell and J.U.M.Jarvis) Springer Verlag, Berlin, pp. 86–100.

Packham, J. R. and Harding D. J. L. (1982) *The Ecology of woodland processes*. Edward Arnold, London.

Packham, J. R. and A. J. Willis (1977) The effects of shading on *Oxalis acetosella*. *Journal of Ecology*, **65**, 619–42.

Parkhurst, D. F. (1986) Internal leaf structure: a three dimensional perspective, in *The economy of form and function in plants* (ed. T. J. Givnish), Cambridge University Press, Cambridge, pp. 215–50.

Parrish, J. T. (1987) Global palaeogeography and palaeoclimate in the late Cretaceous and Early Tertiary, in *The origins of angiosperms and their biological consequences*. (eds. E. M. Friis, W. G. Chaloner and P.R. Crane), Cambridge University Press, Cambridge, pp. 17–50.

Pettitt, J. M. (1985) Pollen tube development and characteristics of the protein emission in conifers. *Annals of Botany*, **56**, 379–97.

Porter, J. R. (1989) Modules, models and meristems in plant architecture, in *Plant canopies: their growth form and function* (eds. G. Russel, B. Marshall and P. G. Jarvis), Cambridge University Press, Cambridge, pp. 143–59.

Postgate, J. (1978) *Nitrogen fixation. Studies in Biology*, no. 92, Edward Arnold, London.

Queller, D. C. (1984) Pollen–ovule ratios and hermaphrodite sexual allocation strategies. *Evolution*, **38**, 1148–51.

Rackham, O. (1980) *Ancient woodland: its history, vegetation and uses in England*. Edward Arnold, London.

Ramsay, H. P. (1979) Anisospory and sexual dimorphism in the Musci, in *Bryophyte systematics* (eds G. C. S. Clarke and J. G. Duckett) Academic Press, London, pp. 281–316.

Raunkiaer, C. (1934) *The Life Forms of Plants and Statistical Plant Geography*, Clarendon Press, Oxford.

Raymond, A. (1985) Floral diversity, phytogeography and climatic amelioration during the Early Carboniferous. *Paleobiology*, **11**, 293–309.

Rayner, M. C. 1927. *Mycorrhiza. New Phytologist* reprint No. 15. Wheldon and Wesley Ltd, London.

Read, D. J. and Stribley, D. P. (1975) Some mycological aspects of the biology of mycorrhiza in the Ericaceae, in *Endomycorrhizas* (eds. F. E. Sanders, B. Mosse and P. B. Tinker), Academic Press, London, pp. 105–17.

Remy, W. (1982) Lower Devonian gametophytes: relation to the phylogeny of land plants. *Science*, **215**, 1625–27.

Remy, W. and Remy, R. (1980) Devonian gametophytes with anatomically preserved gametangia. *Science*, **208**, 295–6.

Retallack, G. J. (1975) The life and times of a Triassic lycopod. *Alcheringa*, **1**, 3–29.

Retallack, G. J. and Dilcher, G. J. (1981) A coastal hypothesis for the dispersal and rise to dominance of the flowering plants, in *Palaeobotany, Palaeoecology and Evolution* vol II (ed. K. J. Niklas), Praeger, New York.

Richards, P. W. (1952) *The Tropical Rain Forest: An Ecological Study*. Cambridge University Press, Cambridge.

Richards, A. J. (1986) *Plant Breeding Systems*. Allen and Unwin, London.

Rivett, M. F. (1924) The root tubercles in *Arbutus unedo*. *Annals of Botany*, **38**, 661–77.

Robinson, D. and I. H. Rorison (1988) Plasticity in grass species in relation to nitrogen supply. *Functional Ecology*, **2**, 249–257.

Robinson, R. K. (1973) Mycorrhiza in certain Ericaceae native to southern Africa. *Journal of South African Botany*, **39**, 123–9.

Rothwell, G. W. (1988) Cordaitales. in *Origin and Evolution of Gymnosperms* (ed. C. B. Beck), Columbia University Press, New York, 273–297.

Rothwell, G. W. and Scheckler S. E. (1988) Biology of Ancestral Gymnosperms, in *Origin and Evolution of Gymnosperms* (ed. C.B.Beck), Columbia University Press, New York, pp. 85–134.

Rothwell, S. and Warner, S. (1984) *Cordaixylon dumusum* n.sp. (Cordaitales) 1. vegetative structure. *Botanical Gazette*, **145**, 275–91.
Rowley, G.D. (1987) *Caudiciform and pachycaul succulents*. Strawberry Press, Mill Valley, California.
Saenger, P. (1982) Morphological, anatomical and reproductive adaptations of Australian mangroves, in *Mangrove ecosystems in Australia: structure function and management* (ed. B. F. Clough), Australian Institute of Marine Science in association with Australian National University, Canberra, pp. 153–192.
Schofield, W. B. (1985) *Introduction to Bryology*. Macmillan, New York.
Schweitzer, H.–J. (1967) Die Oberdevon Flora der Bäreninsel 1. *Pseudobornia ursina* Nathorst. *Palaeontographica*, **120B**, 116–137.
Schweitzer, H.–J. (1972) Die Mitteldevon–Flora von Lindlar (Rheinland) 3. Filicinae – *Hyenia elegans* Krausel und Weyland. *Palaeontographica*, **137B**, 154–75.
Scott, A. C., Chaloner, W.G. and Paterson, S. (1985) Evidence of pteridophyte interactions in the fossil record. *Proceedings of the Royal Society of Edinburgh*, **86B**, 133–140.
Simmonds, N. W. (1979) *Principles of Crop Improvement*. Longman, London.
Slack, A. (1988) *Carnivorous Plants*. 3rd edn, A. and C. Black, London.
Slade, B. F. (1971) Stelar evolution in vascular plants. *New Phytologist*, **70**, 879–84.
Smirnoff, N. and Crawford, R. M. M. (1983) Variation in the structure and response to flooding of root aerenchyma in some wetland plants. *Annals of Botany*, **51**, 237–49.
Smith, A. J. E. (1963) Variation in *Melampyrum pratense* L. *Watsonia*, **5**, 336–67.
Smith, S. S. E. (1980) Mycorrhizas of autotrophic higher plants. *Biological Reviews*, **55**, 475–510.
Sota, E. R. De La (1973) On the classification and phylogeny of the Polypodiaceae, in *The Phylogeny and Classification of the Ferns* (eds. A. C. Jermy, J. A. Crabbe and B. A. Thomas), Academic Press, London, pp. 229–44.
Sporne, K. R. (1974) *The Morphology of Gymnosperms*. 2nd edn, Hutchinson University Library, London.
Sporne, K. R. (1975) *The Morphology of Pteridophytes*. 4th edn, Hutchinson University Library, London.
Stace, C. A. (1976) The study of infraspecific variation. *Current Advances in Plant Science*, **8**, Commentary 23.
Stace, C. A. (1989) *Plant Taxonomy and Biosystematics*. 2nd edn, Edward Arnold, London.
Stebbins, G. L. (1974) *Flowering Plants: Evolution above the Species Level*. Belknap Press of Harvard University Press, Cambridge, Massachussetts.
Steeves, T. A. and I. M. Sussex (1989) *Patterns in Plant Development*. Cambridge University Press, Cambridge.
Stelleman, P. (1978) The possible role of insect visits in pollination of reputedly anemophilous plants, exemplified by *Plantago lanceolata* and syrphid flies, in *The pollination of flowers* (ed. A. J. Richards), Academic Press, London, pp. 41–66.
Stewart, W. N. (1983) *Paleobotany and the Evolution of Plants*. Cambridge University Press, Cambridge.
Stewart, G., Graves, J. and Press, M. (1989) Vampires and witches – the physiology of parasitic flowering plants. *Plants Today*, **2**, 159–65.
Taylor, D. W. and Hickey, L. J. (1990) An Aptian Plant with attached leaves and flowers: implications for angiosperm origin. *Science*, **247**, 702–4.
Taylor, T. N. (1988) Pollen and pollen organs of fossil gymnosperms: Phylogeny and Reproductive Biology, in *Origin and Evolution of Gymnosperms* (ed. C. B. Beck), Columbia University Press, New York, pp. 177–217.
Taylor, T. N. and Brauer, D. F. (1983) Ultrastructural studies of in situ Devonian spores *Barinophyton citrulliforme*. *American Journal of Botany*, **70**, 106–12.
Taylor, T. N. and Taylor, E. L. (1989) The lush vegetation of Antarctica. *Plants Today*, **2**, 116–20.

Thien, L. B. (1980) Patterns of pollination in primitive angiosperms. *Biotropica*, **12**, 1–13.

Thien, L. B., White, D. A. and Yatsu, L. Y. (1983) The reproductive biology of a relict – *Illicium floridanum* Ellis. *American Journal of Botany*, **70**, 719–27.

Thomas, H. H. (1915) On *Williamsoniella* a new type of flower. *Philosophical Transactions of the Royal Society, London*, **207**, 113–48.

Thomas, B. A. and R. A. Spicer (1987) *The Evolution and Palaeobiology of Land Plants*. Croom Helm, London.

Tiffney, B. H. (1984) Seed size, dispersal syndromes, and the rise of the angiosperms: evidence and hypothesis. *Annals of the Missouri Botanical Garden*, **71**, 551–76.

Tiffney, B. H. (1986) Evolution of seed dispersal syndromes according to the fossil record, in *Seed dispersal* (ed. D. R. Murray), Academic Press, Australia, Sydney, pp. 273–305.

Tippett, J. T. (1982) Shedding of ephemeral roots in gymnosperms. *Canadian Journal of Botany*, **60**, 2295–302.

Tivy, J. (1990) *Agricultural ecology*. Longman, London.

Turnbull, C. L. and D. C. Culver (1983) The timing of seed dispersal in *Viola nuttallii*: attraction of diaspores and avoidance of predators. *Oecologia*, **59**, 360–5.

Upchurch, G. R. and J. A. Wolfe (1987) Mid Cretaceous to Early tertiary vegetation and climate: evidence from fossil leaves and woods, in *The origins of angiosperms and their biological consequences* (eds. E. M. Friis, W. G. Chaloner and P.R. Crane), Cambridge University Press, Cambridge, pp. 75–106.

Vincent, J. F. V. (1982) *Structural Biomaterials*. Macmillan, London.

Visser, J. (1981) *South African Parasitic Flowering Plants*. Juta, Cape Town.

Watson, J. (1988) The Cheirolepidiaceae, in *Origin and Evolution of Gymnosperms* (ed. C. B. Beck), Columbia University Press, New York, 382–447.

Watson, E.V. (1967) *The Structure and Life of Bryophytes*. 2nd edn, Hutchinson University Press, London.

Watt, A. S. (1970) Contributions to the ecology of bracken (*Pteridium aquilinum*) VII Bracken and litter, the cycle of change. *New Phytologist*, **69**, 431–49.

Weberling, F. (1989) *Morphology of Flowers and inflorescences*. Cambridge University Press, Cambridge.

White, M. E. (1986) *The Greening of Gondwana*. Reed, Sydney.

White, R. A., Bierhorst, D. W., Gensel, P. G., Kaplan, D. R. and Wagner, W. H. (1977). Taxonomic and morphological relationships of the Psilotaceae. *Brittonia*, **29**, 1–68.

Williams, W. G., Kennedy, G. G., Yamamoto, R. T., Thacker, J. D. and Bordner, J. (1980) 2–Tridecanone: a naturally occurring insecticide from the wild tomato *Lycopersicon hirsutum* f. *glabratum*. *Science*, **207**, 888–9.

Wing, S. L. and Tiffney, B. H. (1987) Interactions of angiosperms and herbivorous tetrapods through time, in *The Origins of Angiosperms and their Biological Consequences* (eds. E. M. Friis, W. G. Chaloner and P. R. Crane), Cambridge University Press. Cambridge, pp. 203–24.

Wolfe, J. A. (1979) Temperature parameters of humid to mesic forests of eastern Asia and their relation to forests of other areas of the Northern Hemisphere and Australasia. *US Geographical Survey Professional Paper*, **1106**, 1–37.

Woodhead, S. and Chapman, R.F. (1986) Insect behaviour and the chemistry of plant surface waxes, in *Insects and the Plant Surface* (eds. B. Juniper and T. R. E. Southwood), Edward Arnold, London, pp. 123–136.

Zak, B. (1973) Classification of Ectomycorrhizae, in *Ectomycorrhizae: their Ecology and Physiology* (eds. Marks, G. C. and T. T. Kozlowski), Academic Press, London, 43–78.

Zavada, M. S. (1984) The relation between pollen exine sculpturing and self–incompatibility mechanisms. *Plant Systematics and Evolution*, **147**, 63–78.

Zimmerman, W. (1952) The main results of the telome theory. *Palaeobotanist*, **1**, 456–70.

Zimmerman, M. H. (1983) *Xylem Vessels and the Ascent of Sap*. Springer Verlag, Berlin.

Zimmerman, M. H. and Brown, C. L. (1971) *Trees Structure and Function*. Springer Verlag, New York.

Index

Abies 201
Abscission 25, 41, 77, 126, 206–7, 297
Acacia 42, 118, 150, 164, 219, 221, 224, 244, 272–4, 285–6
Acaena 163–4
Acanthocalycium 50–1
Acer 200
Achene 155, 163, 171, 186, 297
Achillea 172–3
Aconitum 161
Acriopsis 274
Acrostichum 239
Actinomorphy (ic) 144, 161, 166, 168, 170–1, 174, 176, 297
Actinostele 73–4
Adansonia 200, 204
Adaptation 10, 19
Adenia 204
Adiantum 95
Aechmea 216
Aegiceras 266
Aegiliatis 242
Aerenchyma 32, 236–7, 239–40, 297
Aestivation 149, 150, 297
Agathis 81
Agave (Agavaceae) 224–5, 227
Agrostis 286
Aizoaceae 141, 156, 224–5
Alchemilla 163–4
Algae 4, 34, 58, 61–2, 99, 103, 244, 255
Allelopathy 266–7
Allium (Alliaceae) 175–6, 265–6
Alnus 183, 244–5
Alpinia 204
Alternation of generations 62, 88–90, 105, 127–9, 297
Althaea 151–2
Amaranthus (Amaranthaceae) 44, 141, 156
Amaryllidaceae 176
Amentiferae 160, 182–4
Ammophila 236, 258, 267
Amorphophallus 181
Amphigastrium (a) 215, 297
Amyema 221, 250
Anabaena 244, 246
Anacamptis 177
Andreaea 96, 99, 105, 106
Anemone 162
Anemophily *see* Wind pollination
Angelica 278
Angiopteris 72, 94
Angiosperm
  classification 126, 160, 165
  flower 134–86
  origins 82, 84, 135
Anigozanthos 256
Anisogamy 58
Anisophylly 41, 298
Anisospory 107, 108, 298
Annonaceae 136, 174
Annulus 93, 99, 280, 282, 298
Anothofixus 250
Ant plant 213–14, 272–3
Anther 115–16, 118, 149, 151–2, 166, 171, 176, 178, 282, 298
Antheridium (a) 62, 90, 99, 101–9, 298
Anthoceros (Anthocerotopsida) 91, 96–7, 106, 224, 230
Anthocyanidins 140
Antirrhinum 168–9
Apiaceae 154, 156, 165, 173, 277–8
Apical meristem 28, 40, 43, 45, 47–9, 52–3, 65, 77, 202, 226, 256, 264, 298
Apocarpy (ous) 136, 152, 161, 164, 174, 181, 298

Apocyanaceae 167, 220, 225
Apophysis 97, 98, 234, 285, 298
Apoplasm 35, 298
Aquilegia 161
Arabidopsis 36
Araceae 179, 181, 210, 214
Araliaceae 165, 213
Araucaria (Araucariaceae) 67, 79, 81, 115, 118, 125, 200
Arbutus 248
Archaeopteris 63–4, 76–7, 190
Archaeosperma 78
Archegonium (a) 62, 90, 97, 99–111, 120, 122, 124, 298
Architectural model 12, 198, 200, 205, 218
Arecaceae (-ales) 40, 43, 48, 84, 135, 174, 179, 181, 200–2, 205–6, 210, 218
Arenga 202
Areole 44, 226, 298
Argyroxiphium 256
Aril 113, 125–6, 156, 274, 285, 298
Armeria 241
Arum 180–1
Asclepiadaceae 118, 166, 167, 214, 224, 225
Asplenium 213, 266
Asteraceae (-ales, -anae, -idae) 3, 21, 146, 156, 160, 166, 170–3, 176, 224, 255–6, 276, 279
Asterella 97
Asteroxylon 59–63, 73, 76
Astragalus 21–3
Atactostele 74–5
Atrichum 233–4
Atriplex 239, 284
Aulocomnium 230, 264
Auricle 268, 298
Autochory 282
Avicennia 239, 241–2, 266
Axis *see also* Stele or Branching
  evolution of 74
  form 38
Azolla 238, 239, 246
Azotobacter 244

Baiera 79
Bambusa (Bamboo, Bambusoideae) 186, 203–5, 268
Banksia 165, 249, 256
Banyan 201
Baobab *see Adansonia*
Barinophyton 107
Bark 47, 56, 178, 187, 189, 200, 208–10, 214–15, 218, 255–6, 269–70, 299
Bazzania 215, 217
Begonia 46
Behaviour *see* Tropism or Growth
Bennettitales 78, 119, 136
Berberidaceae 161
Betalains 141
Betula 183, 188, 200, 208–9
Bignoniaceae 169–70, 207, 210, 220, 282
Boraginaceae 167–8
Bordered pit pair 26
Bothrodendron 66
Botrychiopsis 67
Bracken *see Pteridium*
Bract scale 112, 114, 125
Branching *see also* Axis or Growth
  axillary 40
  dichtomous 40, 302
  monopodial 40, 307
  sympodial 40, 314
Brasilodendron 67
Brassicaceae 145, 152, 154–5, 246, 249, 291
Breeding systems 133, 141
Bromeliaceae (Bromeliads) 213, 254 *see also Tillandsia*
Bromus 185
Bruguiera 241–2, 266
Brunoniaceae 171
Bryidae 91, 108, 232
Bryophyte *see also* Moss, Liverwort or *Anthoceros*
  life form 227–8
    ectohydric 215, 217–8, 228–32
    endohydric 228, 232–4
  reproduction 90, 97, 107, 128
Bryum 99, 233
Bud
  scars 39
  scales 14, 255–6, 259
Bulbils 264–5
Bundle sheath 45, 234, 235, 299
Buxbaumia 106
Byblis 254

C4 photosynthesis 234–5 *see also* Kranz anatomy
Cabomba 236
Cactaceae (Cacti) 50, 141, 156, 194–5, 202, 204, 210–14, 224, 226–7, 276
Caesalpiniaceae 150, 164, 244
Calamites 65–8, 108, 109, 200, 205
Calamocarpon 108–9
Calamostachys 108

Calamus 218–19
Calceolaria 168–9
Callistophyton 67, 78
Callitrichales 170
Calluna 248
Calobryales 106, 232
Caltha 240
Calyptra 93, 97, 149, 165, 299
Calyx 133, 136, 140, 148–9, 153, 160–1, 164–5, 170–1, 174, 299
CAM (Crassulacean Acid Metabolism) 212, 214–15, 224, 226, 234, 236
Cambium *see also* Growth secondary
  fascicular 189
  initial 193
  interfascicular 189
Campanula (Campanulales) 146–7, 166, 171
Cannabis (Cannabaceae) 25, 28, 184
Capitulum 156–7, 170–3, 299
Capparaceae 154, 155
Capsella 123, 288
Capsule 91, 96–9, 154, 234, 280–2, 284–6, 299
Carboniferous 57, 63, 65–8, 72, 76–9, 111, 198, 200–1, 205, 269, 285
Carex (sedges, Cyperaceae) 184–6, 235–6, 240
Carpel 13, 113, 115, 136–7, 143, 152–6, 164, 168, 299
Caryophyllales (-anae) 126, 141, 156, 160
Caryopsis 155, 186, 299
Caryota 202
Cassia 139, 150
Cassytha 221–2, 250–1
Casuarina (Casuarinaceae) 184, 194, 214, 221, 244, 250
Catalpa 169
Catkin 112, 115, 160, 182–3, 280, 297, 299
Catopsis 254
Caytonia (Caytonialaes) 78, 112, 126
Cecropia 204, 272, 274
Cedrus 40
Cell
  types 24, 27–8, 47, 187, 194
  wall 24–6, 29, 35–6, 58, 101, 111–12, 189–91, 207, 209, 230, 235, 239, 264
Cellulose 24–6, 56, 191
Centrospermae 141
Cephalanthera 249
Cephalotaxus 81, 112, 113, 115
Cephalotus 252, 254
Cephalozia 282
Ceratozamia 112
Cereus 224
Chaenomeles 163
Chamaecyparis 12–13, 81
Chamaephytes 257
Characters 1, 5, 7, 14, 17–20, 23
Chemical defences 272, 274–9
Chenopodiaceae 141, 156, 224–5, 239
Chiloschista 214
Chlorophyta 58
Citrus 209
Cladode 42, 300
Cladoxylales 63
Classification
  land plants 130
  process of 3–4
Cleistogamy 148, 300
Clematis 161, 162, 284
Climacium 233
Climbers 84, 170, 204, 212, 218, 220
Cline 16–18, 300
Coconut (*Cocos*) 284
Coelogyne 177
Cold adaptations for 255
Coleoptile 283, 300
Collenchyma 26–30, 274, 300
Columella 97–9, 300
Commelina (Commelinales, -anae) 148, 175, 184
Companion cell 29, 300
Competition 65, 267, 288–90
Compositae *see* Asteraceae
Cone
  female 112–14
  male 115–16
Conifer *see* Pinopsida
Conocephalum 106, 229
Convolvulaceae 166–8, 222, 250
Cooksonia 59, 62, 63, 198
Corallorhiza 249
Cordaites 65, 66, 79
Cordyline 202
Corolla 133, 136, 140–3, 148–9, 152–3, 160, 165–74, 184, 300
Cortex 29–30, 47, 49, 75, 190, 208, 210, 220, 240, 300
Corymb 158, 173, 301
Corypha 181
Corystospermales 78
Costaceae 205
Cotyledon 173–4, 287, 301
Crassulaceae 224–5
Crassulacean acid metabolism *see* CAM
Crataegus 163
Cretaceous 57, 68, 79, 82, 84, 135–6, 174, 182, 184, 269–70

Crocosmia 258
Crocus 151
Crop plants 291–6
Cross pollination 133, 138, 148, 179
Crucicalamites 65
Cryptocereus 213
Cryptogam 2
Cryptophytes 257
Cucurbita (Cucurbitaceae) 28–9, 146
Cunninghamia 81
Cuscuta 221–2, 250–1
Cutin 25, 28, 124
Cyanobacteria 212, 244, 246
Cyanogenesis 274
Cyathea (Cyathaceae) 203, 210
Cyathium 146, 147, 301
Cyathodium 97
Cycadopsida (cycads, Cycadales) 23, 44, 48, 78–80, 93, 101–3, 111–13, 115–16, 118–20, 122, 126–7, 139, 198, 201–2, 207, 244, 267, 284–5, 301 *see also Cycas, Stangeria, Encephalartos, Zamia*
Cycas 80, 112–13, 115–16, 202, 284–5
Cyclostigma 63, 64, 67
Cymbidiella 210
Cyme 156–8, 301
Cyperaceae 186, 207, 246, 249 *see also* *Carex*
Cypsela 155, 171, 301
Cytinus 250–1

Dactylis 185
Dactylophyllum 67
Darlingtonia 252
Dawsonia 232
Decaspermium 145
Delphinium 161, 162
Dendrocalamus 205
Dendrochilum 274
Dendrotrophe 252
Desmazeria 185
Development *see also* Embryogenesis
  heteroblasty 12–13
  gametophyte 91, 109, 112
  vascular tissue 31–2, 73
Devonian vegetation 57, 59–64, 72, 78, 107, 198, 228
Dichasium 157–8, 301
Dichondra 168
Dicliny 106, 145–7, 184, 301
Dicotyledon 174, 242
Dicranella 99
Dicranopteris 219
Dicranum 264
Dieffenbachia 272
Digitalis 170
Dillenianae 161, 182
Dinochloa 218, 219
Dioecy 99, 107–8, 145–6, 301
Dionaea 254
Dioscorea 204
Diplophyllum 96, 215, 217
Dipsacus 273, 274
Dischidia 214, 273
Disocactus 213
Dispersal 62, 91–9, 124–7, 161, 165, 170–1, 173, 263–4, 280–7
  animal (zoochory) 270, 274, 285
  explosive 280
  passive release 279–80
  timing of 286
  water 284
  wind (anemophily) 279, 284
Disseminules size of 282
Double fertilization 122, 134, 301
Dracaena 202
Drimys 136–7, 149, 151, 153
Drosera 254
Drosophyllum 254
Drug plants 291–2, 294
Dryas 37
Dryopteris
  archegonium 100, 102
  gametophyte 104
  leaf anatomy 46
  sporangium 95
Duisbergia 64

Ecbalium 281, 282
Eccremocactus 213
Echeveria 225
Echinops 172–3
Echium 167–8
Ecotype 14, 16, 302
Eichornia 239
Elaiosome 274, 285, 302
Elaters 91, 97, 280, 282, 302
Embryogenesis 90, 107, 122–4, 302
Embryo 122–4, 156, 302
Embryosac 112, 122, 134, 173, 302
Encelia 226, 266
Encephalartos 80, 123, 125
Endodermis 32–3, 35, 44, 73, 232, 235, 239, 246, 302
Endogonaceae 248
Endosperm 122, 124, 126, 127, 134, 284, 297, 302
Endosperm nucleus 122, 124, 134, 302
Eospermatopteris 63
Ephedra 82–3, 101, 111–12, 115, 118, 122, 132, 139

Epidermis 27–30, 45, 47, 49, 118, 210, 214, 221, 224, 232, 236, 239, 248, 256, 268, 270–2, 277, 302
Epigyny 153, 154, 163, 303
Epipactis 249
Epiphyll 303
Epiphylum 211, 213
Epiphyte 67, 107, 170, 176, 187, 198, 209–10
  aerophyte 215–16
  ant plant 213–14, 272–3
  bark 214–15
  bryophyte 215, 217–18
  Cactaceae 212
  humus 211–12
  root climbers 220
  stem parasites 221–2, 250, 257
  tank 213, 216
Equisetum
  apical meristem 48
  morphology 70
  origin 68
  reproduction
    antheridium 101–2
    archegonium 99–101
    elaters 91–2, 280
    gametophyte 104–5
    heterothallism 107
    lateral spread 266
    sporangium 92–3
    strobilus 70, 93
  silicified epidermis 270–1
  stele (stem anatomy) 70, 75
Eremolepidaceae 221
Erica (Ericaceae) 7, 118, 151, 213, 247–9
Eriocaulaceae 184
Eryngium 165
Erysimum 151–2
Erythrina 204
Espeletia 255
Establishment 126, 127, 202, 218, 263, 266, 287
Euanthium theory 134
Eucalyptus 45, 149, 165, 227, 246
Euphorbia (Euphorbiaceae) 139, 146–7, 202, 225–5, 271
Eusporangium 91, 94, 303
Evolution
  coevolution of plant and herbivore 277–8
  convergent 20–2
  Darwinian 20
Exine 56, 110, 119, 122, 138–9, 284, 303
Exodermis 211, 214–15
Fabaceae (-ales, Leguminosae) 150, 152, 155, 164, 279, 291 *see also Lathyrus, Lupinus, Trifolium, Vicia*
Fagus 183, 194, 201, 208, 242
Family
  taxonomic rank 3
Fern *see also Dryopteris*
  desert 223
  vascular anatomy 73
  sorus 93
  spore dispersal 281–2
  leptosporangiate 282
  primitive 72
Ferocactus 50, 51
Fertilization 88–90, 101, 103, 109–12, 120–4, 126–7, 134
Fibonacci sequence 50
Fibre 24–9, 45, 189–97, 202, 204, 209–10, 220
Ficus (fig) 184, 200–1, 203, 256
  pollination 180–1
  strangling 213
Filament 115–16, 149–52, 303
Filipendula 163
Fire adaptations for 256
Fissidens 230–1
Fitness 19
Fottonia 207
Floral diagram and formula 142
Flower 131–86
  achlamydeous 135
  ancestral 135–8
  architecture (shape) 141–2
  attractants
    primary 138
    secondary 140
    pigments 140–1
    scent 141
  Bennettitalean
  brood 140
  evolutionary trends 148–86
  explosive 282
  parasite 250
  variation within species 17–19
Follicle 136, 303
Fontinalis 48, 231
Food bodies 273–4
Forsythia 48
Fossils 55–9
Fossombronia (Fossombroniaceae) 45, 224
Fragaria 164
Frankia 244
Fraxinus 206, 242
Freycinetia 219–20

Fruit 127, 133, 136, 153–6, 161, 163–5, 170–1, 180, 182, 186, 221, 266, 269 *see also* Dispersal, Establishment or Pericarp
Frullania 91, 217, 218, 281–2
Funiculus 115, 124, 154, 170, 298–9, 303

Galeola 249
Gametogenesis 88, 303
Gametophyte 88–91, 99–122, 173, 248, 303 *see also* Heterothallism
  development
    Andreaea 105
    angiosperm 112, 118, 121, 133–4
    bryophyte 105–7
    endosporic 91
    female gametophyte 107, 109–112, 120, 122, 126, 134, 173, 269, 302, 306–8
    pteridophyte 103–5, 248
    Sphagnum 105–6
    seed plant 111–12, 114, 118
  distribution of male and female 99
  evolutionary limitations 90
  primitive 62
  nutritive tissue in seed 126–7
Gas relations 223, 234–5
Gemma (ae) 71, 264–5
Genetic drift 18, 23, 303
Genetic engineering 294
Genotype 7, 10, 124, 303
Gentianaceae (-ales) 166–7
Gentianella 10–11
Geological ages 56–7
Geophyte 258–60, 289, 303
Geothallus 224
Gesneriaceae 169–70, 210, 213
Geum 163
Gingers *see* Zingiberiaceae
Ginkgo
  characteristics 132
  female development 100–1, 111–12, 127
  leaves 43, 80, 206
  morphology 80
  male development 103, 115–16, 118
  origin 79
  pollination 120, 122
Ginkgoites 79
Girder 45
Glaucidium 161
Glechoma 148
Globulariaceae 169, 170
Glossopteridales 78
Glume 185–6, 304
Gnetopsida (gnetophytes) *see also* *Gnetum, Ephedra* or *Welwitschia*
  characteristics 132
  origins 82
  morphology 83
  female gametophyte 83, 101, 111–12
  male development 115, 118
  fertilization 122
Gnetum 83, 101, 111–12, 115, 118, 122 *see also* Gnetophytes
  leaves 43
  liane 218, 220
Golden section 50–1
Gondwana
  flora 67
Goodeniaceae 171
Grasses *see* Poaceae
Grevillea 188
Grime's plant strategies 287–90
Growth and behaviour 24, 31–2, 36–7 *see also* Branching
  lateral spread 266–7
  determinate 40
  indeterminate 38
  rhythmic (seasonal) 40, 198, 260
  secondary 31–2, 189–90
    rings 39, 40, 78, 194
Guard cell 27–8, 304
Gunnera 244
Gymnocalycium 50–1
Gymnosperm 2
  living 35, 43, 48, 80–1 *see also* Pinopsida, Gnetopsida, Cycadopsida and *Ginkgo*
  fossil 55, 78–82
Gynodioecy 107, 146
Gynoecium 7, 113, 115, 133, 152–3, 161, 166, 304
Gynostegium 166

Halophyte 45, 224–5, 239–41, 304
Hamamelis (Hamamelidales) 145, 183 *see also* Amentiferae
Haplostele 73–4
Hartig net 246–8
Haustorium 120, 221, 249–51, 304
Hebe 169
Hedera 12, 220
Hedgehog plants 21
Heliamphora 252, 254
Helichrysum 170, 172
Helleborus 46, 136, 151, 153–4, 162
Hepaticites 228
Herberta 217, 230
Herbivory
  chemical defences 173, 274–8

dinosaur 269–70
insect 269
physical defences 270–2
Heterospory 91, 107–9, 126, 304
Heterostyly 143–4, 305
Heterothallism 88, 107–10
Hevea latex production 276
Honey leaves 161
Horneophyton 59–62, 97
Hornwort *see Anthoceros*
Horsetail *see Equisetum*
Hosta 175
Humulus 184
Hyacinthoides 288–90
Hydathode 305
Hydnophyton 213, 214
Hydnoraceae 250
Hydrocotyle (Hydrocotyloideae) 165
Hydroid 233, 305
Hydrophyte 236–40, 259, 260, 305
Hyenia 63–4
Hymenophyllum (Hymenophyllaceae) 93, 95
Hypanthium 153, 161, 163–4, 305
Hypericum 151–2
Hypnum 230–1
Hypocotyl 34, 74, 77, 124, 214, 221, 266, 283, 305
Hypodermis 210, 224, 240, 305
Hypoepigyny 153–4, 305
Hypogyny 153–4, 163, 305

Idioblast 305
Illicium 143, 145, 191
Inbreeding 141, 148, 170, 179, 293
Inflorescence 133–40, 146–9, 156–8, 165–6, 170–6, 179, 181–6
naming of types 157
Insect pollination 133–45, 170, 174, 181–4, 285
Integument 110–15, 119–20, 124, 126–7, 156, 173, 306
Involucre 93, 96, 106, 170, 173, 306
Iris (Iridaceae) 6, 32, 153, 175–6
Isoetes
reproduction
antheridium 101
archegonium 100
gametophyte 104
sporophyll 69, 92
heterospory 108–10
morphology 68–9, 72, 77
corm 69
ligule 72
roots 77
Isogamy 58–9

Jasione 171
Juglans 266
Juncus (rushes, Juncaceae) 184–6, 240, 270
Jungermanniales 215, 228, 230
Jurassic 57, 79, 82, 84, 266, 269, 270

K-strategist 287
Kalanchoe 264
Khaya 200
Kingia 227, 256
Knautia 171
Kranz anatomies 234–5, 268

Labellum 176–8
Laelia 214
Lagurus 185
Lamiaceae (-ales, Labiatae) 21, 141, 152, 166–8, 170
Land plants (Embryobionta)
evolution 85
first 59
Larix 40, 81, 114, 206
Latex 276, 306
Lathyrus 42, 150
Laticifer 271, 276, 306
Launaea 21–2
Laurus 149, 151
Leaf
abscission 206
arrangement 39, 41, 76, 208
consistency 260
diversity 41–2, 205–8
growth and development 12–15, 41, 43–4, 48
lamina 45, 252
shade 207–8
shape 8–17, 222
sheath 41, 70, 205, 268
size 23, 40, 206, 250, 260
venation 43–5, 47, 80, 83
Leaf gap 75–6
Legume 154, 164, 306
Leguminosae 164
Lemma 185–6, 306
Lemna 236, 238
Lentibulariaceae 169, 170, 254
Lenticel 209, 240, 242, 271
Lepidodendron 65–8, 77, 190
Lepidopteris 78
Leptoid 233
Leptosporangia 91, 93–5, 306
Leucadendron 242
Leucobryum 231–2
Liane (a) 67–8, 192–4, 205, 218–20, 306

Life cycles 128–9
Life form
  bryophyte 227–8
  Raunkiaer's 257–62
  xerophyte 223–7
Lignification 72, 260
Lignin 6, 25, 124, 190, 249
Ligule 41, 72, 268, 306
Liliaceae (-ales) 157, 173–6 *see also* Monocots
Limonium 144–5, 153, 239
Lindera 149, 151–2
Linnaeus 2–3
Lithops 224–5
Liverwort
  leafy 215, 217–8 *see also* Calobryales, Jungermanniales or Metzgeriales
  reproduction
    capsule 97, 280
  thalloid 228–9 *see also* Marchantiales or Metzgerialaes
Lobelia 255
Lodicule 185, 306
Lodoicea 284
Lophocolea 215, 217
Loranthaceae 156, 210, 221, 282
Lumnitzera 221, 240, 242
Lunularia 264
Lupinus 245, 281
Lychnis 273, 274
Lycopodites 72
Lycopodium 48, 61, 68, 71–3, 76–7, 102–4, 248, 264, 306
  morphology and anatomy 71–2
    gemmae 71, 264
    roots 77
    stele 73
  reproduction
    gametophyte 103–4, 248
    sporophyte 71, 92
Lyginopteris 67, 78
Lygodium 12–13, 43, 218, 220
Lyonophyton 59, 60, 62, 106
Lythrum 143–4

Macaranga 272, 273
Macromitrium 108
Magnolia 3–4, 113, 136–7, 143
Magnoliaceae (-ales) 135–6, 143, 157, 160–1, 173–4 *see also* Dicots
Malus (Maloideae) 163–4
Malvales 194
Mamillaria 50, 51
Mangroves 65, 201, 221, 239–42, 266
Marantaceae 205
Marattia 72, 94
Marchantia (Marchantiales) *see also* *Conocephalum, Marchantia* or *Preissia*
  morphology and anatomy 228–30, 264–5
  reproduction 97, 100, 102, 106
Marsilea 95, 110, 236–7
Medullosa 65, 78
Megaphyll 76, 260, 306
Megaspore 92, 95, 108–10, 307 *see also* Ovule
Megasporophyll 112–13, 115 *see also* Cone female
Melampyrum 16–19
Melastoma (Melastomataceae) 139, 207, 213
Melastomataceae 207, 213
Mentha 141, 167–8
Menyanthes 284
Mercurialis 146–7
Meristem 30, 40–1, 43–5, 65, 77, 202, 256, 307 *see also* Cambium or Phellogen
  apical 40–1, 47–50, 52–3
  intercalary 47
  lateral 47
  marginal 47
  protection of 226, 256
Mesophyll 30, 44–6, 141, 207, 234–5, 240, 260, 307
Metasequoia 206
Metaxylem 28, 30–2, 307
Metroxylon 202
Metzgeriales 34, 228
Microphyll 76, 260, 307
Micropyle 110, 113, 119, 122, 274, 298, 307
Microscope 6
Microsporophyll (-phore) 115–17, 307 *see also* Stamen
Mimicry 221
Mimosa (Mimoscaeae) 36–7, 150, 164, 244 *see also Acacia*
Miscanthus 185
Mnium 100, 102
Module 7, 10, 14, 38, 52, 201, 307
Monocaul 201–3, 206, 307
Monochasium 158
Monoclea (Monocleales) 97, 228
Monocots (monocotyledonidae, Liliidae, Liliopsida) 43–4, 48, 73, 75, 77, 84, 122, 135, 156–7, 173–84 *see also* Poaceae, Liliaceae, Orchidaceae, Araceae, Arecaceae
  differences from dicots 174
  wind pollinated 184–6

Monoecy 99, 145–47, 307
Monotropa 34, 249
Montera 43, 220
Moraceae 181, 184, 213
Moresnetia 78
Morus 10–11, 180
Moss *see also Mnium, Fontinalis, Polytrichum, Sphagnum, Leucobryum*
  morphology and anatomy 106, 230, 233
  reproduction
    sporophyte 97–9, 280
    gametophyte 100, 102
Musa 203, 204, 276
Muscari 258
Mycorrhiza 35, 52, 77, 178, 244–9
Myosotis 167–8
Myosurus 161–2
Myrica 244
Myrmecodia 213–14, 273
Myrrhis 278
Myrtales 164

Names
  use of 1–3
  hierarchy of 86
Narcissus 174–5
Nardus 290
Nathorstiana 68–9
Nectar 138–40, 145–6, 165, 179, 182, 236, 272, 291
Neottia 249
Nepenthes 252–4
Nerium 166
Nicotiana 167
Nilsonia 204
Nolanaceae 168
Nopalea 226
Nopalxochia 213
Normapolles 182
Nostoc 244, 246
Nothia 59, 60, 62
Nothofagus 183, 246
Notholaena 223
Nutrient relations 223, 242–57
Nymphaea (Nympheales) 14, 120, 145

Olea (Oleaceae) 46, 169–70
Oogamy 58
Operculum 97–9, 280
Ophrys 177, 179
Orchidaceae (orchids) 127, 176–9
  epiphytic 210, 212–14
  mycorrhizae 247–8
  pollination 178–9
Orestovia 61
Orobanche (Orobanchaceae) 34, 169–70, 250–1, 284
Orthostichy 51
Orthotropy 38–41
Ovary 152–6 *see also* Pistil or Carpel
Ovule 79, 110–15, 119–22, 134, 155–6, 308
Ovuliferous scale 112, 114, 124–6

Pachycaul 190, 200, 204, 308
Pachypodium 225
Paeonia (Paeoniaceae) 136, 148, 161–2
Palea 185–6, 308
Palisade layer 30, 45–6, 206–7, 227, 308
Palms (Palmae) *see* Arecaceae
Pancratium 152, 174–5
Pandanus (Pandanaceae) 179, 202–3, 220
Panicle 157–8, 179, 308
Papaver (Papaveraceae, -ales) 149, 153–4, 161
Paphiopedalum 176, 177
Pappus 171, 281, 284
Parasites 221–2, 249–52
Parasponia 244
Parastichy 50–1
Parenchyma 26–30, 35, 44, 187, 189–90, 194, 196–7, 202, 204, 209–10, 220, 235, 240, 285, 309
  in xylem 194, 196–7
  in succulents 226
Parnassia 14, 16
Passage cell 32–3
Passiflora 154
Pebble plant 224
Pellia 91, 96, 100–1, 228–30
Peltaspermales 78
Pentstemon 169
Peperomia 135
Pereskia 202, 226
Perianth 14, 106, 112, 133, 135–6, 145–6, 148–50, 152, 160, 174, 178, 182, 184, 186, 280, 309
Pericarp 127, 155–6, 164, 284, 286, 309
Periderm 187, 208–10, 227, 256, 257, 271, 309
Perigyny 153, 163, 309
Perisperm 126, 156, 309
Peristome 98–9, 280, 286, 309
Permian 57, 67–8, 78–9, 269
Persoonia 249
Pertica 63
Petal 58, 149, 176, 178, 282, 309
Phanerophytes 257, 260

Phaseolus 283
Phellem 208–10, 309
Phelloderm 47, 208–10, 309
Phellogen 7, 10, 12, 208–10, 309
Philodendron 220
Phloem 26, 29–33, 35, 43–4, 58, 75, 189–90, 208–9, 220, 232, 250, 309
Photosynthesis 206–7, 212, 214–15, 224, 226, 234–5
Phragmites 258, 288, 290
Phreatophyte 224, 309
Phyllanthus 42
Phyllocladus 42, 81, 112–13
Phylloglossum 68–9
Phyllotaxis 41, 50, 309
Phymatodes 12–13, 273
Phytoalexins 279
Picea 201
Pillostylis 95, 110, 236–7, 250
Pilularia 95
Pinguicula 42, 254, 309
Pinopsida (conifers) 12, 40, 43–4, 79, 81–2, 84, 93, 101, 103, 111–12, 115, 118–20, 122, 126–7, 138, 187–93, 197–8, 200, 206, 246, 267, 284 *see also Pinus* or *Larix*
Pinus
  reproduction
    female 111, 281
    male 114, 115–16, 116, 284
  leaf 27, 43, 81
  mycorrhizae 246
  resin canal 135, 157, 210, 213, 271
Pistia 238–9
Pistil 112, 115, 127, 133–4, 136, 140, 152–6, 161, 164–81, 184–6, 309 *see also* Carpel
Pisum 283
Pitcher plant 42, 220, 252–4
Pith 6, 29, 30, 47, 73, 75, 202, 204, 256, 274
Pit 26, 189–91, 193, 204, 226, 236
Placentation 133, 154–5, 169–70, 178, 310
Plagiothecium 230, 231
Plagiotropy 38–41
Plantago (Plantaginaceae) 281–2
Plants
  adaptations for land 59
  aquatic 236–9
  as sources of drugs 291–2
  carnivorous 252–4
  differences from animals 24
  environment of 33
  homoiohydric 223
  poikilohydric 223
  succulent 224
Plasmalemma 25, 35–6, 248, 310
Plasmodesmata 24, 26, 35–6, 310
Plasticity 10–12, 22–3, 54, 207
Plastochron 48, 310
Platanus 152–4
Platycerium 210–11, 213
Platyzoma 104, 108
Plectostele 73
Pleurococcus 58
Pleuromeia 68–9
Plicosepalus 221, 252
Plumbaginaceae 143, 239
Plumule 34, 124, 283, 310
Pneumatodes 214–15, 310
Pneumatophores 241–2, 310
Poa 266, 290
Poaceae (grasses) *see also* Bambusoideae
  inflorescence 185–6
  origin 204, 270
  vegetative characters 34, 53, 226–7, 235–6, 267–8
Podocarpus (Podocarpaceae) 79, 81, 115, 117–18, 125–6, 246
Podostemaceae 257
Polemoniaceae 167–8
Pollen 114–16, 118–22, 135–6, 138–41, 143, 145–6, 148–9, 152–3, 279–86
  chamber 111, 114, 119, 120, 310
  tube 1034, 112, 114–15, 118, 120–2, 136, 143
Pollination 103, 110–1, 124, 126–7, 133–36, 138–41, 143–9, 152, 157, 165, 170, 174, 176–86
  siphonogamy 118–22, 312
Pollinium 118, 166, 177–9, 285, 310
Polyembryony 122, 124, 138
Polygonaceae 249
Polygonatum 258
Polygonum 42, 257
Polypetalae 160–6
Polytrichum (Polytrichidae) 47, 99, 232–4
Populus 182, 206, 208
Posidonia (Posidoniaceae) 236, 284
Potamageton 237
Preissia 229–30
Primula (Primulaceae) 143–4, 155
Procambium 28, 30, 47, 189, 194, 310
Proembryo 122–4, 310
Prolepsis 39–40, 310
Propagules 62, 91, 209, 264–5
Protandry 99, 106, 146–8, 157, 172, 311
Proteaceae (-ales) 164–5, 249
Prothallus 103, 264, 311

Protoderm 47, 311
Protogyny 99, 106, 146–8, 311
Protosalvinia 59
Protoxylem 30, 31
Prunella 167
Prunus (Prunoideae) 146–7, 163–4
Psaronius 65, 72
pseudanthium theory 134
Pseudobornia 63–4
Pseudobulb 214
Pseudosporochnus 63, 64
Pseudotsuga 119–20, 125
Psilophyton 63
Psilotum (Psilophytina)
  relationships 72, 77
  gametophyte 90, 105, 248
  sporophyte 93–4
  morphology and anatomy 48, 61, 71–3, 76, 212, 264, 266
Pteridium (bracken) 3, 63, 75, 266
Pteridophyte 86
  life cycle 128–9
  living *see Equisetum, Psilotum, Tmesipteris, Selaginella, Isoetes, Lycopodium* and ferns
  fossil 59–69, 72
Peridosperm 78, 311
Puccinellia 239
Puya 255
Pyrrosia 211

Quaternary 57
Quercus 14, 15, 120, 188, 194, 200, 209

R-strategist 287
Raceme 157–8, 176, 311
Rachis 21, 42–3, 77, 79, 311
Radicle 34, 77, 124, 156, 174, 221, 242, 266, 287, 311
Rafflesia 250
Ranunculus (Ranunculaceae, -ales) 10–11, 31–2, 136, 161–2, 236, 240
Raunkiaer's life forms 259–61
Rays 141, 194, 196–7, 202, 204, 226
Reaction wood 10, 190, 311
Regnellidium 236–7, 276
Resin canal 271, 276
Resurrection plane 223
Rhacomitrium 230–1
Rhacopteris 67
Rhipsalis 211
Rhizobium 244–5
Rhizoctonia 248
Rhizoids 34, 61, 72, 77, 103, 105, 108–9, 215, 218, 228–30, 232, 246, 264
Rhizome 17, 35, 63, 71–2, 90, 105, 204–5, 232, 236, 240, 258, 260, 265–6, 311
Rhizophora (Rhizophoraceae) 206, 221, 239, 241, 242, 266
Rhizophore 65, 70, 77, 312
Rhodophyta 58
Rhynia 59–63, 72–3, 97, 106, 198, 267
Rhyniophyte 61–2
Rhypsalis 212–13
Ribes 32
Riccardia 264
Riccia 97
Ricinus 125, 146–7, 151–2, 182, 273–4
Root 77–8
  adventitious 52, 77, 210, 243, 264
  apical meristem 53
  cap 48–9, 52–3, 77–8, 312
  contractile 35, 257, 258
  development 10, 32
  hair 30, 52–3, 242, 246, 312
  lateral 32, 243
  nodules 244–5
  pressure 35
  system 34–5, 52–4, 73, 77, 198, 201, 227, 236, 241–4, 246, 249, 262, 267
  primary 77, 242–3
  see also Mycorrhizae, Rhizome
Rosaceae (-anae, -oideae) 153, 156, 160–1, 163–5
Rostellum 177–8, 312
Roystonea 203
Rubiaceae 213
Rubus 163–4
Ruderal 246, 249, 288–90
Ruscus 41–2
Rushes *see Juncus*

Sago 202
Salicaceae 182
Salicornia 225, 239
Salix 182
Salt gland 239, 241
Salvia 151–2, 167–8, 267
Salvinia 43, 238–9
Samara 155, 281, 312
Sanango 170
Saniculoideae 165
Santalaceae (-ales) 221, 250, 252
Sapindaceae 220
Saprophytes 34, 249, 312
Sargassum 58
Sarracenia 42, 154–5, 252, 254
Satureja 151–2
Saxifraga (Saxifragaceae) 153–4, 264

Scabiosa 171
Scapania 217–18
Schefflera 213
Schizanthus 168
Schlumbergera 211, 213
Schomburgkia 214
Sclerenchyma 26–8, 30, 44–5, 270, 274, 312
Sclerophyll 224, 257, 312
Scrophulariaceae (-ales) 166, 168–70, 178, 194
Secale 185
Secondary growth 28, 43, 47, 76, 135, 187, 189–90, 202, 220
Secondary wall 26, 28, 189, 191
Sectioning 5–6
Sedges *see* Cyperaceae
Sedum 289
Seed 124–7, 280–8, 312
  size 107, 125, 127, 155, 284
Seedling 125, 283, 288
Selaginella
  gametophyte 103–4, 109–10
  morphology and anatomy 46, 70, 72–3, 77, 223
  sporophyte 70, 72, 108–10
Selaginellites 67, 72, 91–2
Selenicereus 212, 219
Self-incompatibility 119–20, 122, 138, 141, 143–6, 148, 171, 173
Senecio 171, 173, 255
Sepal 149, 312
Setaria 185
Shade
  adaptations for 10, 17, 21, 23, 33, 67, 107, 127, 207, 220, 223, 289
Shoot system 34–5, 52
  architecture 38–9, 197–201, 201–5
  growth 39–40
  long and short 40, 78, 200
Sieve cells 35, 189, 312
Sieve tube elements 26, 312
Sigillaria 65, 66, 68
Silene 146
Silique (a) 154, 312
Silurian 57, 59, 62, 63
Siphonogamy *see* Pollination
Smilax 21–3
Solanaceae (-ales) 166–8, 279, 291
Solanopteris 214
Solanum 257
Solidago 172–3
Sonneratia (Sonneratiaceae) 239, 241
Spadix 141, 157, 179–81, 313
Spartina 239
Spathe 67, 149, 174–5, 179–81, 313
Sperm 58, 62, 88–90, 101–3, 106–7, 109, 118, 120, 122, 127
Sphaerocarpus (Sphaerocarpales) 228–9
Sphagnum 97, 99, 105–6, 228, 231–2, 280–1
Sphenophyllum 66–8, 190
Spikelet 185–6, 313
Spiroideae 164
Splachnum 99, 285
Sporangium 91–7, 313
Spores 62, 91, 97, 282–2
  sporogenesis 90
Sporophyte 62, 88, 90, 92–3, 96–7, 99, 105–7, 109–10, 127
Stamen 14, 115–16, 133, 140, 146–52, 160–1, 164–6, 168, 170, 176, 178, 182, 184, 186, 313
Staminodes 140, 152, 166, 178, 313
Stangeria (Stangeriaceae) 79–80, 116
Stapelia 167, 225
Stele 29–33, 35, 36, 55, 70, 73–6, 220, 232, 240, 313
Stem *see* Stele or Shoot
Stenochlaena 220
Sterculia 226
Stigma 113, 115, 119–20, 122, 136, 143, 145–6, 152–5, 161, 164, 166, 170, 176, 178, 182, 250, 284, 313
Stigmaria 56, 77
Stigmarian axis 56, 65
Stipule 41, 256, 313
Stolon 35, 265
Stoma (ta) 28, 30, 33, 36, 45, 61, 68, 97, 206, 210, 221, 223–4, 226–7, 234–6, 239, 313
Strategies
  ecological 287–9
  regenerative 263, 288–9
Streptocarpus 169
Striga 250–1
Strobilus 70, 72 *see also* Cone
Style 113, 115, 120, 122, 136, 140, 143, 146, 152–5, 161, 164, 166, 168, 170–1, 178, 186, 282, 313
Stylites 68, 72, 77, 108
Suaeda 46, 240
Suberin 25, 209, 304
Subularia 148
Succulent 195, 224–6
  anatomy 202, 204, 224, 226
  arborescent 195, 202
  halophyte 225, 239, 240
Suspensor 122, 124, 248, 313
Swietenia 200
Syconium 156, 179–81, 313
Syllepsis 39, 40, 313

Sympetalae 160, 165, 167–71, 173
Symplasm 24, 35, 239, 313
Synangium 93, 115, 149, 313
Systematics 1
  taxonomic groups 2, 3, 20
  taxonomic hierarchy 1, 3–5, 17

Tapetum 91, 116, 118, 122, 145, 314
Taxodium 115, 117, 201, 206, 242
Taxonomy *see* Systematics
Taxus 112–13, 115, 117, 119, 126
Teleology 20
Telome theory 74, 76
Tepals 148–9, 160, 174, 176
Tertiary 23, 26, 44, 57, 65, 82, 84, 133, 201, 270
Tetraphis (Tetraphidae) 99, 232
Thalictrum 161–2, 280
Thallus *see* Liverworts thalloid
Thesium 252
Thorns 21, 43, 218, 270–1, 274
Thuja 123
Thyrse 157–8, 314
Tillandsia 213, 215, 216
Tissues 24
Tmesipteris 71–3, 76, 77, 212, 264
Tortula 230
Tracheary elements 189, 190, 194, 197
Tracheid 28, 35, 44, 61, 187, 189–94, 197, 200, 220, 232, 240, 314
Tradescantia 175
Transfer cell 44, 314
Transfusion tissue 44
Transpiration 33–6, 191, 206, 208, 221, 226–7, 249, 314
Tree
  adaptability 200
  architecture 187, 197–200
  Carboniferous 65, 67, 77
  Devonian 63, 64
  fern 63, 203, 210
  heartwood 279
  monocaul 201–3
  pachycaul 200
  stability 201
  trunk 38, 198, 200–2, 204–5, 207, 209–10, 212–13, 218, 269
Trentepohlia 59
Triassic 57, 68, 78, 79, 82, 270
Trichocereus 204
Trichocolea 215, 217
Trichomanes 95
Trichome 24, 26, 28–30, 45, 213, 215, 216, 254, 271–2, 276–7
Triglochin 184
Trimerophytes 63, 76
Triticum 173, 283, 291–2, 294–5
Trochodendron 184
Trollius 140, 162
Tropism 37, 314
Tuber 35, 68, 103, 213, 224, 248, 258, 260
Tubercle 51
Tulipa 257
Tunica 48
Twigs 200, 206, 208, 276
Tyloses 190, 196
Typha (Typhaceae) 179–80

Ulex 282
Ulmus (Ulmaceae) 184, 244
Umbel 157, 158, 165, 175, 266
Umbelliferae *see* Apiaceae
Urtica (Urticaceae, -ales) 182, 184, 194, 272, 277, 282
Utricularia 36–7, 43, 213, 254

Valeriana (Valerianaceae) 171, 281, 284
Variation 14, 18
Vascular bundle 30–1, 43–4, 75, 235, 315
Vascular cambium 30, 47, 189, 202, 204
Vacular system 33, 61, 228 *see also* Stele
Velamen 211, 214–15, 315
Venus fly trap *see Dionaea*
Verbascum 168, 169
Verbenaceae 168, 239
Veronica 169
Verticillaster 157–8, 315
Vessels 187, 191, 193–4, 197, 200, 220
Vesturing 189
Vicia 42, 125, 151
Victoria 236
Vinca 166
Viscaceae 156, 221
Vivipary 266
Voltziales 67, 79

Water relations 223, 228
Water transport 35–6
Wax 6, 28, 179, 209, 272, 277, 279, 285
Welwitschia
  morphology and anatomy 43, 83
  reproduction 83, 111–12, 115, 118, 122
Wheat *see Triticum*
Wild fire 256
Williamsoniella 136, 137
Wind pollination 119, 133, 136, 170, 174, 179, 182–6

Winteraceae (-ales) 136, 145, 149
Wolffia 236, 238
Wood anatomy 184, 187–8, 190–1

Xanthium 284
Xanthorrhoea 203, 256
Xerophyte 79, 82, 212, 217, 223–5, 236, 239, 289, 315
Xylem 25–6, 28–32, 35–6, 44, 52, 73, 75, 187–97, 315 *see also* Hadrom
  arid habitats 194–5
  growth rings 194
  liane 220
  specialization 190–5, 197
  vessel diameter 191, 193
  vessel element 27–9, 190–1, 195
Xyridaceae 184

Yucca 44, 140, 151, 152, 175, 176, 202, 227

Zamia (Zamiaceae) 79, 115, 202
Zea 31, 283
Zingiberiaceae 178, 204
Zizyphus 270
Zostera (Zosteraceae) 236–7, 284
Zosterophyllum 59, 61, 62
Zygomorphy 149, 157, 166, 168–70
Zygote 90, 97, 122, 124, 134, 315